LINNAEUS'
Philosophia Botanica

STEPHEN FREER

Stephen Freer, born at Little Compton in1920, was a classical scholar at Eton and Trinity College Cambridge. In 1940, he was approached by the Foreign Office and worked at Bletchley Park and in London. Later, Stephen was employed by the Historical Manuscripts Commission, retiring in 1962 due to ill health. He has continued to work since then, first as a volunteer for the MSS department of the Bodleian Library with Dr William Hassall, and then on a part-time basis at the Oxfordshire County Record. In 1988, he was admitted as a lay reader in the Diocese of Oxford. His previous book was a translation of Wharton's *Adenographia*, published by OUP in 1996.

A fellow of the Linneau Society of London, Stephen lives with his wife Frederica in Gloucestershire. They have a daughter, Isabel.

COVER ILLUSTRATION

Rosemary Wise, who designed and painted the garland of flowers on the book cover, is the botanical illustrator in the Department of Plant Sciences in the University of Oxford, associate staff at the Royal Botanic Gardens, Kew, and a fellow of the Linneau Society of London. In1932 Carl Linnaeus made an epic journey to Lapland, the vast area across arctic Norway, Sweden, and Finland. In 1988, to mark the bicentenary of the Linneau Society of London, a group from Great Britain and Sweden retraced his route. Rosemary, was the official artist and the flowers featured here are taken from ones painted at that time, plants with which Linnaeus would have been familiar.

The garland of flowers surrounds an image of the medallion portrait of Linnaeus by C. F. Inlander, 1773, reproduced with kind permission from the Linnean Society of London.

CAROLUS LINNAEUS. M.D.
Sᵣₒ Rⁱᶜᵉ Mᵗⁱˢ Sueciæ Archiater; Medic: et Botan: Profeſſ.
Upsal: ordin. Horti Academ: Præfectus; nec non Acc:
Imper: Nat. Curiof. DIOSCORIDES 2ᵈᵘˢ Upsal:
Stockh: Berol: Monſp: et Pariſ: Soc.
Natus 1707. Maj. ²³⁄₁₃.   Delin. 1748.

J. M. Bernigeroth ſc Lips 1749.

Portrait of Linnaeus, engraved by Johann Martin Bernigeroth.

LINNAEUS'
Philosophia Botanica

TRANSLATED BY

Stephen Freer

NATURÆ DISCERE MORES

OXFORD

UNIVERSITY PRESS

OXFORD
UNIVERSITY PRESS

Great Clarendon Street, Oxford OX2 6DP
United Kingdom

Oxford University Press is a department of the University of Oxford.
It furthers the University's objective of excellence in research, scholarship,
and education by publishing worldwide. Oxford is a registered trade mark of
Oxford University Press in the UK and in certain other countries

This work was first published in Latin in 1751 in Stockholm and Amsterdam
This English translation first published 2003

Reprinted 2013

British Library Cataloguing in Publication Data
Data available

Library of Congress Cataloging in Publication Data
Data available

ISBN 978-0-19-856934-3

FOR FREDERICA

CONTENTS

PREFACE

Linnaeus' *Philosophia Botanica* (*The Science of Botany*) was first published in 1751. In 1736 he had produced *Fundamenta Botanica* (the *Foundations of Botany*), consisting of 365* aphorisms concerning all aspects of the subject, divided into 12 chapters. These are repeated as the first paragraphs of each of the sections of the *Philosophia*, and are followed by detailed explanatory matter, probably based on his lecture notes. The book is an important stage in the development of binominal nomenclature, which was carried further in his *Species Plantarum* (*Plant Species*) of 1753, and which has become universal in botany and zoology.

The following is a bibliographical description of the first edition:

CAROLI LINNÆI/ARCHIATR. REG. MEDIC. ET BOTAN. PROFESS. UPSAL./ACAD. IMPERIAL. MONSPEL. BEROL. TOLOS. UPSAL./STOCKH. SOC. ET PARIS. CORRESP./ PHILOSOPHIA/BOTANICA/ IN QVA/EXPLICANTUR/*FUNDAMENTA BOTANICA*/CUM/*DEFINITIONIBUS PARTIUM,*/*EXEMPLIS TERMINORUM,*/*OBSERVATIONIBUS RARIORUM,*/ADJECTIS/*FIGURIS*/*ÆNEIS* [sic]/Medallion with design of *Linnaea Borealis* plants and motto TANTUS AMOR FLORUM/Rule/CUM PRIVILEGIO/Rule/STOCKHOLMIÆ apud GODOFR KIESEWETTER,/AMSTELODAMI apud Z. CHATELAIN./1751.
8º; [Π]³, A-Y⁸, Z⁵; 41†ll; 362 pp.

---

* It is probably not accidental that this corresponds to the number of days in the year.
† This figure is from pages containing uniform matter, such as lists and indexes; the number of lines is often variable on pages where several different founts are used.

Contents: [Π]1ᵃ title, 1ᵇ blank, 2ᵃ dedication, 2ᵇ blank, 3ᵃᵇ preface; pp. 1–362 text; engravings opposite even–numbered pages 288–304, and on pp. 307 and 309.

Copies: Linnean Society (several, including one annotated by the author); Bodleian Library. Some copies include as a frontispiece an engraving, of a portrait of Linnaeus, by Johann Martin Bernigeroth.

Later editions were published at Vienna in 1755, 1763, and 1770. Revised editions were produced at Berlin in 1780 and at Halle an der Saale in 1809. In 1775 there appeared an English work based on the *Philosophia*, namely *Elements of Botany* by Hugh Rose. A complete French version, Charles Linné's *Philosophie botanique*, translated by François-Alexandre Quesné from the editions of 1770 and 1780, was published at Paris and Rouen in 1788. In the same year, a Spanish version of the *Fundamenta*, Carlos Linnéo's *Fundamentos Botánicos*, translated by Angel Gomez Ortega, was published (with the Latin text) at Madrid. In 1966 a facsimile of the first edition was published at Lehre in Germany, as Vol. 48 of *Historiae Naturalis Classica*, edited by J. Cramer and H.K. Swann, and issued by Wheldon and Wesley Ltd, Stechert Hafner Agency Inc., at Codicote, Hertfordshire, and New York.

Linnaeus was precise in his use of technical terms, with some very fine distinctions. For instance, a male plant is described as *mas*, whereas a male flower is *masculus*; a single plant bearing both male and female flowers is androgynous, but a bisexual flower is hermaphrodite. Yet some words that he regularly uses in a technical sense can also appear with a more general application; thus *pistillum* can be the pistil of a flower or the clapper of a bell; *calyx* the botanical calyx or a common cup; *hybernaculum* may refer to a winter bud or a greenhouse!

I have used the name Linnaeus rather than Linné, because it is the form that is familiar in English.* With other names, I have generally used the vernacular form, if ascertained, except where the Latin form is generally used in English (e.g. Dillenius rather than Dillen). In some cases, the Latin is a translation rather than an adaptation of the vernacular, (e.g. Tragus,* Bock), and here I have given both forms.

---

* As was usual in Scandinavia, the family had no hereditary surname (but used patronymics) until the time of the matriculation of Carl's father Nils Ingemarsson at the university, where a distinctive surname was required; (it was derived from a prominent lime tree on land belonging to the family). When Carl was ennobled in 1761, a vernacular version was needed, and this is the name by which he is usually known in Sweden and elsewhere.

In the English rendering of the title, I have put 'Science' rather than 'Philosophy', since in this context, the latter term might give a misleading impression in modern English. Again, *ars* may mean 'art' in some passages and 'technique' or 'technical skill' in others; *scientia* may be 'science' or 'knowledge'; *Ringens* may be 'gaping' or 'ringent' (without any distinction of meaning); *loculus*, rendered 'chamber', and *loculamentum*, 'compartment' or 'space', appear to be synonyms (see Sections 93 and 94). Linnaeus was the first to give the terms 'corolla' and 'petal' their now accepted botanical meanings.

For the spelling of the names of the genera, *The Plant Book* by D. J. Mabberley has been taken as the standard; so we have 'Liriodendron' (with Greek ending), where Linnaeus wrote 'Liriodendrum'; conversely 'Haematoxylum' for 'Haematoxylon'; and 'Hippophaë' rather than 'Hippophaës'.[†] But in some places (e.g. in the lists of genera at the end of Section 209, and in Section 241), it seemed better to retain the spelling used by the author to whom the name is attributed. In the English plurals of Latin words and names, I have put 'Campanulas' rather than 'Campanulae', and 'Lychnises' rather than 'Lychnides', but 'Ranunculi', not 'Ranunculuses'. Obvious misprints have been silently corrected. Some other errors noted in the errata have been recorded in the Notes.

The work contains a large number of Greek names of plants. In transcribing these, I have eliminated some obsolete forms of single letters and conjoined letters; they are printed here without accents or breathings.

Linnaeus gives many details concerning the derivations of plant names. Some of these are extremely fanciful and far-fetched; it must be remembered that, in his time, the scientific study of etymology was at a very early stage, and its results not widely available.

There are some unresolved problems. The meaning of 'kyber, lex.150' (Section 200) is not clear; (in the French version it is preceded by *avec*). In Section 69b, (nos. 12 and 13) the sense of $1\frac{1}{2}$ and $1\frac{3}{4}$ is not immediately obvious; (the French translator interprets no. 12 as meaning that the number of stamens is greater by half than the number of petals; and no. 13, that the number of stamens is greater than that of the petals, and five are longer). I have not been able to ascertain the significance of the crosses against some names in

---

[*] The Greek τραγος, goat, in Latin letters.
[†] Both forms occur in the text; the latter agrees with the original Greek ιπποφαες.

Section 6. No explanation is given of Linnaeus' use of the exclamation mark; it is said to mean that he had actually observed the thing described.

Linnaeus frequently uses round brackets, which are reproduced in the translation. I have used square brackets to indicate that words have been added for the sake of clarity; and angular brackets when the original Latin word is included after the translation, e.g. to show the connexion between a name and the derivation given, as in Section 234.

Some chapters contain summaries of his previous work on classification and taxonomy; and there is much new material concerning the fruit-body, the characters of plants, and the medicinal and other properties of various kinds of vegetable. He gives a list of the natural orders, and also develops his own system, according to the number of stamens and pistils. He deals with the effects of different climates, soils, and environments.

Ten engravings contain 167 figures of the shapes of leaves and other parts of the plant. At the end are 7 short memoranda concerning the botanist, his pupils and their botanical excursions, travelling, the layout of a garden, and the construction of a herbarium. Linnaeus attached great importance to careful observation and measurement (for the latter, he recommended the use of parts of the body, e.g. the foot, rather than instruments); but he did not confine his interest to the details of natural history, and advised the traveller to study all aspects of the countries visited, including art and architecture, and the manners and customs of the people.

Linnaeus provided three indexes of Terms, Genera, and Contents. These have not been reproduced here; but two modern indexes have been compiled by Claire Stenson: the first combining Contents and Terms; the second of the Genera.

My wife has given incalculable moral and material help throughout, as has my daughter, Isabel. Others have been generous in giving or lending reference books, and providing much valuable information and advice; I must mention especially my brother Tom Freer of Guiting Power, Gloucestershire, and Dr Lulu Stader, formerly of the Oxford University Press, Cologne, without whose encouragement the whole project might have been abandoned. Miss Janet Barber, a neighbour and friend, has given many knowledgable suggestions. The staff of the Oxford University Press, the Bodleian Library, the Linnean Society of London, and the Cheltenham public library have shown unfailing patience and care. The Linnean Society has made a grant towards the costs of publication, from the Appleyard Fund. Mrs Mandy Pullen and Ms Vicky Culling have dealt very efficiently with an untidy draft of a very difficult text. Thanks are also due to the late Dr Arthur Cain, Professor of Zoology at

Liverpool, who first suggested the *Philosophia* as the most appropriate of Linnaeus' works for translation and; to that most learned botanist, the late Professor William Stearn, who gave much encouragement and useful advice, and provided a photographic copy of the fourth (Vienna) edition; he had also intended to write the Introduction to this translation. Professor Thomas Elmqvist of Stockholm has kindly checked the manuscript for any errors and corrected some of the plant names in accordance with modern practice, Professor Paul Cox of the Tropical Botanical Garden, Hawaii, and editor-in-chief of *Plant Talk*, has provided a most valuable Introduction.

S. F.
Caudle Green
Gloucestershire
July 2001

For this paperback edition, Mr. Philip Oswald has provided some useful corrections, for which I am most grateful.

S. F.
Caudle Green
Gloucestershire
February 2005

❧ INTRODUCTION

Paul Alan Cox
National Tropical Botanical Garden
Hawaii

One of the great achievements of the last century has been the political and economic unification of Europe. In this increased climate of openness and cooperation, botanical research has flourished, and Europe as a whole has focused on plant conservation, ranging from strong leadership in international treaties such as the Convention on International Trade in Endangered Species (CITES), to the creation of successful non-governmental organizations like the International Union for the Conservation of Nature (IUCN), and even to the establishment of popular conservation journals such as *Plant Talk*. Given these trends, organismal and conservation biology in the twenty-first century may indeed belong to Europe, particularly if new efforts complement significant progress which has been made by the great natural history museums, botanical gardens, and universities which dot the European landscape.

As with all progress, though, there is invariably loss and one of the unheralded casualties of European unification might be the Swedish 100 kroner note. This piece of currency (Figure 1), currently worth roughly 13 US dollars, bears not the image of a military strategist, nor a visionary statesman, nor even a sports hero, but instead the portrait of a botanist—Carl Linnaeus, pictured in front of his beloved garden in Uppsala. No current design for the new euro comes close to capturing the scientific detail and intellectual achievement of Sweden's favourite son as portrayed on the 100 kroner note; indeed it might very well be possible to teach an entire botany course based on this single banknote.

Linnaeus, of course, is renowned throughout the scientific world for his invention of binomial nomenclature. Few personalities outside the realm of religion, save perhaps Freud, Darwin, or Marx, have had such a singular impact on the history of modern human thought. Indeed, the publication date of his *Species Plantarum* demarcates taxonomic history into the pre- and post-Linnaean periods, with names published prior to that date being regarded as invalid. His singular unification of biological nomenclature has remained, through two and half centuries of international conflict and two world wars, the universally accepted standard by which all plants, from Spitsbergen to Patagonia, are named and discussed by scientists. For this achievement alone Carl Linnaeus truly deserves his memorial on the currency of a nation known for international diplomacy and peace-making. But Linnaeus, despite his volumes of published scholarly works and international fame, is best remembered in his native Sweden for his impact as a teacher.

I had the great privilege to live for a year in that Mecca of the Botanical world, Uppsala, and to walk the same paths trodden centuries ago by Linnaeus. I was a frequent visitor to his wonderful botanic garden (which is pictured behind Linnaeus on the 100 kroner note), supped at his nearby summer home—Hammarby—and even strolled though the halls of the Stockholm palace where Linnaeus was honoured by the king as a Knight of the Polar Star. Yet to my mind, the place that best resonates with the spirit of this most famous of all Swedes is not his beloved garden, nor the rooms of his summer home, but instead the paths of the little state forest in the centre of Uppsala, where Linnaeus once led the assembled students and citizens of Uppsala on grand field trips, complete with trumpeters, wine, and gourmet food. Linnaeus saw in plants a grand adventure—'God creates, but Linnaeus names,' he was fond of saying. Carl Linnaeus raised botany to an elegance and a social acceptability both within Sweden and internationally that it has seldom since attained.

Those of us who are botanists and who read Linnaeus' scholarly works often forget that Carl Linnaeus was first and foremost a teacher. Indeed, much of the fame that he acquired was due, in no small part, to the dozens of students that flocked from throughout the globe to learn at his feet. His teaching about plants was so persuasive, and the devotion to the science his pedagogy engendered so complete, that the students he dispatched to distant climes to collect plants became known in Sweden not as scholars, but as 'apostles', with the Messianic antecedent of that term unstated but well understood. For those of us fortunate enough to have fallen under the benevolent influence of a charismatic teacher of biology, we can only imagine what a lecture given by Linnaeus must have been like. Imagine an elegant Uppsala lecture hall in the presence of a professor who is the undisputed world authority, and even the inventor of entire disciplines (for example, systematic botany, ethnobotany), and yet who was so engaging, direct, and entertaining in his classroom lectures that even the townspeople would queue up to hear him. Unfortunately, the direct magic of that most ephemeral of arts—teaching—evaporates soon after the lecture hall empties, and can often be inferred only from its impact on the listeners. Of one thing we can be certain: boring was not an adjective that described Linnaeus' presentations in the classroom or in the field. We understand Linnaeus as a scholar, but what accounts for his power as a teacher? As we sift through the history and writings and lecture notes, it appears that teaching ability was not something that Linnaeus acquired during his studies, but something he brought with him to Uppsala.

When Linnaeus arrived in that academic centre of Scandinavia on 5 September 1728, as an impoverished student from Lund, he carried with him a thirst for botanical knowledge that was unquenchable. He was born on 23 May 1707 in the charming village of Råshult, in the province of Småland, southern Sweden. His mother, Christina Brodersonia, soon found that this, the first child she bore to her husband, the village priest Nils Linnaeus, had a unique affinity for plants: he preferred plants to toys and would cry when flowers were removed from his crib. The origin of such a unique botanical eye is a discussion perhaps better left to child development experts, or even theologians, but I suspect that one answer might lie within the magic of Swedish spring—and if you have never experienced a Swedish spring, I would definitely add it to any biologist's life-list of key experiences. Swedish spring is almost a religious experience—out of what was darkness and gloom and cold is called into existence light, and if on cue, there is a sudden explosion of leaf, and flower, and fragrance. Linnaeus was born just as the flowers began to bloom; while Swedish spring offers grand panoramas, there is much beauty and wonder to be found within a child's grasp. I tested this hypothesis by taking my own infant daughter to the Råshult cottage where Linnaeus was born, with its sod roof and rustic red timbers (a note to foreigners—Henry Ford's innovation of monochromatic black model Ts may not have been original, as ox-blood red appears to be the only colour of house paint available in Sweden). There on the lawn, little Jane became completely entranced with the tiny crocuses peeking up through the blades of grass, and as she played I could easily imagine little Carl as a small boy frolicking around the floral table setting that his father planted in front of the house. On one thing, though, we need not speculate—the young Linnaeus, as soon as he discovered that each flower had a name, determined to learn them all, with the result that tried even the pastoral patience of his father.

Soon there was a promotion—Nils was appointed Rector of the parish of Stenbrohult, which is just a stone's throw from Råshult, and so in many ways Stenbrohult became for Linnaeus 'min ljuva natale'—'my sweet birthplace'. The fondness of Linnaeus' memory of the Stenbrohult was augmented by his father's ambitious gardening around the rectory, an activity strongly influenced by Nils' uncle Sven Tiliander, who was continually importing rare plants from Germany, and using Latin rather than provincial names to refer to his new acquisitions. As an amateur botanist, Nils learned the Latin and had established a herbarium of 50 species while he was a university student. At the new rectory, Nils Linnaeus determined to create a garden that would surpass any in the province. Mimicking his father, little Carl laid out his own toy garden, which continually expanded.

Young Linnaeus' botanical eye was not only developed in Stenbrohult, 'by his mother's milk with a never-extinguished love of flowers,' but was also deeply affected by his aesthetic experience of the Swedish landscape, particularly the beautiful scenery surrounding Lake Möckeln in front of his father's church. 'I doubt if there is a spot in the whole world,' Linnaeus wrote in his autobiography, 'set out in a more pleasant fashion, so that it is not surprising if I had cause to complain *Nescio qua natale solum dulcedine cunctos ducit et immemores non sinit esse sui* (All men are drawn to their native country by some attraction or other, and it does not allow them to forget it). Ovid, *Epistulae ex Ponto* (Letters from the Black Sea) (I 3 ll.35–6) Linnaeus was far from unique among his countrymen in this sentiment, and even today, the Swedish conception of heaven seems to be solitude on a lake in the Swedish woods.

The point of this digressive analysis of Linnaeus' childhood is quite simple: his love for plants far preceded his scientific study of them. He experienced plants in a deep aesthetic way—Linnaeus saw flowers as unique (I know of no other botanist who portrayed new species of flowers as naked women in his student notebook)—and this aesthetic was what truly animated his extraordinary career as a botanist. How else can we account for Olof Celsius, Dean of Theology at Uppsala, being so deeply moved by his first meeting with the young student that he brought him to reside in his own household?

> Linnaeus was sitting one day in the neglected Botanic Garden when an elderly clergyman came up to him and asked him what he was looking at, and whether he knew about plants and had studied botany … He finally asked him how many plants he had collected and pressed. When Linnaeus told him he had more than six hundred native flowers, he invited the young man to come back with him to his house. (Blunt, 1971)

After Celsius, Olof Rudbeck, and other Uppsala professors were amazed when they heard Linnaeus speak about plants. After they read his revolutionary thesis on plant sexuality, they appointed him to be a demonstrator in botany. Shocking as it was to appoint a young student over more senior candidates for the position, their choice could not have been more prescient: instead of the 30 to 70 students who attended botany lectures, 400 or more would crowd in to hear Linnaeus. There has never been a lecturer, before or since, who could whip up a crowd in Uppsala like Carl Linnaeus.

Paradoxically, the ever-popular Linnaeus had so many qualities that are anathema to Sweden's culture: he was boastful, he was flamboyant, he was loud and unorthodox, and all this in a consensus culture that anthropologists compare to Japan, where quiet competency, self-effacement, and contribution to the group is valued above all. But his brilliance burned so brightly that Linnaeus stood out wherever he went, and the Swedes, particularly students, flocked to see this self-proclaimed 'King of the Flowers' sometimes decked in Lappish costume, speaking with such intensity and passion about plants that he generated in eighteenth-century Uppsala the sort of celebrity wielded today only by rock musicians or movie stars. Not all Swedes were captured like moths by Linnaeus' flame, and a few, such as Nils Rosén, even took it upon themselves to extinguish it. However, the intellectual duel between Rosén and Linnaeus, in which Rosén had Linnaeus kicked off the Uppsala faculty, and for which Linnaeus nearly ran Rosén through with a sword, is a story for another place and time.

Instead, the key issue here is that this essential passion, his flamboyance, the extraordinary excitement of Linnaeus that encouraged an entire generation of students, is largely missing from his scholarly botanical works. Can any reader of *Species Plantarum* derive from that work a true image of Linnaeus' charismatic classroom personality? Of course we, as did all of the eighteenth-century botanists before us, can see the brilliance of the author of *Species Plantarum*, but Linnaeus' scientific prose is anything but flamboyant. His early botanical writing is cold, precise, and reads with about as much excitement as source code for the computer design of a toaster. Linnaeus even took great pains to hide the good stuff from his readers: he refused to allow the extraordinary diary of his journey to Lapland to be published, and just as Newton concealed his invention of calculus which enabled the writing of the *Principia*, Linnaeus failed to articulate the sexual system he used to categorize the plants in both *Flora Lapponica* (1737) and *Hortus Cliffortianus* (1737). Was this because Linnaeus was, after all, just another self-effacing, quiet, but competent Swede? Such a hypothesis is simply untenable—Linnaeus was always Linnaeus, and many of his very good ideas occurred to him while he was still a student. In these early scientific works, we must realize that Linnaeus, particularly before he gained international fame and notoriety, knew all too well the unwritten rules of eighteenth-century scholarly publication, and played the game well. That is precisely why *Philosophia Botanica* in an English translation by Stephen Freer is such a gift to the world. Essentially a compilation of Linnaeus' lectures on botany, *Philosophia Botanica* allows us to imagine what it must have been like to sit in that old Uppsala lecture hall, and to hear without the filter of culture or society Carl Linnaeus, the greatest student of plants ever to grace this earth, discuss with excitement and enthusiasm his view of plants.

This translation will be of great importance to scholars, because *Philosophia Botanica*, originally published two years before *Species Plantarum*, gives a clear insight into the evolution of Linnaeus' thought prior to writing his *magnum opus*, indisputably one of the most important biological books ever published. *Philosophia Botanica* should rekindle careful consideration of the general principles of botanical nomenclature pioneered by Carl Linnaeus, particularly at a time when support for those principles and for plant systematics in general seems to be waning. But *Philosophia Botanica* speaks to an audience far broader than historians of science or plant systematists. I believe that all people who are interested in plants should read *Philosophia Botanica*, for the same reason that all people who are interested in physics should read Richard Feinberg's *Five Easy Pieces*, for in both cases we have individuals of extraordinary gifts and historical importance summarizing their field for a lay audience.

Linnaeus begins *Philosophia Botanica* with a historical overview of the major figures in botany from ancient to his then present time. The position of Linnaeus as the central figure of botany emerges slowly in this discussion: we can only wonder if Linnaeus held up a copy of his *Hortus Cliffortianus* for his students when he told them that Ehret was an outstanding artist of plant form (Linnaeus lists both *Hortus Cliffortianus* (illustrated by Ehret) and *Flora Zeylanica* as 'characteristic works' in *Philosophia Botanica*). Primacy of plant form and the visual aspects of illustrative botany are stressed; but herbaria, Linnaeus warns us early on, are far more important than mere drawings: 'A herbarium is better than any picture, and necessary for every botanist.' Ever practical, Linnaeus tells his students how to prepare a herbarium sheet, right down to the type of glue to use and where on the sheet to write the identification. It is precisely this juxtaposition of the philosophical with the pedestrian that I think makes *Philosophia Botanica* such a good read today, even though the reader is jarred by continual notes of Linnaeus' achievements as a plant physiologist, botanical legislator, systematist, taxonomist, and so on. This may seem proof of affectation to readers whose own student days are distant, but if *Philosophia Botanica* is read as a series of lectures, rather than a scholarly tome, the effect is muted: introduction of a lecturer's own research into an introductory class remains today a very fine pedagogical technique. Those who seek to seize upon Linnaeus's laudatory mention of his own work as evidence of hubris may in fact be misinterpreting evidence of a skilled teacher. Furthermore, given the egalitarian nature of Swedish society today, it is easy to forget that eighteenth-century Sweden was once as class-stratified as any European kingdom. While there is no evidence that Linnaeus was ever ashamed of his humble beginnings in the little village of Råshult, his lectures in schoolboy Latin or parochial Swedish spoken with a Småland accent made him no match for the social mavens of Stockholm and their pampered children who appeared regularly in his lecture hall. A gentler reading is that Linnaeus' focus on his own work in the lecture hall bespeaks not conceit but instead enthusiasm. And Linnaeus was particularly certain that nature would reveal her secrets to one who sought them diligently.

Part of the way through *Philosophia Botanica*, Linnaeus reveals his own holy grail: a search for the elusive natural method of classification, a method which if successful would bring together related organisms in the true natural order of things:

> The fragments of the NATURAL METHOD are to be sought out studiously. This is the beginning and the end of what is needed in botany. Nature does not make leaps. All plants exhibit their contiguities on either side, like territories on a geographical map. (p. 40)

This biological cartography that Linnaeus proposed to map was not restricted to plants; Linnaeus had previously noted in his Lapland diary that humans and great apes share many similarities. We can only conjecture how Darwin's later understanding of common descent would have been handicapped had Linnaeus not launched his students on a search for the 'fragments of a natural method'. Yet, concerning evolution, Linnaeus was still a continent apart from Darwin, as he regarded species as immutable: 'We reckon the number of species as the number of different forms that were created in the beginning' (p. 113).

In seeking a natural classification of plants, Linnaeus thought that the floral organs were most conservative, and even today, in our age of molecular biology, classifications based on floral characters are seldom wildly wrong. In *Philosophia botanica* (p. 72) he presents a little quiz-diagnosis of a plant based solely on floral characters—and asks his students if they had diagnosed a *Linum*.

For Linnaeus, floral characters were paramount, but his posit of the most 'natural' number of calyx lobes, the most 'natural' placement of stamens, etc. might be confusing at first, particularly if we mistakenly assume in our reading that he is making an evolutionary argument. Let us not forget that in Linnaeus' day biology was still strongly Aristotelian, with the search for a Platonic *eidos* or natural form being the goal of every biologist. I think Linnaeus here though is less interested in positing an *eidos* than he is in introducing young botanists to a mnemonic device, having them learn a simple ground state, and then remember each plants species by the way they differed from this 'natural flower'. Such an interpretation is strengthened by pages (which quite frankly in the setting of a lecture hall must have represented a dazzling intellectual *tour de force*) in which Linnaeus illustrates each possible variant of shape or form or placement by a different genus of plants. Variations on a theme seemed to be the way Linnaeus remembered floral characters.

Linnaeus realized that a workable taxonomic system had to be simple, understandable, and universal in its scope. Thus, in *Philosophia botanica* we hear, as if from Sinai, the commands from him who first distilled order from botanical chaos:

> A plant is completely named, if it is provided with a generic name and a specific one. (p. 219)

> The generic name must be fixed unalterably before any specific name is devised. (p. 171)

> No sane person introduces primitive generic names. (p. 172)

> Generic names made from two entire and separate words are to be banished. (p. 172)

> Generic names, that do not have a root derived from Greek or Latin are to be rejected. (p. 175)

> Generic names that have been formed to perpetuate the memory of a botanist, who has done excellent service, should be religiously preserved. (p. 185)

> The shorter the name, the better. (p. 246)

> Names used by the ancients, that are $1\frac{1}{2}$ feet long, and which constitute descriptions instead of definitions, are to be abhorred. (p. 246)

Of all of *Philosophia Botanica*, perhaps the most exciting to an eighteenth-century audience would have been his analysis of plant sexuality. It is easy to forget today how stunning Linnaeus' imputation of sex to plants must have been then, particularly in his titillating anthropomorphic approach to the topic: 'The calyx is the bed-room, the corolla the curtain, the filaments the spermatic vessels, the anthers the testicles, the pollen the sperm, the stigma the vulva, the ovary the ovary, the pericarp the fertilized ovary, and the seed is the egg' (p. 105). Yet in his careful emphasis on plant reproductive biology, and particularly on the variety and lability of gender in plants, Linnaeus once again laid a careful foundation for later workers, including Darwin, who would discover in the analysis of plant-breeding systems much of the mysteries of evolution. The echoes of Linnaeus' sexual system can even be found today in the renaissance of understanding of the relationship between plant-breeding systems and speciation.

What else can we learn today from *Philosophia Botanica*? Is it merely a historical text, of interest only to those interested in the arcana of antique botany? I think not. As we enter a new century, those of us who care about the conservation of biodiversity are continually disheartened by the paucity of trained taxonomists: with all the amazing new machinery that fills the laboratories of modern botanists, we are stung by Linnaeus' reproach, 'If you do not know the names of things,' he tells us in *Philosophia Botanica*, 'the knowledge of them is lost.' Today, we might broaden his dictum: 'If you do not know the name of things, the things themselves may be lost.' At the time of writing this introduction, one out of eight plant species in the world, by IUCN estimates, are threatened with immediate extinction, yet most plant species pass from this world unnamed and unmourned. And despite the advent of high-speed computers, satellite reconnaissance, and DNA sequencing, the conservation crisis seems only to get worse.

I think that is why an English translation of *Philosophia Botanica* is so timely today, because the solution to the mass extinction of species may not lie in technology but in teaching. Those teachers who communicate to their students an enthusiasm for life and a reverence for it may eventually move our society beyond ever thicker *Red Book* lists of extinct species, into actually doing what is needed to save them in the first place. Here in *Philosophia Botanica* are the lectures of a master teacher who inspired scores of students to risk their lives in the search for plants. As the last Swedish 100 kroner notes are collected and destroyed in anticipation of the introduction of the euro, and this image of Carl Linnaeus begins to disappear, more than ever before we need teachers who love both their topic and their students. *Philosophia Botanica* is our best source for the teaching of the greatest professor of botany who has ever lived. Linnaeus' passion, enthusiasm, and achievement are recorded in its pages. For that reason alone it merits close study.

Paul Alan Cox
Poipu, Kauai

A final note: It was hoped that the great Linnaean scholar of our day, Professor William Stearn of the Natural History Museum, London, would have written this introduction, but due to his illness I was pressed into service. While I am only a poor substitute for Professor Stearn, I do share his love of Linnaeus and therefore wish to dedicate this Introduction to him.

P. A. C.

CAROLI LINNÆI

ARCHIATR. REG. MEDIC. ET BOTAN. PROFESS. UPSAL.
ACAD. IMPERIAL. MONSPEL. BEROL. TOLOS. UPSAL.
STOCKH. SOC. ET PARIS. CORRESP.

PHILOSOPHIA BOTANICA

IN QVA
EXPLICANTUR

FUNDAMENTA BOTANICA

CUM

DEFINITIONIBUS PARTIUM,
EXEMPLIS TERMINORUM,
OBSERVATIONIBUS RARIORUM,

ADJECTIS
FIGURIS ÆNEIS.

CUM PRIVILEGIO.

STOCKHOLMIÆ apud GODOFR KIESEWETTER,
AMSTELODAMI apud Z. CHATELAIN.
1751.

✾ TITLE PAGE

PHILOSOPHIA BOTANICA, in which the *foundations of botany* are explained, with *definitions of the parts*, *examples of the technical terms* and *observations of rarities*; with the addition of *copper-plate illustrations*;

by

CARL LINNAEUS, Chief Physician to the King, Professor of Medicine and Botany at Uppsala, Fellow of the Imperial Academy,[1] and of the Academies of Montpellier, Berlin, Toulouse, Uppsala, and Stockholm, and corresponding member of the Academy of Paris.[2]

COPYRIGHT

Gottfried Kiesewetter, Stockholm;
Z. Chatelain, Amsterdam.
1751.

S:æ R:æ M:tis
SUMMÆ FIDEI VIRO
ILLUSTRISSIMO EXCELLENTISSIMOque
L. BARONI

Dn. ANDR.
von HÖPKEN

REGIS REGNIque SVECIÆ
SENATORI
ORDINIS SERAPHINI
EQUITI

Opusculum hoc, Philosophiam Botanices comple-
ctens, sacrum voluit
ILLUSTRISSIMI EXCELLENTISSIMIque
NOMINIS

Upsal. d. 16. Septembr. 1750.

cultor devotissimus
CAR. LINNÆUS.

To the most illustrious and excellent right trusty subject of His Royal Majesty,

FRIHERRE Sir ANDERS von HÖPKEN,
Counsellor to the King and Realm of Sweden, Knight of the Order of the
Seraphim,

This little work, containing the Science of Botany, has been dedicated by one
who devotedly reveres that most illustrious and excellent name,

CARL LINNAEUS,
Uppsala, 16 September 1750.

LECTORI BOTANICO.

FUNDAMENTIS BOTANICIS Theoriam atque Inſtitutiones Rei Herbariæ ſub paucis Aphorismis olim comprehendi, quorum Explicationem per Exempla, Obſervationes & Demonſtrationes, diſtinctis riteque definitis plantarum Partibus & Terminorum vocibus, Philoſophiam Botanicam dixi, cum in his conſiſtant Præcepta Artis.

PHILOSOPHIÆ ejusmodi BOTANICÆ varias Partes dudum emiſimus, utpote in caput I. *Bibliothecam Botanicam*, II. *Claſſes Plantarum*, III. *Sponſalia Plantarum*, VII. VIII. IX. X. *Criticam Botanicam*, XII *Vires Plantarum ;* Reliquas Sectiones Fundamentorum, ſc. Cap. IV. VI. XI. conjunctim cum prioribus in unum opus compingere & auctas novis exemplis, obſervationibus, demonſtrationibus, ſub Philoſophiæ Botanicæ titulo edere, diu animo volvi, in quem etiam finem bene multa collegi; interea territum copia dicendorum ſatietas, imo tædium, operoſæ ſcriptionis me cepit, adeo ut hæc in tempora commodiora ſeponere viſum fuerat, quæ cum annis, auctis indies curis publicis & privatis, muneris negotiis & peregrinationibus Hiſtoriæ Naturalis cauſa inſtitutis, ita ſeſe ſubtrahebant, ut de ſucceſſu operis ipſe dubitare cœperim.

Interea urgente Bibliopola, poſt diſtracta priora exemplaria, novam Fundamentorum Botanicorum editionem, a me quoque enixe petiere diſcipuli, ut *Partes plantarum, artisque Terminos*, a me uſitatos, rite definitos adderem, methodo qua eos in prælectionibus tradere ſolitus ſum; huic eorum petitioni acceſſere amicorum, de Re herbaria optime meritorum, adhortationes, ut Terminos artis explicarem, Partesque plantarum definirem; ut amborum deſideriis adſurgerem, incepi collectanea mea in compendium redigere & typis mandare : at dira Anthritis, vix incepto

TO THE BOTANICAL READER

Long ago in *Fundamenta Botanica* I comprised, under a few aphorisms, the theory and principles of botany; and I called the explanation of them by means of examples, observations, and demonstrations, distinguishing and correctly defining parts of plants and technical terms, the Science of Botany;[1] since the rules of the art consist in these things.

In time I published various parts of this kind of Science of Botany; namely *Bibliotheca Botanica* (Chapter I), *Classes Plantarum* (Chapter II), *Sponsalia Plantarum* (Chapter III), *Critica Botanica* (Chapters VII–X), and *Vires Plantarum* (Chapter XII). For some time I considered combining together the remaining sections of the *Fundamenta* (to wit Chapters IV, VI, and XI) with those mentioned above, into a single work, and publishing it under the title of *Philosophia Botanica*, with the addition of fresh examples, observations, and demonstrations; and for this purpose I made some extensive collections. Meanwhile I was put off by the quantity of things that would have to be said; and satiety, or rather weariness with the labour of writing, overtook me to such an extent that I had decided to lay these matters aside for more convenient opportunities; and as the years passed, with the continual increase in my public and private responsibilities, and in the business and travelling, connected with my office, that I undertook in the cause of natural history, these opportunities disappeared to such an extent that I began to have doubts about the achievement of the work.

Meanwhile my publisher was pressing for a new edition of *Fundamenta Botanica*, after the distribution of the copies of the first issue; and my pupils also vehemently demanded of me that I should add the *parts of plants* and *technical terms* as used by me; and that they should be accurately defined, according to the method by which I habitually propounded them in my lectures. To this request of theirs were added the exhortations of my friends that were eminent in their services to botany, urging me to explain the technical terms and define the parts of plants. In order to meet the wishes of both parties, I began to reduce my collections into a digest and commit them to print. But when I had only just begun the work, a terrible attack of gout incapacitated my mind and spirit, as well as my bodily strength, so much so that it was almost suffocated before it could flower.[2] But now that I have recovered my strength after a fashion, I here present the digest of the Science of Botany; it is small in extent, inasmuch as it comprises the preliminary outlines and rudiments of botany, published only for the sake of my pupils; but, if fate and leisure allow, it will appear in a much larger form at some future time.

At the end of the work I have added some notes about instruction in the subject, to avoid leaving the alternate pages blank.

As I am now occupied in collecting *species of plants*,[3] I vehemently request and implore the most eminent botanists throughout Europe to send me plants complete with flowers, if they have in duplicate any of the comparatively rare plants that I have not described; this would make it possible for me to refer them to their genera, with adequate definitions; in exchange on my part I will express in the work my gratitude to those who have sent them.

CARL LINNAEUS,
Uppsala 16 Sept. 1750

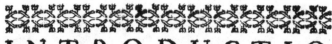

INTRODUCTIO.

1. OMNIA, quæ in Tellure occurrunt, *Elementorum* & *Naturalium* nomine veniunt.

Syſt. nat. 6. Obſ. in regna tria §. 6. 7.
Elementa ſimplicia ſunt, *Naturalia* compoſita arte divina.
Phyſica tradit Elementorum proprietates.
Scientia Naturalis vero Naturalium.

2. NATURALIA (1) dividuntur in Regna Naturæ tria: *Lapideum*, *Vegetabile*, *Animale*.

Syſt. nat. 6. p. 211. §. 14. 8. 9. Neceſſitas cognitionis.
Faun. ſuecic. præfat. 4. Actio primi hominis.
Act. ſtockh. 1740. p. 411. Uſus.

3. LAPIDES (2) creſcunt. VEGETABILIA (2) creſcunt & vivunt (133). ANIMALIA (2) creſcunt, vivunt, & ſentiunt.

Syſt. nat. 6. p. 211. §. 15. idem.
Syſt. nat. 6. p. 219. §. 2. Lapides creſcere.
Sponſal. plant. §. 1-14. Vegetabilia vivere.
Jung. iſagog. c. 1. Planta eſt corpus vivens non ſentiens, ſ. certo loco aut certæ ſedi affixum, unde nutriri, augeri, denique ſe propagare poteſt.
Boerh. hiſt. 3. Planta eſt corpus organicum, alteri cuidam corpori cohærens per aliquam partem ſui, per quam Nutrimenti & Incrementi & Vitæ materiam capit & trahit.
Ludwig. veget. 3. Corpora naturalia eadem ſemper Forma & Loco-motivitate prædita appellantur *Animalia*; eadem ſemper forma, ſed loco-motivitate deſtituta *Vegetabilia*; & quæ diverſam formam obtinent *Mineralia* dicuntur.
Obſ. Petrificata & *Cryſtalli* figura conveniunt omnino in eadem ſpecie. Locomotivitas in *Balano*, *Lernea*; uti in *Mimoſa*.

4. BOTANICE eſt Scientia Naturalis, quæ *Vegetabilium* (3) cognitionem tradit.

Boerh. hiſt. 16. Botanica eſt Scientiæ naturalis pars, cujus ope feliciſſime & minimo negotio plantæ cognoſcuntur & in memoria retinentur.
Ludwig. aphor. 1. Botanica eſt ſcientia vegetabilium, ſ. cognitio corum, quæ per plantas & in plantis fiunt.

A   I. BIBLIO-

INTRODUCTION

1. All things that are found on the earth go by the names of *elements* or *natural [bodies]*.[1]

> Systema naturae 6.[2] Observations on the three kingdoms, sectt. 6 and 7.
> 'The *elements* are simple; the *natural [bodies]* are constructed by divine skill.'
> '*Medicine* treats of the properties of the elements. But *natural science* treats of those of natural [bodies].'

2. The NATURAL [bodies] (1) are divided into the three kingdoms of nature: *mineral, vegetable, and animal.*

> Systema naturae 6. p. 211 sectt.14, 8 and 9.[3] The need for investigation.
> [My] *Fauna Suecica*, preface p. 4. The activity of the first man.
> *Acta Stockholmensia,*[4] 1740, p. 411. Uses.

3. MINERALS (2) have growth. VEGETABLES (2) have growth and life (133). ANIMALS (2) have growth, life, and feeling.

> Systema naturae 6. p. 211 sect. 15. The same
> Systema naturae 6, p. 219 sect. 2. 'Minerals have growth.'
> [My] *Sponsalia plantarum* sectt. 1–14. 'Vegetables have life.'
> *Jung, Isagoge Ch.* 1. 'A plant is a living body without feeling, fixed in a particular place or abode from which it can receive nourishment and growth and eventually propagate itself.'
> *Boerhaave, Historia* 3. 'A plant is an organic body, adhering to some other body by some part of itself, through which it draws the matter for nourishment, growth, and life.'
> *Ludwig, Vegetabile* 3. 'Natural bodies,[5] that are always provided with the same shape and with locomotion, are called *animals*; those that keep the same shape but are without locomotion, are called vegetables; and those that are subject to variations in shape, *minerals*.'

Observations. *Fossils and crystals* always conform in shape, within the same species. There is locomotion in *Balanus* and *Lernea*; as there is in *Mimosa*.

4. BOTANY is a natural science, that treats of the investigation of *vegetables* (3).

> *Boerhaave, Historia* 16. 'Botany is the part of natural science, by means of which plants are investigated and remembered most agreeably and with the least trouble.'
>
> *Ludwig, Aphorismi* 1. 'Botany is the science of vegetables, to wit, the investigation of the things that are done by plants and in plants.'

I. BIBLIOTHECA.

5. BIBLIOTHECA Botanica continet *Libros de Vegetabilibus* scriptos.

Docet Detecta, Fata, Progressus, Loca, Methodum.

Sciat Botanicus quos evolvat auctores de planta quæstionis.

Europæa. *C. Bauhinum* (cum systematico) *Rajus 3, Vaillant.*

Alpina. *Rajus, Linnæus, Hallerus, Gmelinus.*

Capensia. *Hermannus, Burmannus, Commelinus, Linnæus, Rajus, Plukenetius, Petiverus.*

Indica. *Hermannus, Rheede, Plukenetius, Rajus, Burmannus, Linnæus.*

Americ. septentr. *Gronovius, Catesbæus, Rajus, Morisonus, Plukenetius, Cornutus.*

· · · · austral. *Plumierus, Sloaneus, Rajus, Plukenetius, Hernandez, Marcgravius, Piso, Feuillæus.*

Bibliothecæ. *Seguierus, Linnæus, Bumaldus, Scheuchzerus.*

Historiæ. *Tournefortius, Hotto, Royenus.*

6. PHYTOLOGI vocantur Auctores, opere aliquo (5) de vegetabilibus clari, sive *Botanici*, sive *Botanophili* sint.

Fund. Bot. §. 7. cum 43. distinctio.

Botanici primarii secundum Tempus:

Theophrastus	XVII. SECULO.		XVIII. SECULO.	
Plinius	*Robinus*	1601	Sherard	1701
Dioscorides	*Spigelius*	1608	Rudbeck f.	- - -
	Boot	1609	Meriana	1705
	Renealme	1611	Jussiæus A.	1709
	Swertius	1612	Boerhaave	1710
	Bry	- - -	*Petitus*	- - -
	Beslerus	1613	*Heucherus*	1711
	Jungermannus	1615	Kæmpferus	1712
XV SECULO.	*Brossæus*	1628	*Helving*	- - -
Gaza	Parkinson	1629	*Tita*	1713
Barbarus	*Donatus*	1631	Feuillée	1714
	Laurembergius	1632	Knaut f.	1716
	Ferrarius	1633	*Bradlæus*	- - -
	Cornutus	1635	Isnard	- - -
	Veslingius	1636	VAILLANT	1717
	Stapelius	1644	*Blair*	1718
				XVI.

❧ I. THE LIBRARY

5. The botanical LIBRARY contains *books about vegetables.*

    It gives information about discoveries, events, proceedings, locations, and method. The botanist must know which authors he should consult about the plant that is the subject of his enquiry.

    European [vegetables]. *C. Bauhin* (with systematic [arrangement]) *Ray 3, Vaillant.*

    Alpine. *Ray, Linnaeus, Haller, Gmelin.*

    South African. *Hermann, Burman, Commelin, Linnaeus, Ray, Plukenet, Petiver.*

    Indian. *Hermann, Rheede, Plukenet, Ray, Burman, Linnaeus.*

    North American. *Gronovius, Catesby, Ray, Morison, Plukenet, Cornut.*

    South American. *Plumier, Sloane, Ray, Plukenet, Hernandez, Marcgraaf, Piso, Feuillée.*

    Libraries. *Seguier, Linnaeus, Bumaldus,*[1] *Scheuchzer.*

    Histories. *Tournefort, Hotto, Royen.*

6. Authors who are famous for any work (5) on vegetables, whether they are *botanists* or *amateurs of botany* are called PHYTOLOGISTS.

    For the distinction, compare [my] *Fundamenta botanica* section 7 with section 43.
    The principal botanists, chronologically:

Theophrastus	**16TH CENTURY**		Matthioli	1548
Pliny	Brunfels	1530	*Lonitzer*	1550
Dioscorides	Tragus[2]	1532	Dodoens	1552
	Cordus	1535	*Belon*	1553
	Ruel	1536	*Guilandini*[3]	1557
	Dorstenius	1540	*Anguillara*	1561
15TH CENTURY	GESNER	1541	*Calceolari*	1566
Gaza	Fuchs	1542	*Pena*	1570
Barbaro	*Brassavola*	1545	Lobel	"

continued

Garcias	1574	Turner	1651	**18TH CENTURY**	
Monardes	"	Pancovius	1654	Sherard	1701
Clusius[4]	1576	Loesel	"	Rudbeck the younger	"
Carrichter	"	Jung	1657	Meriana	1705
Costeo	1578	Ambrosini	"	Jussieu, A.	1709
Acosta	"	Joncquet	1658	Boerhaave	1710
CESALPINO	1583	Rudbeck	1659	Petit	"
Durante	1584	Ray	1660	Heucher	1711
Dalechamp	1587	Hoffmann, M.	1662	Kaempfér	1712
Camerarius[5]	"	Schaeffer	"	Helving	"
Tabernæmontanus[6]	1588	Elsholtz	1663	Tita	1713
Thalius	"	Vallot	1665	Feuillée	1714
Alpini P.	1591	Chabrée	1666	Knaut the younger	1716
Bauhin, J.	"	Merret	1667	Bradley	"
Cortusi	"	Boccone	1668	Isnard	"
Colonna	1592	Aldrovandi	"	VAILLANT	1717
BAUHIN, C.	1593	MORISON	1669	Blair	1718
Pona	1595	Munting	1672	Pontedera	1718
Gerard	1597	Barrelier	+ 1673	Ruppius	"
Imperato	1599	Tillands	"	Dillenius	1719
Swenckfeld	1600	Sterbeck	1675	Monti	"
		Zanoni	"	Buxbaum	1721
		Amman	"	Tillius	1723
17TH CENTURY		Dodart	1676	Jussieu, B.	1725
Robin	1601	Breynius	1678	Martyn	1726
Van de Spiegel	1608	Rheede	"	Micheli	1729
Boot	1609	Mentzel	1682	Catesby	1731
Renéaulme	1611	Commelin	1683	Geoffroy	"
Swertius	1612	Trionfetti	1685	Celsius	1735
Bry	"	Magnol	1686	Walther	"
Besser	1613	Hermann	1687	Zannichelli	"
Jungermann	1615	Rivinus[8]	1690	Linnæus	"
De la Brosse	1628	Plukenet	1691	Haller	"
Parkinson	1629	Petiver	1692	Miller	1736
Donatus	1631	Cupani	"	Burman	1737
Laurenberg	1632	Plumier	1693	Ludwig	"
Ferrari	1633	TOURNEFORT	1694	Moehring	"
Cornut	1635	Bromel	"	Weinman	"
Vesling	1636	Zwinger	1696	Blackwell	1739
Stapel[7]	1644	Sloane	"	Amman the younger	"
Hernandez	1647	Commelin	1697	Gronovius	"
Marcgraaf	1648	Bobart	1699	Royen	1740
Piso	"	Volkamer	1700	Gleditsch	"

continued

Seguier	"	Wachendoff	"	French	1699
Barère	1741	Lecheus	1748	of Uppsala	1700
Gerberus	1743	Kalm	1750	Russian	1728
Gesner	†1744	Hasselqvist	1750	of Nuremberg	1731
Stellerus	†1746	SOCIETIES		of Stockholm	1739
Gmelin	1747	German	1670		
Guettard	"	English	1682		

7.   The (true) BOTANISTS (6) have a real basic understanding of botany, and should know how to name all vegetables with intelligible names; they are either *collectors* (8) or *methodizers* (18).

> [My]*Fundamenta botanica* 151, 164–7, and 152.
> [My]*Systema naturae* 6. p. 211 sect. 13, 12, 11; and p. 215 p. 2.
> [My]*Genera plantarum*, preface sect. 1–3.
> [My]*Classes plantarum*, preface sect. 1.

8.   The COLLECTORS (7) have been concerned primarily with the number of species of vegetables: they include the fathers (9), commentators (10), illustrators (11), describers (12), monographers (13), the meticulous (14), Adonises (15), compilers of Floras (16), and travellers (17).

9.   The FATHERS established the first rudiments of botany.

> Events: The Greeks learnt from the Egyptians; the latter from the Chaldaeans.
>   The *Romans* after the defeat of Pompey at Actium.[9]
>   The *Goths* sack Rome, 4th Cent.; the *Lombards*, 5th Cent.
>   The *Arabs* invade Egypt, 6th Cent.; and Spain, 7th Cent.
>   Calipha, King of *Morocco*, 11th Cent.

> The GREEKS flourished before the birth of Christ.
>   *Hippocrates* the father of medicine, 5th Cent. BC.
>   *Aristotle* chief of the peripatetics, 4th Cent.
>   *Theophrastus* the father of botany, 3rd Cent.
>   *Xenophon* of Athens, 2nd Cent. [*sic*]

> The ROMANS flourished in the earliest centuries of the Christian era.
>   *Cato* under Julius Caesar, 149 BC.
>   *Varro* under Augustus, 62 BC.
>   *Virgil* under Augustus, 70 BC.
>   *Columella* under Claudius.

Pliny from Tiberius to Titus.

Dioscorides under Nero.

Rufus under Trajan.

Palladius under Antoninus Pius.

ASIATICS:

Galen of Pergamum lived at Rome, 133.

Oribasius of Pergamum, Julian's accountant.

Aetius of Amida, under Constantine and Theodosius.

(*Paulus*) *Aegineta*, under Constantine Pogonatus.

ARABS:

Mesueh of Nishapur, end of the eighth century.	
Serapio, son of John, physician.	
Rhazes, a native of Rayy in Persia,	*c.*920.
Avicenna, from Bukhara in Persia, physician,	*c.*1030.
Averroës, from Cordoba in Spain,	*c.*11th Cent.
Avenzoar,	*c.*1150
Abenguefit,[10] physician, Arab,	12th Cent.

OBSCURE authors of the 12th and immediately following centuries, as *Myrepsus* Praepositus, *Quiricius* Tertonensis, Abbess *Hildegard*, *Sylvaticus*, *Dondi*, *Suardus*, *Villa Nova*, *Cuba*, *Platearius*, and others.

The SKETCHES 325, 326, and 332 ignored them all.

In Greek:	*Hippocrates,*	*Aristotle,*	THEOPHRASTUS
	Bassus,	*Nicander,*	*Xenophon,*
	Apuleius,	DIOSCORIDES,	*Rufus,*
	Galen,	*Oribasius,*	*Aëtius,*
	Aegineta,	*Myrepsus.*	
In Latin:	*Cato,*	*Varro,*	*Virgil,*
	Columella,	PLINY,	*Palladius*
In Arabic:	*Mesueh,*	*Serapio,*	*Rhazes,*
	Avicenna,	*Averrhoës,*	*Abenguefit,*[10]
	Abenzoar.		
In vernacular:	*Cuba,*	*Hildegard,*	*Myrepsus,*
	Sylvaticus,	*Quiricius,*	*Bosco,*
	Suardus,	*Platearius,*	*Dondi,*
Agriculture:	*Bassus,*	*Xenophon,*	*Palladius,*
	Cato,	*Varro,*	*Columella.*
Medicine:	*Hippocrates,*	*Galen.*	
Natural [bodies]	*Aristotle,*	*Pliny.*	
Botany:	*Theophrastus,*	*Dioscorides,*	*Pliny.*

10. The COMMENTATORS (8) have clarified the writings of the fathers (9).

> *Events* of the mid-15th century: Turkish [power] separated the Greek or eastern
> empire from the western; refugees brought literature into Italy. Discoveries:
> *printing*, 1440; *gunpowder; America*, 1492.
> ARISTOTLE was translated by *Constantin* and others.
> Commentaries were supplied by *Scaliger* and others.
> THEOPHRASTUS was translated by *Gaza* and others.
> Commentaries by *Stapel, Scaliger*, and others.
> PLINY was edited by *Dalechamps, Hardouin, Gronovius*, and others.
> Commentaries by *Barbaro, Guilandini, Saumaise*, and others.
> DIOSCORIDES was translated by *Cornario, Sarracenus*, and others:
> Commentaries by *Barbaro, Fuchs, Matthioli, Cordus, Gesner, Ruel*, and others.
> CATO: Commentaries by *Meurs, Ursinus*, and others.
> AVICENNA was translated by *Alpago, Costeo, Plempius*, and others; and by *Amathi*
> into Hebrew.
> Commentary by *Lonitzer*.
> MESUEH: commentaries by *Campeggio, Monardes*, and others.

11. The ILLUSTRATORS (8) have represented the figures of vegetables in
pictures.

> This craft was uncommon among the ancients, like representation in a mirror.
> Requirements: a *botanist*, a *draughtsman*, and an *engraver*.
> These are outstanding; *Dillenius, Colonna, Ehret*.
> All parts should be recorded in their *natural* position and *size*, including the most
> minute parts of the *fruit-body*.
> They are either
> WOODCUTS, in which *Gesner* and *Rudbeck* excelled.
> COPPER-PLATES (Dondi 1536), *Ferrari, Dodart, Breynius, Commelin, Loesel,*
> *Rheede, Hermann, Rivinus.*
> TIN-PLATES, *Dillenius.*
> BASIC without shadow: as *Brunfels, Fuchs, Clusius, Plumier.*
> TINTED with lively colours: *Martyn, Blackwell, Weinmann.*
> ORIGINALS, from actual leaves instead of print. *Hessel* in America 1707,
> *Kniphof* in Germany 1733. ['Nature printing']
> VALUABLE: *Rheede, Sloane, Dillenius.*
> POOR: *Brauner, Miller, Brunswig, Cuba, Lonitzer, Nieuwlandt, Palmberg, and*
> *Hernandez.*
> The COMMONEST: *Matthioli, Camerarius, Lobel, Dodoens, Fuchs, Clusius, Besler,*
> *Sweerts, Robert, Morison, Rivinus, Vaillant.*
> For EXOTICS: *Morison, Plukenet, Petiver, Dodart, Martyn, Alpini, Barrelier,*
> *Boccone, Catesby, Dillenius, Hermann.*

A HERBARIUM[11] is better than any picture, and necessary for every botanist.

1. *Plants* should be gathered when they are not wet.
2. No *parts* should be removed.
3. They should be slightly *spread out*,
4. but not *bent*.
5. The *fruit-body* should be present.
6. They should be *dried* between dry pieces of paper,
7. *as quickly as possible*, short of using a hot iron,
8. in a *press*, with slight pressure.
9. They should be *glued* with fish-glue.
10. They should always be kept on *folio* [sheets].
11. Only *one* to a page.
12. The sheet should not be *bound*.
13. The *genus* should be written above.
14. The *species* and explanation on the back.
15. Those of *the same genus* should be placed within a band of linden bark.
16. They should be arranged *methodically*.

12.   The DESCRIBERS (8) have produced sketches (325) of vegetables.

OBSOLETE:	*Brunfels, Tragus, Dorstenius, Lonitzer, Ruel, Durante, Carrichter, Thurneiser, Turner, Salomon.*
COMMONLY USED:	*Matthioli, Cordus (Val.), Fuchs, Clusius, Dodoens, Lobel, Tabernaemontanus, Gerard, Parkinson, Dalechamps.*
UNIVERSAL:	*J. Bauhin* with *Cherler, Morison, Ray.*
CHARACTERISTIC:	*Hortus Cliffortianus, Haller,* [my] *Flora Zeylanica.*
TO SCALE:	*Tournefort, Feuillée.*
SPECIALISTS:	in Mosses: *Dillenius.*
	Grasses: *Scheuchzer, C. Bauhin.*
	American ferns: *Plumier.*
	Medicinal: *Pomet, Valentini, Godofredus.*
EXOTIC:	in the Indies: *Rheede, Rumpf.*
	American: *Plumier, Sloane, Hernandez.*

13.   The MONOGRAPHERS (8) have described a *single* vegetable at length in a separate work.

So that everything in a particular case is determined with greater accuracy.
Instruction in the CURIOSITIES of NATURE is praiseworthy.
The SELECT list of these is short.

Dillenius' *Mesembryanthema*,   Boerhaave's *Proteae*,
Kaempfer's *Thea*.

The *succulent* plants are worthy of distinction; so are the largest genera, e.g. *Euphorbia*. The chief of this kind are

Haller's	*Allium*	Our	Musa.
	Alpine Veronicas		Betula nana.
	Pediculares		Ficus.
Breynius	*Ginseng*		Passiflora.
Lafitau	*Pana*		Anandria.
Bradley	*Aloe*		Acrosticum.
Hellwing	*Pulsatilla*		Senega.
			Lignum colubrinum.
			Splachnum.

14. The METICULOUS (8) have drawn attention to the rarer vegetables.

The authors to be consulted about the rarer plants are

EUROPEAN:	Gmelin, *Siberian.*	Amman, *Russian.*
	Linnaeus, *Lapland.*	Haller, *Swiss.*
	Buxbaum, *Oriental.*	Tournefort, *Oriental.*
	Loesel, *Prussian.*	Mentzel's *Pugillus.*
	Vaillant, *Parisian.*	Magnol, *Montpellier.*
	Ray, *English.*	Tilli, *Pisan.*
	Colonna's *Ecphrasis.*	Barrelier's *Icones.*
	Boccone's *Museum.*	Alpini's *Rariora.*
	Aldini, Trionfetti	

INDIAN:	Dillenius, *Hortus Elthamensis.*[12]	Martyn's *Centuriae.*
	Hermann's *Hortus Lugd. Bat.*[13]	Commelin's *Hortus Amstelodamensis.*[14]
	Breynius, *Centuriae* and *Prodromus.*	Hermann's *Paradisus.*
	Dodart's *Historia Plantarum.*	Plukenet's *Phytographia.*

MUSEUMS that are made up from the threefold Kingdom of nature:

	Seba's, *Thesaurus.*	Catesby, *Carolina.*
	Petiver's *Gazophylacium.*	Anon., *St. Petersburg.*

15. The ADONISES[15] (8) present the *cultivated* vegetables of a particular garden.

The UNIVERSITY GARDENS in Europe are:

Padua	1540.	*Paris*	1636.	*Uppsala*	1657.
Pisa	1547.	*Oxford*	1683.	*Berlin*	
Bologna	1547.	*Leiden*	1677.	*Leipzig*	1580.
Montpellier	1598.	*Amsterdam*	1686.	*Göttingen.*	
		Utrecht	1638.	*Wittenberg.*	

Eminent among the PUBLIC gardens are:

Magnol's Montpellier.

Hermann's Leiden.

Boerhaave's Leiden.

Royen's Leiden.

Tilli's Pisa.

Volkamer's Nuremberg.

Haller's Göttingen.

Linnaeus' Uppsala.

The chief PRIVATE gardens are

Linnaeus' Hortus Cliffortianus.

Morison's Hortus Blesensis [of Blois].

Thraus' Catalogus Carolsruhensis
[of plants at Karlsruhe].

Moehring's own garden.

Walther's Designatio plantarum
[of plants at Idstein].

16. The compilers of FLORAS (8) list the vegetables that grow *naturally* in any particular place.

The list should be systematic, so that notice is taken even of those that are absent; those that are present should be recorded with the location, the [quality of] the soil, the time, and the vernacular names.

The principal Floras are

Gmelin, Siberian.

Haller, Swiss.

Linnaeus, Swedish
Lapland.

Ray, English.

Vaillant, Paris.

Dalibard, Paris.

Guettard, Étampes.

Magnol, Montpellier.

Dillenius, Giessen.

Gottesched, Prussian.

17. The TRAVELLERS (8) have gone to remote regions to investigate plants.

The OUTSTANDING [examples] are found in botanical [works].

Scheuchzer's Itinera Alpina [Alpine journeys].

Calceolari and *Pona*, Monte Baldo.

Ray's Travels and voyages.

Tournefort's Voyage du Levant.

Shaw's Calendarium Plantarum Africae.

Alpini, Egypt

Belon, Rauwolf,

Kaempfer's Amoenitates[16]

Marcgraaf and *Piso*, Brazil

Feuillée, Peru

Hernandez, Mexico
Cornut, Canada
The more SELECTIVE botanical [works] are:
Indian: *Rheede*, Malabar.
            Burman's Thesaurus Zeylanicus [of Ceylon].
            Linnaeus' Flora Zeylanica.
            Rumpf, Amboina.
            Kaempfer, Japan.
American: *Sloane*, Jamaica.
            Plumier, American.
            Gronovius' Flora Virginica [of Virginia].

18. The METHODIZERS (7) have worked primarily on the arrangement (VI) of vegetables and the nomenclature that is derived from that; and they are *philosophers* (19), *systematists* (24), *or nomenclators* (38).

19. The PHILOSOPHERS (18), by demonstration from rational principles, have reduced botanical knowledge to the form of a science: for instance, *rhetoricians* (20), *controversialists* (21), and *legislators* (23).

> These botanists are to be called theoreticians.
> The rules and regulations in botany are due to them.
> Empirical investigation was the botany of the ancients.

20. The RHETORICIANS (19) have expounded all things that are learned ornaments of science.

> BRIEF ORATIONS.
>   *Hellwig's* Botanices nobilitas.
>   *Trionfetti's* Prolusio.
>   *Commelin's* Oratio.
> NATURAL HISTORY.
>   *Biberg's* Oeconomia naturae.
>   *Söderberg's* Curiositas naturalis.
> EMBLEMS.
>   *Camerarius'* Emblemata.
>   *Mylius'* Hortus Philosophicus.

21. CONTROVERSIALISTS (19) have disputed in published writings.

WARS ABOUT SYSTEMS.

TOURNEFORT'S *Elementa* ..

.... *Colet's* Litterae criticae.

Chomel's Responsum.

* *

RAY'S *Sylloge.*

.... *Rivinus'* Epistola.

Ray's Epistola.
Ray's Variae methodi.

.... *Tournefort's* Optima methodus.

Dillenius' Judicium.

.... *Rivinus'* Responsio.

Dillenius' Examen.

* *

LINNAEUS' *Methodus.*

.... *Siegesbeck's* Epicrisis.

Browallius' Examen.
Gleditsch's Consideratio.

.... *Siegesbeck's* Vaniloquia.
Heister's Schedulae.

THE WARS OF KINGS about plants have been bloodier, being written not with a pen, but with cannons and swords.
The *fir-tree* called the *cedar* provoked Hadrian to the destruction of Jerusalem.
 Logwood provoked a war between the Spaniards and the English, 1736–1743.[17]
 A *fig* provoked Xerxes against the Athenians: and Rome against Carthage, as exhorted by Cato.
 The Jews and the Romans fought about a *balsam-bush.* Pliny Bk. XII.
 The cashew is often the cause of battles among the Brazilians.
 The *date-palm* has frequently been the cause of disputes among the orientals.
 Nutmeg has incited the Dutch to arms in the [East] Indies.

22. The PHYSIOLOGISTS (9) have discovered the laws of vegetation and the mystery of sex (V).

Millington 1676.
Camerarius, (Rudolf), Epistola.
Vaillant's Sermo.
Wahlbom, our Sponsalia plantarum.[18]

23. The LEGISLATORS (19) have compiled the rules and regulations.

> *Jung's* Isagoge phytoscopica.
> *Our own* Fundamenta botanica, this work.[19]
> *Ludwig's* Regnum vegetabile.
>   Aphorismi botanici.

24. The SYSTEMATISTS (18) have arranged the plants in particular ranks; and they are either *orthodox* (26) or *heterodox* (25).

25. The HETERODOX systematists (24) have divided the vegetables on a principle other than that of the fruit-body (164): such are *the alphabetists, the root-choppers, those that study the leaves, the physiognomists, the chronologists, those that study the locations, the empirics,* and *the dealers in unguents.*

> The *alphabetists* divide according to an alphabetical method.
> The *root-choppers* according to the structure of the root: such are the gardeners.
> *Those that study the leaves,* according to the shapes of the leaves.
> The *physiognomists* according to the external appearance.
> The *chronologists* according to the time of flowering.
> *Those that study locations* according to the place where they are indigenous.
> The *empirics* according to their use in medicine.
> *Those that deal in unguents,* according to the order followed by the druggists.

26. The ORTHODOX systematists (24) have taken their systems from the true foundation, that of the fruit-body (164) and they are *universal* (27) or *partial* (32).

> They keep to the natural genera.
> They arrange the genera according to some part of the fruit-body.
> They point out the things that are present; then those that are absent make themselves obvious.

27. The UNIVERSAL orthodox systematists (26) have established all the classes of vegetables according to a genuine system: such are the *fructists* (28), the *corollists* (29), the *calycists* (30), and the *sexualists* (31).

28. The FRUCTISTS (27) have arranged the classes of the vegetables by the pericarp (80), the seed (86), or the receptacle (86); such are *Cesalpino, Morison, Ray, Knaut, Hermann*, and *Boerhaave*.

> *Cesalpino* (54) Professor at Padua 1583.
> *Morison* (55) Professor at Oxford 1680.
> *Ray* (59) English priest 1682, 1700.
> *Knaut* (57) Physician at Halle 1687.
> *Hermann* (56) Professor at Leiden 1690.
> *Boerhaave* (58) Professor at Leiden 1710.

29. The COROLLISTS (27) have divided the classes by the corolla (86) with its petals: such are *Rivinus* and *Tournefort*,

> *Rivinus* (61) Professor at Leipzig 1690.
>   *Heucher* Professor at Wittenberg 1711.
>   *Ruppius* (61) student at Jena 1718.
>   *Hebenstreit* Professor at Leipzig 1721.
>   *Ludwig* (63) Professor at Leipzig 1737.
>   *Knaut* (62) Librarian at Halle 1716.
> *Tournefort* (64) Professor at Paris 1694.
>   *Plumier* a monk 1703.
>   *Pontedera* (65) Professor at Padua 1720.

30. The CALYCISTS (27) have divided the classes by the calyx (86): such as *Magnol* and *ourself*.

> *Magnol* (66) Professor at Montpellier, posthumously, 1720.
> *Ourself* (67) 1737.

31. The SEXUALISTS (27) have founded their systems on sex: such as *Myself*.

> *Myself* (68) (in the Netherlands) 1735.

32. The PARTIAL orthodox systematists (26) have arranged a system consisting of one class only: e.g. the *compounds* (77, order 21), the *umbellates* (77, order 22), the *grasses* (77, order 14), the *mosses* (77, orders 65 and 66), and the *funguses* (77, order 67).

33. The classes of the COMPOUNDS (32 and 117) have been expounded by *Vaillant* and *Pontedera*.

> *Vaillant* (70) demonstrator at Paris 1718.
> *Pontedera* (70) Professor at Padua 1720.

34. The class of the UMBELLATES (32 and 118) has been arranged by *Morison* and *Artedi*.

> *Morison* (71) Professor at Oxford 1672.
> *Artedi* (71) student of medicine, a Swede, (1735).

35. The classes of the GRASSES (32) have been arranged by *Ray, Monti, Scheuchzer, Micheli,* and *ourself*.

> *Ray* (72) English priest 1703.
> *Monti* (72) Professor at Bologna 1719.
> *Scheuchzer* (72) Professor at Zürich 1719.
> *Micheli* (71) botanist of Tuscany 1729.
> *Ourself* (72) in Genera plantarum 1737.

36. The class of the MOSSES (32) was the subject of the work of *Dillenius*.

> *Dillenius* (73 and 74) Professor at Oxford 1741.

37. The class of the FUNGUSES (32) has been arranged by *Dillenius* and *Micheli*.

> *Dillenius* (75) physician at Giessen 1719.
> *Micheli* (75) botanist of Tuscany 1729.

38. The NOMENCLATORS (18) have been concerned with the naming of the vegetables: such are the *synonymists* (39), the *taxonomists* (40), the *etymologists* (41), and the *lexicographers* (42).

39. The SYNONOMISTS (38) have collated the different names for the vegetables, that have been given to them by botanists in the past.

> *C. Bauhin's* Pinax is the most important.
> *Haller* in various works.

40. The TAXONOMISTS[20] (38) have determined the truly proper names for genera and species.

> Linnaeus' Critica botanica.

41. The ETYMOLOGISTS (38) dig up the roots and sources of the generic names.

> Falugi's Prosopopoeia.

42. The LEXICOGRAPHERS (38) collate the names in different languages.

> Mentzel's Index multilinguis or Lexicon polyglotton.

43. The AMATEURS OF BOTANY (6) are those who have produced various [works] about botany, though they do not properly pertain to botanical science: such as *anatomists* (44), *gardeners* (45), *physicians* (46), and *miscellaneous* [*writers*].(52)

44. The ANATOMISTS (43) have examined the internal structure of the vegetables.

Anatomical [works]	*Malpighi.*
	Grew.
Physiological [works]	*Feldman.*
The laws of vegetation,	*Hales.*
	Gesner.
	Ludwig.

45. The GARDENERS (41) have treated of the cultivation of vegetables.

> The AGRICULTURALISTS have applied it to common use
> *Lauremberg's* Horticultura. *Ferrari's* Flora.
> *Elsholt's* Horticultura. *Bradley's* various English [works].
> *Ligier* in French *Miller's* dictionary.
> The UNIVERSITY garden and the arrangement of it. See *Naucler's Dissertation concerning the garden at Uppsala.*
> Parts: the *hot-house*, the *warm-house*, the *cool-house*,
>   the *steam-house*, the *sun-house*, the *sunny corner*,

the *area* skilfully divided into *hot-beds,*
perennial, annual, spring, and *autumn,*
open-air *walks,* uncovered and covered,
hedges, pergolas, and *alleys.*

Tools: *mattock, hoe, trowel,*
fork, rake, crowbar,
barrow, trolley, scoop,
knife, grapple, two-pronged hook,
shears, watering-can with spout,
harrow, sun-bell,[21] *shading-pot.*

Jobs: *digging, watering, weeding,*
mucking, hoeing, transplanting,
clipping, pruning, hardening off,[22]
banking up, flooding, dividing,
notching, barking, grafting.

The AGRICULTURALISTS have the care of the fields and meadows.
Writers about *country matters,* sect. 9. They are concerned with
the ploughshare, harrow, roller, hoe, rake,
bank, ditch, ridge, furrow, strip,
turning, hoeing, trenching, harrowing, ploughing,
thrashing, winnowing, sieving, and grinding.

46. The PHYSICIANS (43) have sought out the potencies and uses of
vegetables for the human body: such as *astrologers* (47), *symbolists* (47),
chemists (48), *observers* (49), *mechanists* (49), *dieticians* (50), and *botanical*
systematists (51).

Dioscorides' Materia medica.　　*Zorn's* Botanologia.
Simon Paulli's Quadripartitum.　　*Dale's* Pharmacologia.
Koenig's Regnum vegetabile.
Pomet's Histoire des drogues.
Valentini's Museum museorum.

47. The ASTROLOGERS (46) have divined the influence on plants of
powers derived from the stars, the SYMBOLISTS[23] (46) the
potencies derived from the resemblance between a plant and a part of
the body.

Bodenstein.
Pappen.

48. The CHEMISTS (46) have come to believe that they can draw out the efficient properties of vegetables by means of analysis.

> *Geoffroy, Tournefort, Tawry.*

49. The OBSERVERS (46) have deduced the potencies of vegetables from chance and experience, the MECHANISTS (46) from the principles of physiological mechanics.

> *Observers: Geoffroy's* treatise on Materia medica.
> *Hermann's* Cynosyra.
> *Boerhaave's* Historia plantarum.
> *Haller's* Synopsis Helvetica.
> *Linnaeus'* Materia medica.

50. The DIETICIANS (46) have discerned the potencies of things to be consumed, by the taste and smell.

> *Duchesne's* Diaeteticon.
> *Nonnius'* Res cibaria.
> *Behren's* Selecta diaetetica.
> *Lister* in Apicium [against Apicius].

51. The BOTANICAL SYSTEMATISTS (46) have carefully distinguished the potencies of medicines according to the natural classes.

> *Camerarius'* Convenientia plantarum.
> *Hasselqvist's* Vires plantarum.

52. The MISCELLANEOUS [writers] (46) are those who have written various [works] for the use of others: such as *economists, biologists, theologians,* and *poets.*

> The economists treat of the uses of plants in ordinary life.
> [My] *Flora* oeconomica.
> [My] *Pan* Svecicus.
> [My] Öländska resa.
> [My] Gothländska resa.
> [My] Wästgöta resa.
> [My] Skånska resa.

The biographers have mostly given vent to *panegyrics*.
The theologians have explained plants mentioned in the Bible.

    Celsius' Hierobotanicum.

Poets: *Macer*, spurious

    [Walahfrid] *Strabo's* Hortulus [Little garden]

    Rapin, Evelyn, *Gardener*.

    Neefs' Poëmaticum.

    Pictorius' Pantopolon.

    Santolinus' Pomona.

    Falugi's Prosopopoeia.

    Cowley's Sex libri plantarum.

II. SYSTEMATA.

53. SYSTEMATICIS (24) Orthodoxis (26) nitor & certitudo scientiæ Botanices debetur.

Syst. nat. obs. veget. 3. idem.
Syst. nat. obs. veget. 4. Systematici qui.

Cæsalpinus.	Rivinus.	Vaillantius.	Linnæus.
Morisonus.	Knautius.	Jussiæus.	Royenus.
Rajus.	Ruppius.	Scheuchzerus.	Gronovius.
Hermannus.	Ludvigius.	Dillenius.	Gmelinus.
Magnolius.	Tournefortius.	Michelius.	Guettardus.
Boerhavius.	Plumierus.	Hallerus.	Wachendorffius.
		Gesnerus.	Gleditschius.
		Burmannus.	Dalibard.

54. CÆSALPINUS (28) est Fructista & primus verus Systematicus, secundum Corculi (86:6) & Receptaculi (86:7) situm distribuens.

Arbores corculo ex apice seminis. - - - **1.**
e basi seminis - - - **2.**
Herbæ solitariis seminibus - - - - **3.**
baccis - - - - **4.**
capsulis - - - - **5.**
Binis seminibus - - - - **6.**
capsulis - - - **7.**
Triplici principio fibrosæ - **8.**
Bulbosæ - **9.**
Quaternis seminibus - **10.**
Pluribus seminib. *Anthemides* **11.**
Cichorac. s. *Acanaceæ* **12.**
flore *Communi* - **13.**
folliculis - **14.**
flore fructuque *carentes* **15.**

55. MORISONUS (28) est Fructista cum Physiognomis (25) & Corollistis (29) conspirans.

Lignosæ *Arbores* - - **1.**
Frutices - - **2.**
Suffrutices - - **3.**

Herba-

❧ II. SYSTEMS

53. To the Orthodox (26) SYSTEMATISTS (24) we owe the clarity and accuracy of botanical science

> [My] *Systema naturae, observations on the vegetable kingdom* 3; the same.
> *Systema naturae, observations on the vegetable kingdom* 4; who the systematists are.

Cesalpino.	Rivinus.	Vaillant.	Linnaeus.
Morison.	Knaut.	Jussieu.	Royen.
Ray.	Ruppius.	Scheucher.	Gronovius.
Hermann.	Ludwig.	Dillenius.	Gmelin.
Magnol.	Tournefort.	Micheli.	Guettard.
Boerhaave.	Plumier.	Haller.	Wachendorff.
		Gesner.	Gleditsch.
		Burman.	Dalibard.

54. CESALPINO (28) is a fructist and the first true systematist; he divides according to the positions of the corcle (86: VI) and the receptacle (86: VII).

Trees with corcle	from the apex of the seed					…	1
	from the base of the seed					…	2
Herbs with solitary seeds	…		…	…	…	…	3
with solitary berries				…	…	…	4
with solitary capsules				…	…	…	5
with pairs of seeds	…		…	…	…	…	6
of capsules	…		…	…	…		7
with a triple beginning	fibrous			…	…	…	8
	bulbous			…	…	…	9
with seeds in fours	…		…	…	…	…	10
with numerous seeds, *Anthemises*					…	…	11
cichoraceae or *acanaceae*				…	…	…	12
with a *common* flower					…	…	13
with follicles				…	…	…	14
lacking flower and fruit		…	…	…	…	15

55. MORISON (28) is a fructist in league with physiognomists (25) and corollists (29).

Woody	trees	...	...	...	...	...	...	1
	shrubs	...	...	...	...	...	...	2
	sub-shrubs		...	...	...	...	...	3
Herbaceous	climbing	...	...	...	...	...	...	4
	leguminous		...	...	...	...	...	5
	siliquous	...	...	...	...	...	...	6
	with three capsules			...	...	...	...	7
	described by the number of capsules						...	8
	bearing clusters		...	...	...	...	...	9
	milky or pappose		...	...	...	...	...	10
	bearing stalks		...	...	...	...	...	11
	umbelliferous		...	...	...	...	...	12
	with three kernels			...	...	...	...	13
	helmet-shaped		...	...	...	...	...	14
	with many capsules			...	...	...	...	15
	bearing berries		...	...	...	...	...	16
	capillary	...	...	...	...	...	...	17
	irregular	...	...	...	...	...	...	18

56. HERMANN (28) is a fructist who reckons according to the gymnosperm (200) and angiosperm (200) fruits.

Herbs, Gymnosperm	with one seed,	simple	4.
		compound	3.
	with two seeds,	stellate	5.
		umbellate	2.
	with four seeds,	rough-leaved	6.
		whorled	7.
	with many seeds,	many-seeded gymnosperms	1.
Angiosperm	bulbous,	with three capsules	16.
	with capsule,	with one seed-vessel	8.
		with two seed-vessels	9.
		with three seed-vessels	10.
		with four seed-vessels	11.
		with five seed-vessels	12.
	with silique,	siliquous	14.
	with pod,	leguminous	15.
	with many capsules,	with many capsules	13.
	fleshy,	bearing berries	17.
		bearing fruits	18.

Without petals	with epicalyx,	*without petals*	19.
	glumose,	*staminal*	21.
	without covering,	*mossy*	20.
Trees	incomplete,	*bearing catkins*	22.
	fleshy,	*umbilicate*	23.
		not umbilicate	24.
	not fleshy,	*with dry fruit*	25.

57. KNAUT (Christoph) (28) adopted the system of Ray (59), turned upside down.

Herbs with petals	and with fleshy fruit,	*bearing berries*	1.
	with membraneous fruit,	*with one petal*	2.
		with four petals, regular	3.
		with four petals, irregular	4.
		with five petals	5.
		with six petals	6.
		with many petals	7.
		with many capsules	8.
	with uncovered fruit,	*gymnosperms*	9.
		solid	10.
		pappose	11.
Herbs without petals		*without petals*	12.
		staminal	13.
		inconspicuous	14.
		imperfect	15.
Woody		*trees*	16.
		shrubs	17.

58.

BOERHAAVE (28) reconciled the system of Hermann (56) with those of Ray (59.b) and Tournefort (64).

under-water [plants]	1.	gymnosperms with two seeds,		5.
			umbelliferous	
land [plants]	2.		*stellate*	11.
capillary [plants]	3.	gymnosperms with one seed,		6.
			simple	
gymnosperms with many seeds	4.		*with flat petals*	7.
gymnosperms with four seeds, *whorled*	12.		*with rays*	8.
rough leaved	13.		*uncovered*	9.
with four petals	14.		*capitate*	10.
with one capsule	15.	*bearing berries*		25.
with two capsules	16.	*bearing pomes*		26.

continued

with three capsules	17.	*with no petals*		27.
with four capsules	18.	*with one colyledon,*	*with bracts*	28.
with five capsules	19.		*without petals*	29.
with many capsules	20.	*trees*	*with one cotyledon*	30.
with many siliques	21.		*with no petals*	31.
siliquous	22.		*with catkins*	32.
with four petals, cruciform	23.		*with one petal*	33.
leguminous	24.		*rosaceous*	34.

59. RAY (28) formerly a fructist (28) eventually became a corollist (29).

a. HIS PECULIAR METHOD

trees		1.
shrubs		2.
herbs	*imperfect*	3.
	without flowers	4.
	capillary	5.
	staminal	6.
	gymnosperms with one seed	7.
	umbellate	8.
	whorled	9.
	rough-leaved	10.
	stellate	11.
	bearing pomes	12.
	bearing berries	13.
	with many siliques	14.
	with one petal, regular	15.
	with one petal, irregular	16.
	with four petals, siliquous	17.
	with four petals, siliculous	18.
	butterfly-shaped	19.
	with five petals	20.
	cereals	21.
	grasses	22.
	with grass-like leaves	23.
	bulbous	24.
	allied to the bulbous	25.

b. THE REVISED METHOD

herbs	under-water	1.
	funguses	2.
	mosses	3.
	capillary	4.
	with no petals	5.
	with flat petals	6.
	disc-shaped	7.
	bearing clusters	8.
	capitate	9.
	with a solitary seed	10.
	umbelliferous	11.
	stellate	12.
	rough-leaved	13.
	whorled	14.
	with many seeds	15.
	bearing pomes	16.
	bearing berries	17.
	with many siliquae	18.
	with one petal	19.
	with two or three petals	20.
	siliquous	21.
	leguminous	22.
	with five petals	23.
	bearing flowers	24.
	staminal	25.
	anomalous	26.
	reed-like	27.
trees	with no petals	28.
	with fruit umbilicate	29.
	with fruit not umbilicate	30.
	with dry fruit	31.
	with siliquous fruit	32.
	anomalous	33.

60. KAMEL tried to arrange plants according to the valves (86) of the pericarp.

Pericarps	with no aperture	with four apertures
	with one aperture	with five apertures
	with two apertures	with six apertures
	with three apertures	

61. RIVINUS (29) is a corollist according to the regularity and number of the petals, with three kinds of fruit.

Regular	with one petal	1.	Irregular	with one petal	11.
	with two petals	2.		with two petals	12.
	with three petals	3.		with three petals	13.
	with four petals	4.		with four petals	14.
	with five petals	5.		with five petals	15.
	with six petals	6.		with six petals	16.
	with many petals	7.		with many petals	17.
Compound [consisting] of florets	regular	8.	Incomplete	imperfect	18.
	regular and irregular	9.			
	irregular	10.			

Three kinds of fruit: to wit, *uncovered*, *pericarp* (dry), and *fleshy*.
RUPPIUS improved RIVINUS' system in the compound [flowers].

62. KNAUT (Christian) (29) adopted the system of Rivinus (61) in reverse, to wit, preferring the number [of petals] to their regularity.

with one petal,	regular	1.	irregular	2.
clustered	regular	3.	irregular	4.
	regularly irregular			5.
with two petals	regular	6.	irregular	7.
with three petals	regular	8.	irregular	9.
with four petals	regular	10.	irregular	11.
with five petals	regular	12.	irregular	13.
with six petals	regular	14.	irregular	15.
with many petals	regular	16.	irregular	17.

He denied the existence of flowers with no petals and of uncovered seeds.

63. LUDWIG combined the method of Rivinus (61) with that of Linnaeus (68).

The *classes* of Rivinus [are arranged] by the regularity and number of the petals; the *orders* by the sexual method within a single class; likewise *Wedel*.

with one anther	*with one style*
with two anthers	*with two styles*
with three anthers	*with three styles*
with five anthers	*with four styles*
with ten anthers etc.	*with many styles etc.*

64. TOURNEFORT (29) is a corollist, according to the regularity and shape, with two situations for the receptacle (86) of a flower.

Simple,			compound,		
	bell-shaped	1.		*with florets*	12.
	funnel-shaped	2.		*with demi-florets*	13.
	anomalous	3.		*rayed*	14.
	labiate	4.	with no petals,	*with no petals*	15.
	cruciform	5.		*without flowers*	16.
	rosaceous	6.		*without flowers or fruit*	17.
	umbellate	7.	trees,	*with no petals*	18.
	like a pink	8.		*with catkins*	19.
	liliaceous	9.		*with one petal*	20.
	butterfly-shaped	10.		*rosaceous*	21.
	anomalous	11.		*butterfly-shaped*	22.

Orders from the pistil or the calyx going off into fruit.

65. PONTEDERA attempted to combine Tournefort's [system] (64) with Rivinus' (61).

Uncertain,					
	uncertain	1.		*anomalous*	13.
	without flowers	2.		*butterfly-shaped*	14.
Without buds,	*imperfect*	3.		*liliaceous*	15.
	anomalous	4.		*like a pink*	16.
	labiate	5.		*cruciform*	17.
	bell-shaped	6.		*umbellate*	18.
	salver-shaped	7.		*with filaments*	19.
	wheel-shaped	8.	bearing buds,	*with filaments*	20.
	funnel-shaped	9.		*with no petals*	21.
	with florets	10.		*anomalous*	22.
	tongue-shaped	11.		*bell-shaped*	23.
	rayed at the heads	12.		*wheel-shaped*	24.
				funnel-shaped	25.
				butterfly-shaped	26.
				rosaceous	27.

66. MAGNOL (30) is a calycist, associated with the fructists (28).

Herbs with an *external* calyx enclosing a flower,	*unknown*	1.
	staminal	2.
	with a single petals	3.
	with many petals	4.
	compound	5.
upholding a flower	*with a single petal*	6.
	with many petals	7.
internal only		8.
external and *internal*, with a flower *with a single petal*		9.
with 2–3 petals		10.
with 4 petals		11.
with many petals		12.
Trees with an *external* calyx only		13.
internal only		14.
both *internal* and *external*		15.

67. WE (30) have worked out a calycine system according to the forms of the *calyx* (86).

with spathe	1.	*irregular*	…	…	…	…	…	…	10.
with husks	2.	*caducous*	…	…	…	…	…	…	11.
with catkins	3.	*persistent,*	*regular,*		*with one petal*			12.
umbellate	4.				*with many petals*			13.
compound	5.		*irregular,*		*with one petal*			14.
doubled	6.				*with many petals*			15.
flowering profusely	7.	*incomplete*	…	…	…	…	…	…	16.
crowned	8.	*with no petals*	…	…	…	…	…	17.
anomalous	9.	*uncovered*	…	…	…	…	…	…	18.

68. I (31) have worked out a sexual system according to the number, relative size, and position of the *stamens*, together with the pistils.

with one stamen	1.
with two stamens	2.
with three stamens	3.
with four stamens	4.
with five stamens	5.
with six stamens	6.
with seven stamens	7.
with eight stamens	8.
with nine stamens	9.
with ten stamens	10.

with twelve stamens	11.
with twenty stamens	12.
with many stamens	13.
didynamia	14.
tetradynamia	15.
with stamens united	16.
with stamens in two sets	17.
with stamens in many sets	18.
with anthers united	19.
gynandrous	20.
monoecious	21.
dioecious	22.
polygamous	23.
cryptogamous	24.

69. The NATURAL method has been sought in the cotyledons, the calyx, the sex[ual] and other [parts]; by Royen, excellently; by Haller, eruditely; and by Wachendorff, in Greek.[1]

    a.   ROYEN, professor at Leiden, 1740

Palms	1.	*calyx-flowered*	11.
lilies	2.	*gaping*	12.
grasses	3.	*siliquous*	13.
with catkins	4.	*bearing columns*	14.
umbellate	5.	*leguminous*	15.
compound	6.	*with few anthers*	16.
clustered	7.	*with double anthers*	17.
with three kernels	8.	*with many anthers*	18.
incomplete	9.	*with concealed anthers*	19.
fruit-flowered	10.	*mineral-vegetables*	20.

    b.   HALLER, professor at Göttingen, 1742

Funguses	1.	*with double stamens*	9.
mosses	2.	*with equal stamens*	10.
with seeds on the leaves	3.	*with smaller stamens*	11.
with no petals	4.	with stamens *of* $1\frac{1}{2}$	12.
grasses	5.	with stamens *of* $1\frac{1}{3}$	13.
allied to the grasses	6.	*with four gaping* stamens	14.
monocotyledons, *petal-like*	7.	*clustered*	15.
with many stamens	8.		

    c.   WACHENDORFF, professor at Utrecht, 1747

Gymnosperms	1.	*with cylindrical anthers*	11.
with two perianths of similar shape	2.	*with one perianth*	12.
with two perianths of dissimilar shape	3.	*monophythanthae* [sic]	13.
palloplostemonopetalae [sic]	4.	*diphythanthae* [sic]	14.

with petals unequal to the stamens	5.	*with no calyx*	15.
cylindrobasiostemones [sic]	6.	*with developed calyx*	16.
with two long stamens	7.	*with spathe*	17.
with four long stamens	8.	*with husk*	18.
distemonopleantherae [sic]	9.	*with concealed flowers*	19.
with free anthers	10.		

70. The COMPOUND flowers (77:21) are distinguished by VAILLANT by means of the calyx (86.I), the receptacle (86, VII), and the coronule of the seeds (86, VI); PONTEDERA (33) agrees theoretically with Vaillant concerning compound [flowers] in [his arrangement of] the orders.

> VAILLANT 1718
> CLASSES *Artichoke-headed,[2] bearing clusters, like succory, like teasel.*
> ORDERS Calyx: *simple, overlapping, with epicalyx.*
> Receptacle: *uncovered, hairy, chaff-like.*
> Coronule: *none, hairy, feathery.*

71. The UMBELLATE [flowers] (77:22) were arranged by MORISON (34) according to the shape of the seeds; but ARTEDI (34) was the first to distribute them into three [classes] according to the involucres.

> MORISON regarded the resemblances of the seeds.
> ARTEDI presented only the involucres.
> Involucre 1. *complete* and *partial*
> 2. *partial* only.
> 3. *none.*

72. The GRASSES (77:13,14) are arranged by RAY (35) according to their affinity with the cereals, and MONTI agrees with him; and SCHEUCHZER worked it out excellently; MICHELI (36) classified them according to the husks, simple and compound, and WE (35) have done so according to the sex[ual parts].

The classes of RAY, MONTI, and SCHEUCHZER are:

With ears:			
	like wheat		*finger-shaped*
	like barley	Tufted:	*simple, docked*
	like rye		*awned*
	like rye-grass		*compound*
	like millet	Allied:	*Linagrostis*
	shaped like canary grass		*rush-shaped*

shaped like a fox-tail	*Juncus [rush]*
shaped like a bulrush	*Canna [reed]*
shaped like a mouse-tail	*Scirpus*
prickly	*Cyperus [sedge]*
crested	*sedge-shaped.*
spices	

73. The MOSSES (77:65) were discovered and sorted out with stupendous industry by DILLENIUS (36).

> *with veil.*
> *without veil.*

74. The ALGAE (77:66) were arranged by DILLENIUS (37) by means of the texture, by MICHELI according to the flowers.

75. The FUNGUSES (77:67) were classified by DILLENIUS (37) according to the caps, by MICHELI according to the fruit-bodies.

> DILLENIUS' division according to the caps, thus:
> > *with gills,*
> > *porous,*
> > *like a hedgehog*

76. The MINERAL-VEGETABLES which were once left to *Pluto*[3] [i.e. the mineral kingdom], were subjected to the empire of *Flora* by MARSIGLI; but PEYSONNEL restored them to the kingdom of *Fauna*.

> [My] *Amoenitates academicae* 80, Peysonnel's scheme 1727, B. Jussieu's 1741.

77. The fragments of the NATURAL METHOD are to be sought out studiously.

> This is the beginning and the end of what is needed in botany.
> Nature does not make leaps.
> All plants exhibit their contiguities on either side, like territories on a geographical map.
> These are the fragments proposed by me:

1. PIPERITÆ
   [PUNGENT].
   Arum
   Dracontium
   Calla
   Acorus
   Saururus
   Pothos
   Piper
   Phytolacca.
2. PALMÆ [PALMS].
   Corypha
   Borassus
   Coccus
   Chamærops
   Phœnix
   Coix.[4]
3. SCITAMINA
   [DELICACIES].
   Musa
   Thalia
   Alpinia
   Costus
   Canna
   Maranta
   Amomum
   Curcuma
   Kæmpferia.
4. ORCHIDEÆ
   [ORCHIDS].
   Orchis
   Satyrium
   Serapias
   Herminium
   Neottia
   Ophrys
   Cypripedium
   Epidendrum
   Limodorum
   Arethusa.
5. ENSATAE
   [SWORD-SHAPED].
   Iris
   Gladiolus
   Antholyza

Ixia
Sisyrinchium
Commelina
Xyris
Eriocaulon
Aphyllanthes.
6. TRIPETALODEÆ
   [WITH THREE
   PETALS]
   Butomus
   Alisma
   Sagittaria.
7. DENUDATÆ
   [UNCOVERED].
   Crocus
   Gethyllis
   Bulbocodium
   Colchicum.
8. SPATHACEÆ
   [WITH SPATHES].
   Leucojum
   Galanthus
   Narcissus
   Pancratium
   Amaryllis
   Crinum
   Hæmanthus.
9. CORONARIÆ
   [CORONARY].
   Ornithogalum
   Scilla
   Hyacinthus
   Asphodelus
   Anthericum
   Polianthes.
10. LILIACEÆ
    [LILIES ETC].
    Lilium
    Fritillaria
    Tulipa
    Erythronium.
11. MURICATÆ
    [MURICATE].
    Bromelia
    Renealmia

Tillandsia
Burmannia.
12. COADUNATÆ
    [UNITED].
    Annona
    Liliodendron
    Magnolia
    Uvaria
    Michelia
    Thea.
13. CALAMARIÆ
    [REEDS ETC].
    Bobartia
    Scirpus
    Cyperus
    Eriophorum
    Carex
    Schœnus

Flagellaria?
Juncus?
14. GRAMINA
    [GRASSES].
    Zea
    Coix
    Cornucopiae
    Nardus
    Saccharum
    Zizania
    Phalaris
    Phleum
    Alopecurus
    Panicum
    Milium
    Agrostis
    Lagurus
    Dactylis
    Calla
    Acorus
    Holcus
    Melica
    Aira
    Poa
    Briza

continued

Uniola
Cenchrus
Cynofurus
Bromus
Festuca
Avena
Arundo
Lolium
Triticum
Ægilops
Secale
Hordeum
Elymus
Anthoxanthum
Oryza.

15. CONIFERÆ
[CONIFERS].
Abies
Pinus
Cupressus
Thuja
Juniperus
Taxus
Ephedra.

16. AMENTACEÆ
[WITH CATKINS].
Pistacia
Myrica
Alnus
Betula
Salix
Populus
Platanus
Carpinus
Corylus
Juglans
Quercus
Fagus.

17. NUCAMENTACEÆ
[WITH FIR-CONES].
Xanthium
Ambrosia
Parthenium
Iva
Micropus
Artemisia?

18. AGGREGATÆ
[CLUSTERED].
Statice
Protea
Leucadendros.
Hebenstretia
Brunia
Cephalanthus
Globularia
Scabiosa
Knautia
Dipsacus
Valeriana
Morina
Boerhaavia
Circæa?

19. DUMOSÆ
[BUSHY].
Viburnum
Tinus
Opulus
Sambucus
Rondeletia
Bellonia
Maurocenna
Cassine
Rhus
Cotinus
Celastrus
Euonymus
Ilex
Tomex
Prinos
Callicarpa
Lawsonia.

20. SCABRIDÆ
[SOMEWHAT
ROUGH].
Ficus
Dorstenia
Parietaria
Urtica
Cannabis
Acnida
Humulus
Morus.

21. COMPOSITI
[COMPOUND].
a. *Semiflosculosi*
[with demi-florets]
Prenanthes
Lactuca
Chondrilla
Hieracium
Crepis
Andryala
Hypochæris
Pitris
Hyoseris
Leontodon
Scorzonera
Tragopogon

continued

Scolymus
Sonchus
Lapsana
Cichorium
Catananche
Elephantopus
b.   Capitati [Capitate].
Echinops
Sphæranthus
Gundelia
Arctium
Serratula
Onopordum
Carduus
Cynara
Carthamus
Carlina
Cnicus
Atractylis
Centaurea
Corymbium.
c.   Corymbiferi
[bearing clusters].
Stœbe
Santolina
Chrysocoma
Tanacetum
Kleinia
Stæhelina
Xeranthemum
Gnaphalium
Carpesium
Tarchonanthus
Baccharis
Erigeron
Tussilago
Doronicum
Solidago
Senecio
Inula
Aster
Gerberia
Othonna
Chrysanthemum
Matricaria
Buphthalmum

Anacyclus
Cotula
Anthemis
Achillea
Eriocephalus
Helenia
Arctotis
Bellis
Tagetis.
d.   Oppositifolii
[with flowers that are
opposite the leaves].
Helianthus
Rudbeckia
Coreopsis
Bidens
Verbesina
Sigesbeckia
Milleria
Silphium
Eupatorium
Ageratum
Osteospermum
Calendula?
Chrysogonum?
Melampodium?
Tridax?
Tetragonotheca?

22.   UMBELLATÆ
[UMBELLATE].
Eringium
Arctopus
Hydrocotyle
Sanicula
Astrantia
Tordylium
Caucalis
Artedia
Daucus
Ammi
Bunium
Conium
Selinum
Athamanta
Peucedanum
Chrithmum

Cachrys
Ferula
Laserpitium
Ligusticum
Angelica
Sium
Bubon
Sison
Oenanthe
Phellandrium
Cicuta
Coriandrum
Aethusa
Bupleurum
Scandix
Chærophyllum
Seseli
Imperatoria
Heracleum
Thapsia
Pastinaca
Smyrnium
Anethum
Carum
Pimpinella
Ægopodium
Apium
Anisum
Lagœcia.

23.   MULTISILIQUÆ
[WITH MANY
SILIQUAS].
Pæonia
Aquilegia
Aconitum
Delphinium
Garidella
Nigella
Isopyrum
Helleborus
Caltha
Ranunculus
Myosurus
Adonis
Anemone
Hepatica

continued

Pulsatilla
Atragene
Clematis
Thalictrum.

24. BICORNES
[TWO-HORNED].
Ledum
Azalea
Andromeda
Clethra
Erica
Myrsine
Memecylon
Santalum
Vaccinium
Arbutus
Royena
Diospyros
Melastoma
Pyrola.

25. SEPIARIÆ
[HEDGE-PLANTS].
Nyctanthes
Jasminum
Ligustrum
Brunfelsia
Olea
Chionanthus
Fraxinus
Syringa

26. CULMINIÆ
[WITH UPWARD
PROJECTIONS].
Tilia
Theobroma
Sloanea
Bixa
Heliocarpus
Triumfetta
Bartramia
Muntingia
Clusia
Dillenia
Kiggelaria
Grewia
Corchorus

27. VAGINALES
[SCABBARD-
SHAPED].
Laurus
Helxine
Polygonum
Bistorta
Persicaria
Atraphaxis
Rheum
Rumex

28. *CORYDALES
[CRESTED].*
Melianthus
Epimedium
Hypecoum
Fumaria
Impatiens
Leontice
Monotropa?
Utricularia?
Tropaeolum?

29. CONTORTI
[TWISTED].
Rauwolfia
Tevetia
Cerbera
Plumiera
Tabernæmontana
Cameraria
Nerium
Vinca
Apocynum
Cynanchum
Creopegia
Asclepias
Stapelia.

30. RHOEADES
[POPPIES].
Papaver
Argemone
Chelidonium
Bocconia
Sanguinaria
Actæa
Podophyllum

31. PUTAMINEA
[WITH STONES IN
FRUIT].
Capparis
Breynia
Morisona
Crateva
Marcgravia.

32. CAMPANACEI
[BELL-SHAPED].
Convolvulus
Ipomœa
Polemonium
Campanula
Roëlla
Phyteuma
Trachelium
Jasione
Lobelia
Viola.

33. LURIDÆ
[SALLOW].
Capsicum
Solanum
Physalis
Hyoscyamus
Nicotiana
Atropa
Mandragora
Datura
Verbascum
Celsia
Digitalis

34. COLUMNIFERI
[BEARING
COLUMNS].
Camellia
Xylon
Gossypium
Urena
Hibiscus
Turnera
Malope
Lavatera
Althæa
Alcea

continued

Malva
Melochia
Sida
Napæa
Waltheria
Mentzelia
Hermannia
Helicteres
Stewartia.

35. SENTICOSÆ
[THORNY].
Rosa
Rubus
Fragaria
Potentilla
Tormentilla
Sibbaldia
Dryas
Geum
Comarum
Aphanes
Alchemilla.

36. COMOSÆ
[WITH TOP KNOTS].
Spiræa
Filipendula
Aruncus

37. POMACEÆ
[WITH POMES].
Punica
Pyrus
Cratægus
Mespilus
Sorbus
Ribes.

38. DRUPACEÆ
[WITH STONE-
FRUIT].
Amygdalus
Prunus
Cerasus
Padus.

39. ARBUSTIVA
[WOODY].
Philadelphus
Eugenia

Psidium
Myrtus
Caryophyllus.

40. CALYCANTHEMI
[WITH CALYCINE
FLOWERS].
Epilobium
Oenothera
Jussiæa
Ludwigia
Oldenlandia
Isnarda
Ammannia
Peplis
Lythrum
Glaux
Rhexia.

41. HESPERIDEÆ
[CITROUS].
Citrus
Styrax
Garcinia

42. CARYOPHYLLEI
[PINK-LIKE].
Dianthus
Saponaria
Drypis
Cucubalus
Silene
Lychnis
Coronaria
Agrostemma
Frankenia
Alsine
Cerastium
Holosteum
Arenaria
Spergula
Sagina
Moerhingia

43. ASPERIFOLIÆ
[ROUGH-LEAVED].
Tournefortia
Cerinthe
Symphytum
Pulmonaria

Anchusa
Lithospermum
Myosotis
Heliotropium
Cynoglossum
Asperugo
Lycopsis
Echium
Borrago.

44. STELLATÆ
[STAR-SHAPED].
Anthospermum
Rubia
Aparine
Galium
Valantia
Spermacoce
Sherardia
Asperula
Crucianella
Hedyotis
Phyllis
Houstonia
Spigelia
Lippia
Diodia
Knoxia
Cornus?
Coffea.

45. CUCURBITACEÆ
[MARROWS ETC.]
Passiflora
Fevillea
Momordica
Trichosanthus
Cucumis
Cucurbita
Bryonia
Sicyos
Melothria
Gronovia?

46. SUCCULENTÆ
[SUCCULENT].
Cactus
Mesembryanthemum
Tetragonia

continued

Aizoon
Sempervivum
Sedum
Cotyledon
Rhodiola
Crassula
Tillea
Anacampseros
Portulaca
Claytonia
Chrysosplenium
Heuchera
Saxifraga
Mitella
Penthorum
Geranium
Linum
Oxalis
Zygophyllum
Fagonia
Tribulus
Neurada?
Averrhoa.

47. TRICOCCA
[WITH THREE
KERNELS].
Cambogia
Euphorbia
Dalecampia
Clutia
Andrachne
Phyllanthus
Osyris
Croton
Tragia
Acalypha
Cneorum
Jatropha
Ricinus
Cliffortia
Mercurialis
Hernandia
Sterculia
Carica
Hura.

48. INUNDATÆ
[AQUATIC].
Hippuris
Elatine
Proserpinaca
Myriophyllum
Ceratophyllum
Potamogeton
Zanichellia
Ruppia
Zostera
Sparganium
Typha.

49. SARMENTACEÆ
[WITH RUNNERS].
Cissus
Vitis
Hedera
Panax
Aralia
Ruscus
Asparagus
Medeola
Uvularia
Convallaria
Gloriosa
Rajania
Dioscorea
Smilax
Tamus
Menispermum
Cissampelos
Asarum
Aristolochia
Hippocratea?

50. TRIHILATÆ
[WITH THREE HILA].
Cardiospermum
Paullinia
Stapindus
Staphylea
Malpighia
Bannisteria
Begonia
Acer
Triopteris
Æsculus

Berberis?

51. PRECIÆ.[5]
Primula
Androsace
Diapensia
Cortusa
Dodecatheon
Soldanella
Cyclamen.

52. ROTACEÆ
[WHEEL-SHAPED].
Gentiana
Exacum
Chironia
Swertia
Lysimachia
Anagallis
Trientalis
Centunculus
Hottonia?
Samolus?

53. HOLERACEÆ
[CULINARY
VEGETABLES ETC].
Spinacia
Blitum
Beta
Galenia
Atriplex
Chenopodium
Rivina
Petiveria
Herniaria
Illecebrum
Polycnemum
Axyris
Achyranthes
Amaranthus
Gomphrena
Celosia
Ceratocarpus
Corispermum
Callitriche
Salsola
Salicornia
Anabasis.

continued

54. VEPRECULÆ
[BRIARS].
Rhamnus
Sideroxylon
Chrysophyllum
Lycium
Ceanothus
Philyca
Cestrum
Catesbæa
Daphne
Gnidia
Passerina
Stellera
Lachnaea.

55. PAPILIONACEÆ
[BUTTERFLY-
SHAPED].
Erythrina
Anagyris
Cytisus
Robinia
Achyronia
Genista
Spartium
Ulex
Borbonia
Colutea
Crotalaria
Ononis
Lupinus
Galega
Securidaca
Glycine
Phaseolus
Dolichos
Clitoria
Pisum
Lathyrus
Vicia
Orobus
Lotus
Dorycnium
Psoralea
Anthyllis
Trifolium

Ervum
Cicer
Coronilla
Ornithopus
Scorpiurus
Hippocrepis
Æschynomene
Hedysarum
Glycyrrhiza
Medicago
Trigonella
Arachis
Phaca
Glycine[6]
Biserrula
Tragacantha
Indigofera
Amorpha
Dalea.

56. LOMENTACEÆ
[WITH LOMENTS].
Sophora
Cercis
Bauhinia
Parkinsonia
Cassia
Poinciana
Tamarindus
Guilandina
Adenanthera
Hæmatoxylum
Cæsalpina
Mimosa.

57. SILIQUOSÆ
[SILIQUOUS].
Myagrum
Anastatica
Subularia
Lepidium
Cochlearia
Iberis
Thlaspi
Biscutella
Clypeola
Alyssum
Lunaria

Draba
Vella
Bunias
Cheiranthus
Hesperis
Raphanus
Dentaria
Cardamine
Brassica
Sinapis
Arabis
Turritis
Erysimum
Sisymbrium
Crambe
Isatis
Bunias[7]

58. VERTICILLATÆ
[WHORLED].
Ajuga
Teucrium
Trichostema
Thymus
Satureja
Clinopodium
Origanum
Lavandula
Hyssopus
Melissa
Horminum
Salvia
Rosmarinus
Ziziphora
Monarda
Lycopus
Amethystea
Glechoma
Mentha
Ocymum
Dracocephalum
Nepeta
Betonica
Sideritis
Cunila
Lamium
Galeopsis

continued

Stachys
Ballota
Marrubium
Moluccella
Leonurus
Orvala
Phlomis
Brunella
Scutellaria
Prasium.

59. PERSONATÆ
[PERSONATE].
Cymbaria
Antirrhinum
Rhinanthus
Pedicularis
Bartsia
Euphrasia
Melampyrum
Obolaria
Orobanche
Lathræa
Chelone
Mimulus
Dodartia
Gesneria
Swalbea
Duranta
Columnea
Gerardia
Craniolaria
Torenia
Martynia
Scrophularia
Sesamum
Gratiola
Capraria
Ruellia
Justicia
Besleria
Browallia
Erinus
Buchnera
Tozzia

Verbena
Veronica
Acanthus
Vitex
Volkameria
Clerodendrum
Cornutia
Lantana
Petrea
Bignonia
Citharexylum
Bontia
Halleria
Gmelina
Ovicda?
Æginetia?

60. PERFORATÆ
[PERFORATED].
Hypericum
Ascyrum
Cistus
Telephium.

61. STATUMINATÆ
[RIBBED].
Ulmus
Celtis
Bosea.

62. CANDELARES
[TAPER-LIKE].
Rhizophora
Mimusops
Nyssa.

63. CYMOSÆ
[WITH CYMES].
Diervilla
Lonicera
Mitchella
Loranthus
Ixora
Morinda
Cinchona?

64. FILICES
[FERNS].
Ophioglossum

Osmunda
Pteris
Trichomanes
Adiantum
Lonchitis
Asplenium
Hemionitis
Polypodium
Acrostichum.

65. MUSCI
[MOSSES].
Lycopodium
Porella
Fontinalis
Sphagnum
Splachnum
Phascum
Mnium
Polytrichum
Bryum
Hypnum.

66. ALGÆ.
Marchantia
Jungermannia
Anthoceros
Lichen
Blasia
Riccia
Ulva
Tremella
Spongia
Conferva
Chara
Fucus.

67. FUNGI
[FUNGUSES].
Agaricus
Boletus
Hydnum
Phallus
Elvela
Clavaria
Clathrus
Peziza

continued

Lycoperdum
Byssus
Mucor.

68. INDETERMINATE
and *as yet of no fixed
abode*.

Pinguicula
Collinsonia
Buffonia
Hirtella
Montia
Mollugo
Siphonanthus
Pavetta
Avicennia
Penæa
Polypremum
Budleja
Plantago
Scoparia
Ptelea
Trapa
Elæagnus
Brabejum
Hamamelis
Cuscuta
Coldenia
Menyanthes
Hydrophyllum
Strychnus
Theophrasta
Paragonula
Plumbago
Phlox
Genipa
Conocarpus
Mirabilis
Coris
Cupania
Itea
Cressa

Nama
Basella
Parnaffia
Suriana
Pontederia
Tradescantia
Aloë Yucca
Hemerocallis
Richardia
Cordia
Triglochin
Dodonæa
Grislea
Jambolifera
Adoxa[8]
Guajacum
Cynometra
Anacardium
Dictamnus
Toluifera
Melia
Schinus
Hydrangea
Cherleria
Mesua
Mammea
Calophyllum
Elæocarpus
Microcus
Ochna
Sauvagesia
Vateria
Chysobalanus
Plinia
Nymphæa
Calligonum
Tetracera
Lœselia
Limosella
Ovieda
Cleome

Hugonia
Connarus
Pentapetes
Polygala
Nepenthes
Pistia
Cynomorium
Liquidambar
Najas
Osyris
Viscum
Hippophaë
Antidesma
Pisonia
Zanonia
Coriaria
Melanthium
Veratrum
Empetrum
Lemna
Marsilea
Calaumaria[9]

A.[10] Hydrocharis
Stratiotes
Valisneria.

B. Ruta
Peganum.

C. Sanguisorba
Poterium.

D. Reseda
Datisca.

E. Ceratonia
Gleditsia.

F. Veratrum
Melanthium.

G. Selago
Camphorosma.

H. Ophiorrhiza
Mitreola.

The absence of things not yet discovered has acted as a cause of the deficiencies of the natural method; but the acquisition of knowledge of more things will make it perfect; for nature does not make leaps.

III. PLANTÆ.

78. VEGETABILIA comprehendunt Familias VII:
*Fungos, Algas, Muscos, Filices, Gramina, Palmas,
Plantas.*

Conſtant Vegetabilia triplicibus vaſis:

   1. VASA SUCCOSA liquorem vehunt.

   2. UTRICUI I alveolis ſuccum conſervant.

   3. TRACHEÆ aërem attrahunt.

   Geſneri Diſſ. de *Vegetabilibus* de his conſulenda.

1. FUNGI;

2. ALGÆ; his Radix, Folium & Caudex in unum.

3. MUSCI; hiſ *Anthera* absque Filamento, remota a flore *Femi-neo:* deſtituto Piſtillo; *Semina* vero propria tunica cotyledo-nibusque carent.

4. FILICES; his *Frondes* ſunt fructificantes averſo latere.

5. GRAMINA; his *Folia* ſimpliciſſima, *Culmus* articulatus, *Ca-lyx* glumoſus; *Semen* unicum.

6. PALMÆ; his *Caudex* ſimplex, apice *frondoſus,* fructificatio-nes in *Spadice* cum *Sputha.*

7. PLANTÆ dicuntur reliquæ, quæ priores intrare nequeunt fa-milias.

   Herbaceæ quotannis ſupra Radicem pereunt, cum Radix omnis herbæ perennis infra terram gemmam proferat.

   Frutices: Caudex adſcendit ſupra terram absque gemmis.

   Arbores: Caudex adſcendit ſupra terram cum gemmis.

      Gemmæ vel diſtinguunt Frutices ab Arbore, vel nulli Limites, cum magnitudo nihil facit. Pontedera ita-que Gemmiparas dixit Arbores.

      Indicæ arbores maximæ frutices dicentur, quum in his raro gemma; adeoque hæc diviſio non eſt na-turalis, cum intra Fruticem & Arborem nullos li-mites poſuit natura, ſed opinio vulgi.

79. Vegetabilium (78) PARTES, primum a Tyrone
diſtinguendæ, ſunt III: *Radix Herba, Fructificatio.*

Conſtat Vegetabile ex *Medulla* 1, veſtita *Ligno* 2, facto ex *Li-bro* 3, ſecedente a *Cortice* 4, inducto *Epidermide* 5.

Medulla creſcit extendendo ſe & *integumenta.*

Fibræ *medullaris* extremitas per Corticem protruſa ſolvitur in *Gemmam* imbricatam ex folioliſ nunquam renaſcituris.

           C 3   Herbæ

❧ III. PLANTS

78. The VEGETABLES comprise seven families: funguses, algae, mosses, ferns, grasses, palms, and plants.

> Vegetables exist with vessels of three kinds:
> 1. JUICY VESSELS that carry fluid.
> 2. LITTLE BAGS that hold fluid in their cavities.
> 3. WINDPIPES that draw in air.
>
> *Gesner's* dissertation on *vegetables* should be consulted about these.
> 1. FUNGUSES;
> 2. ALGAE; these have root, leaf and stem in one.
> 3. MOSSES; these have *anthers* without filaments, distant from the *female* flower, which has no pistil; and the *seeds* lack their own integuments and colytedons.
> 4. FERNS; their *fronds* produce fruit on the lower side.
> 5. GRASSES: these have very simple *leaves*, jointed *culms, calyces* with husks, and solitary *seeds*.
> 6. PALMS: these have simple *stems* with *fronds* at the tip, and fruit-bodies on *stalks* with *spathes*.
> 7. PLANTS is the designation of the rest, which cannot be included in the families listed above.
>
>> *Herbaceous* plants die every year above the root, while the root of every perennial herb produces a bud under the ground.
>>
>> *Shrubs*; the stem springs up above the ground, without buds.
>>
>> *Trees*; the stem springs up above the ground, with buds.
>>
>>> Either the buds differentiate the shrubs from the trees, or there is no distinction; for size is of no importance: therefore Pontedera called [all] the producers of buds trees.
>>>
>>> The largest Indian trees will be called shrubs, because buds rarely occur in them; and so this division is not a natural one, because nature has established no distinction between a tree and a shrub, but popular opinion has done so.

79. The PARTS of vegetables (78), that must be distinguished by the beginner, are three: the *root*, the *herb*, and the *fruit-body*.

> A vegetable is made of *marrow* 1, which is covered by *wood* 2, which is made of *rind* 3, which separates from the *bark* 4, which is overlaid with a *cuticle* 5.
>
> *Marrow* grows by extending itself and its *integuments*.
>
> The end of a *medullary* fibre, protruding through the bark, develops into an imbricated *bud*, from leaflets that will never be renewed.
>
> The bud is a summary of the herb, and extends indefinitely, until the fruit-body sets a limit to the former vigorous growth.
>
> The fruit-body is formed, when leaves that would [otherwise] be separate coalesce into a calyx; this makes the tip of the twig burst into flower in less than a year; then the fruit cannot initiate new life out of the *medullary* substance, unless the *woody* essence of the stamens is first absorbed by the liquid of the pistil. See *Loefling De Gemmis*.
>
> Every *vegetable* is propagated in continuation from the root.
>
> Every *fruit-body* is produced from the root through the herb.
>
> Every vegetable *ends* with the fruit-body, otherwise it would hardly cease to grow.
>
> There is no fresh creation, but *continued* generation, since the *corcle* of the seed consists of a medullary part of the root.

80. The ROOT (79), which draws nourishment in, and produces the herb (81) with the fruit-body (IV), is composed of *marrow, wood, rind*, and *bark*; and it consists of a *stock* and a *radicle*.

> A. The *radicle* is the fibrous part of the root, in which the descending stock terminates and by which the root draws in nourishment for the sustenance of the vegetable.
>
> B. The *descending stock* slowly steals under the ground, and puts forth *radicles* (A), which botanists distinguish by various names because of their varying structure.
>
> > 1. *Perpendicular*, which goes straight down.
> > 2. *Horizontal*, which extends sideways under the ground: *Iris*.
> > 3. *Simple*, fig. 129, not subdivided.
> > 4. *Branching*, fig. 130, divided into lateral branches.
> > 5. *Spindle-shaped*, fig. 129, oblong, thick, and tapering: as *Daucus* and *Pastinaca*.
> > 6. *Tuberous*, fig. 128, consisting of bodies that are almost round, gathered into a bundle: *Paeonia, Hemerocallis, Helianthus, Solanum*, and *Filipendula*.
> > 7. *Creeping*, fig. 131, extending far and sending off radicles hither and thither.
> > 8. *Fibrous*, consisting of fibrous radicles only.
> > 9. *Premorse*, which is cut off below and does not end in a tapering tip: as *Scabiosa, Plantago*, and *Valeriana*.

C. *The ascending stock* slowly rises above the ground, often displaying various forms of the trunk, and it produces the herb (81).

So all *trees and shrubs* are roots above ground.

Therefore, when a tree is turned upside down, it produces leaves from the descending stock, and radicles from the ascending one.

81. The HERB is the part of a vegetable that rises from the root (80), and terminates in the fruit-body; and it comprises the *trunk, the leaves,* the *supports,* and the *winter bud.*

> The *trunk* multiplies the herbage and leads directly to the fruit-body; it is covered with leaves and terminates in the fruit-body.
>
> The *leaves* breathe out and in (like the lungs in animals), and provide shade.
>
> The *supports* are accessories, and the plant rarely perishes, even if deprived of them.
>
> The *winter bud* is a summary of the herb above the root (80) before it grows out.

82. The TRUNK (81) produces leaves and the fruit-body; there are six sorts of it: *stem, culm, scape, peduncle, petiole, frond,* and *stipe;* but a *branch* is a part [of it].

> A. The STEM, the trunk peculiar to the herb, holds up the leaves and the fruit-body.
> a. *Simple,* extending in a continuous line towards the tip.
> 1. *Entire,* the simplest, with hardly any branches.
> 2. *Bare,* without any leaves: *Euphorbia, Cactus, Stapelia, Ephedra,* and *Cuscuta.*
> 3. *Leafy,* furnished with leaves.
> 4. *Zigzag,* bending this way and that according to the joints: *Ptelea.*
> 5. *Revolving,* fig. 115, climbing up spirally on the branch of another [plant].
>   Turning to the left ☾ ; in common parlance, following the sun: *Humulus, Helxine, Lonicera,* and *Tamus.* Turning to the right ☽ : in common parlance, against the motion of the sun: *Convolvulus, Basella, Phaseolus, Cynanchum, Euphorbia, Eupatorium.*
> 6. *Bending back,* in the shape of a bow, towards the ground: *Ficus.*
> 7. *Lying down,* horizontally on the ground.
> 8. *Creeping,* fig. 112, putting out radicles on this side and that, lying down: *Hedera, Bignonia.*
> 9. *As a runner,* fig. 131, creeping, almost bare.
> 10. *Parasitic,* adhering to another plant, and not to the ground: *Epidendron, Viscum,* and *Tillandsia.*
> 11. *Rounded,* cylindrical.
> 12. *Two-edged,* having two opposite corners: *Sisyrinchium.*

13. *Two-cornered, three-cornered, four-cornered, five-cornered,* and *many-cornered* are all sorts of the preceding (12).

14. *Three-edged,* possessing three flat sides.

15. *Triangular, quadrangular, quinquangular,* and *multangular,* from the number of the projecting corners.

16. *Furrowed,* ploughed with wide and deep hollowed-out furrows.

17. *Streaked,* engraved with very thin hollowed-out lines.

18. *Glabrous,* with smooth surface.

19. *Villous,* downy with soft hairs: *Tomex* and *Rhus.*

20. *Scabrous,* roughened with stiffish projecting points.

21. *Prickly,* sprinkled with stiff bristles.

Branchy, furnished with lateral branches.

22. *Ascending,* with branches turning upwards.

23. *Diffuse,* with spreading branches.

24. *Distichous,* putting out branches in a horizontal direction.

25. *With arms,* fig. 117, having branches opposite [each other] cross-wise.

26. *Very branchy,* loaded with many branches, not in any order.

27. *With supports,* descending, with branches to the root: *Ficus.*

28. *Prolific,* sending out branches only from the centre of the tip: *Pinus.*

The rest as in the entire [stem].

b. *Compound,* subdivided and dissolving into small branches as it ascends.

29. *Dichotomous,* always dividing into pairs.

30. *Subdivided,* into branches, not in any order.

31. *Jointed,* geniculate with internodes: *Piper.*

B. The CULM, the trunk peculiar to grass (78), holds up the leaves and the fruit-body; it keeps as many sorts of stalk as possible.

32. *Without knots,* continuous, not interrupted.

33. *Jointed,* fig. 114, connected by various joints.

34. *Scaly,* fig. 111, covered with overlapping scales.

C. The SCAPE, fig. 113, a complete trunk, holding up the fruit-body, but not the leaves: *Narcissus, Pyrola, Convallaria,* and *Hyacinthus.*

D. The PEDUNCLE, a partial trunk, holding up the fruit-body, but not the leaves. The *pedicel* is a partial peduncle. It is defined by place or manner.

By the place in which it is inserted into the basic part of the plant.

1. *Radical,* issuing directly from the root.

2. *Cauline,* situated on the stem.

3. *Rameal,* issuing from the branches.

4. *Axillary,* from an axil, to wit, between a leaf and the stem, or between a branch and the stem.

5. *Terminal,* forming the ends of branches or the stem.

6. *Solitary,* the only one in a place.

7. *Scattered,* when several are produced, not in any order.

By the manner in which it displays flowers and attaches them at the top.

8. *With one flower, two flowers*, etc., or *many flowers* according to the number of fruit-bodies on a single peduncle.

9. A *bundle* constitutes a collection of erect parallel flowers, sloping to a point, and close together: *Dianthus barbatus*.

10. A *head* consists of very many flowers brought firmly together into a globe: *Gomphrena*.

11. A *spike*, fig. 165, sessile flowers sparsely alternate on a simple common peduncle.
    Secund, with flowers turned to one and the same side.
    Distichous, with flowers facing towards both sides.

12. A *cluster*, fig. 163, is made out of a spike, when single flowers are furnished with their own petioles, the height being raised in proportion: *Spiraea with Opulus leaf, Ledum, siliquous* [plants].

13. A *panicle*, fig. 167, the fruit-body scattered over peduncles that are variously divided.
    Diffuse, when the pedicels are wide apart.
    Close set, when the pedicels are close together.

14. A *thyrsus* is a panicle close set in an egg-shaped form: *Syringa* and *Petasites*.

15. A *raceme*, fig. 164, consists of a peduncle with short lateral branches: *Vitis* and *Ribes*.

16. A *whorl*, fig. 166, is made of numerous almost sessile flowers surrounding the stem in a ring.

E. The PETIOLE is a sort of trunk which attaches to the leaf, but not to the fruit-body.
   The petiole, the peduncle, and the pedicel were synonymous to our predecessors, but hardly to us.
   The petiole sends forth a leaf, but the peduncle a fruit-body; very rarely one occurs that does both, as sometimes in *Turnera* and *Hibiscus*.
   A branchy stem always causes opposite leaves.

F. The FROND, fig. 108, a sort of trunk, from a branch, joined to a leaf and often to a fruit-body.
   Peculiar to *ferns* and *palms*.

G. The STIPE is the base of a frond, peculiar to *palms, ferns*, and *funguses*.

83. The LEAF is considered according to its *simplicity, complexity*, or *circumstances*.[1]

A. SIMPLE (Plate I), when the petiole displays a solitary leaf, and the latter varies in outline, corners, edge, surface, tip, and tissue.

a. The *outline* is concerned with the circumference apart from recesses and corners.

1. *Circular*, fig. 1, with diameters length-wise and cross-wise equal, and circumference rounded.
2. *Almost round*, fig. 2; its shape very close to the circular.
3. *Ovate* [egg-shaped], fig. 3; its length-wise diameter is greater than its cross-wise one, with the base circumscribed by a segment of a circle, but the top narrower than that.
4. *Oval or elliptical*, fig. 4; its length-wise diameter is greater than its cross-wise one, with the upper and lower ends narrower.
5. *Parabolic*, fig. 110; its length-wise diameter is greater than its cross-wise one, and from its base upwards it narrows into a semi-ovate [shape.]
6. *Spatula-shaped*, fig. 109; its shape nearly round, with a narrow base, which is linear and elongated.
7. *Wedge-shaped*, fig. 45; its length-wise diameter is greater than its cross-wise one, and it gradually narrows downwards.
8. *Oblong*, fig. 5; its length-wise diameter is several times greater than the cross-wise one, and either end is narrower than a segment of a circle.

b. The corners are projecting parts of a horizontal leaf.
9. *Lanceolate*, fig. 6, oblong (8), becoming gradually thinner on both sides towards the end.
10. *Linear*, fig 7, of even width throughout, sometimes becoming narrower only at the end.
11. *Needle-shaped*, fig. 105, an evergreen linear leaf (10): as in *Pinus*, *Abies*, *Juniperus*, and *Taxum*.
12. *Awl-shaped*, fig. 8; linear (10) below, but becoming gradually thinner towards the tip.
13. *Three-cornered*, fig. 12, when three projecting corners surround the disc.
14. *Four-cornered* and *five-cornered*, fig. 20; these are forms of the preceding.
15. *Deltoid*, fig. 58, rhombic from its four corners, of which the lateral ones are less distant from the base than the others.
16. *Round*, without any corners.

c. Recesses cut the disc of the leaf into sections.
17. *Kidney-shaped*, fig. 9; almost round (2).
18. *Heart-shaped*, fig. 10; ovate (3), hollowed out at the base, without any corners at the back.
19. *Crescent-shaped*, fig. 11; almost round (2), hollowed out at the base, noted for its corners at the back.
20. *Arrow-headed*, fig. 13; three-cornered (13), hollowed out at the base, furnished with corners at the back.
21. *Halberd-shaped*, fig 15; three-cornered (13), with base and sides hollowed out, and corners extended: *Rumex*.
22. *Fiddle-shaped*, oblong (8), broader at the bottom, and narrowed at the sides.
23. *Split*, fig. 16; divided by linear recesses and straight edges: *split into two, three, four, five, etc.*, or *many*, according to the number of this [feature].

24. *Lobed*, fig. 19; divided to the middle into separated parts, with convex edges: *two-lobed*, *three-lobed* fig. 17, *four-lobed*, or *five-lobed*, according to the number of this [feature].

25. *Palm-shaped*, fig. 22; divided length-wise into several almost equal parts towards the base; but there, they cohere into one.

26. *Split like a feather*, fig. 23; divided cross-wise, with oblong horizontal strips.

27. *Lyre-shaped*, fig. 76; divided cross-wise into strips, in such a way that the upper ones are larger, and the lower ones more widely separated.

28. *Slashed in strips*, fig. 24; cut into sections in various ways, the sections likewise subdivided at random.

29. *With recesses*, fig. 25; sustaining wide recesses in its sides.

30. *Cloven*, fig 28; divided right down to the base: *cloven into two, three, four, five*, or *many*, according to the number of this [feature].

31. *Entire*; undivided, without any recess; therefore it is contrasted with the foregoing (17–30).

d. The tip is the end of the leaf, in which it terminates.

32. *Truncated*, ending in a cross-wise line.

33. *Premorse*, fig. 18; having a very obtuse end with uneven incisions.

34. *Retuse*, fig. 46; terminating with an obtuse recess.

35. *Emarginate*, fig. 45; terminating with a notch.

36. *Obtuse*, fig. 40; terminating as it were within a segment of a circle.

37. *Acute*, fig. 41; terminating with an acute angle.

38. *Tapering*, fig. 42; terminating in an awl-shaped tip.

39. *Tendrilled*; ending in a tendril (section 84): *Gloriosa, Flagellaria, Nissolia* T.

e. The edge is the outermost border of the leaf at the sides, when the disc of the leaf is intact.

40. *Thorny*, going out at the edge into points that are relatively hard, stiff, and prickly.

An *unarmed* leaf is contrasted with one that is thorny.

41. *Toothed*, fig. 30; having horizontal points of the same texture as the leaf, separated by spaces.

42. *Saw-toothed*, fig. 31; distinguished by sharp corners, overlapping, and pointing towards the end.

43. *Scalloped*, fig. 38; its edge cut by corners which do not point towards either end:

bluntly [scalloped], fig. 36; sharply, fig. 35; doubly, fig. 33.

44. *Repand*,[2] fig. 29; its edge terminating in corners and the recesses between them, [each] inscribed in a segment of a circle.

45. *Cartilaginous*, fig. 34; its edge stiffened by a cartilage, of a tissue very different from that of the leaf.

46. *Ciliate*, fig. 50; its edge fortified with bristles that are parallel length-wise.

47. *Lacerated*; cut at the edge in various ways, with irregular segments.

48. *Gnawed*, fig. 21; when a leaf with recesses also has other very small recesses at the edge.

49. *Perfectly entire*, fig. 42; its outermost edge entire, without any scalloping.

f. The surface is the covering of the supine (upper) and the prone (lower) discs of the leaf.

50. *Sticky*, smeared with a dampness which is not fluid, but firm.

51. *Tomentose*, fig. 48; covered with villi, which are not very conspicuous; therefore it is usually whitish; as plants of the seaside and plains, which are exposed to the winds.

52. *Woolly*, as if clad with a spider's web: *Salvia* and *Sideritis*.

53. *Shaggy*, fig. 47; when the surface is covered by distinct elongated hairs: *Cortusa*.

54. *Prickly*, fig. 49; when fairly stiff fragile bristles are scattered over the disc of the leaf.

55. *Scabrous*, when fairly stiff protuberances are scattered over the disc of the leaf.

56. *Prickled*; when the disc of the leaf is engrossed by stiff pricking points.

57. *Streaked*; when lines are hollowed out on the surface, and are drawn length-wise and parallel.

58. *Pimply*,[3] fig. 54; covered with vesicular points.

59. *Punctured*; with a scattering of hollowed-out points.

60. *Shining*; bright with glabrosity: *Fenula canadensis* and *Angelica canadensis*.

61. *Pleated*, fig. 37; when the disc of the leaf goes up and down into angles towards the edge.

62. *Wavy*; it becomes so when the disc of the leaf goes up and down convexly towards the edge: *Alchemilla*.

63. *Curled*, fig. 39; when the circumference of the leaf turns out to be greater than the disc allows, so that it becomes wavy: all curled leaves are monstrosities.

64. *Wrinkled*, fig. 51; when the veins of the leaf turn out to be somewhat more contracted than the disc, so that the tissue in between rises: *Salvia*.

65. *Concave*, when the edge of the leaf becomes too constricted to circumscribe the disc, and so the disc is pushed down.

66. *Veiny*, fig. 52; when the vessels that run in different directions turn out to be very much branched, and have junctions that are visible to the naked eye.

67. *Nervous*, fig. 53; when very simple vessels without branches extend from the base towards the tip.

68. *Coloured*; assuming a colour other than green: *Amaranthus tricolor*.

69. *Glabrous*; with a smooth surface without any unevenness.

g. The tissue of the leaf is examined as to its sides.

70. *Rounded*, fig. 62; for the most part cylindrical.

Semi-cylindrical; rounded, but flat length-wise on one of the two sides.

71. *Tubular*; having an internal cavity (if it is cut): *Cepa*.

72. *Fleshy*; full of pulp on the inside: *succulent [plants]*.

73. *Compressed*; pressed together from opposite marginal sides, so that the tissue of the leaf becomes greater than the disc.

74. *Flat*, displaying both surfaces parallel throughout.

75. *Humped*; making both surfaces convex (76), with a fair quantity of pulp in the middle.

76. *Convex*; raised more in the disc.

77. *Depressed*; pressed down more in the disc than at the sides.

78. *Grooved*, fig. 61; hollowed out from a deep furrow along its whole length, almost into a half-cylinder.

79. *Sword-shaped*; two-edged (83), becoming thinner from the base towards the tip.

80. *Sabre-shaped*, fig. 56; compressed and fleshy, with one edge convex and narrow, the other straighter and thicker: *Mesembryanthemum. Dill[enius]*.

81. *Chopper-shaped*, fig. 57; compressed, almost round, and obtuse; humped (75) towards the outer side, with a sharp edge; rather more round below: *Mesembryanthemum. Dill[enius]*.

82. *Tongue-shaped*, fig. 55; linear and fleshy, obtuse, and convex underneath; usually with a cartilaginous edge: *Mesembryanthemum. Dill[enius]*.

83. *Two-edged*; sustain two projecting length-wise corners, with a somewhat convex disc.

84. *Three-edged*, fig. 59; with three flat sides length-wise in an awl-shaped leaf.

85. *Furrowed*, fig. 60; ploughed up length-wise by numerous corners and as many recesses placed between them.

86. *Keeled*; when the prone side of the disc projects length-wise.

87. *Membranous*; not filled with any perceptible pulp between the two surfaces.

B. The COMPOUND [LEAF] (Plate II) provides for several leaves on one petiole, according to the structure or the degrees [of it].

  h. The structure is concerned with the insertion of the leaflets.

88. *Compound*; when a single petiole puts forth more than one solitary leaf.

89. *Jointed*, fig. 100; when one leaf grows out of the top of another.

90. *Fingered*, fig. 66; when a single petiole has several leaflets attached to its tip.

91. *Binate*, fig. 63; fingered (89), with two leaflets.

92. *Ternate*, figg. 64 and 65; fingered (89), with three leaflets.

93. *Quinate*, fingered (89): with five leaflets.

94. *Feather-shaped*: when a single petiole has several leaflets attached to its sides.

With an odd [leaflet], fig. 68; feather-shaped, terminating with an odd leaflet.

Tendrilled, fig. 72; feather-shaped, terminating with a tendril (section 84).

Broken off, fig. 69; feather-shaped, terminating neither with a tendril nor with a leaflet.

In opposite fashion, feather-shaped, with opposite leaflets (111).

In alternate fashion, fig. 70; feather-shaped, with alternate leaflets (112).

In interrupted fashion, fig. 71; feather-shaped, with alternate leaflets of smaller size.

In jointed fashion, fig. 75; feather-shaped, jointed in a common petiole.

In decurrent fashion, fig. 74; feather-shaped, with leaflets running downwards (121) along the petiole.

95. *Coupled*; when a feather-shaped [leaf] (94) consists of only two leaflets and no more.

i. The degrees are concerned with the subdivision of the common petiole.

96. *Decompound*; when the petiole, [already] divided once, has several leaflets attached to it.

97. *Doubly coupled*, when a petiole that is divided into two (section 82) has four leaflets attached to its tips.

98. *Biternate* (doubly ternate), fig. 77; when the petiole has three ternate leaflets attached to it: *Epimedium*.

99. *Bipinnate* (doubly pinnate), fig. 78; when the petiole has feather-shaped leaflets attached to its sides (94).

100. *Pedate* (branchy), fig. 67; when a petiole that is split into two has leaflets attached to it only on the inner side: *Passiflora* and *Arum*.

101. *Supra-decompound*; when a petiole, that has been divided any number of times, has very many leaflets attached to it.

102. *Triternate* (triply ternate), fig. 79; when the petiole has three biternate (98) leaflets attached to it.

103. *Tripinnate* (triply pinnate), fig. 80; when the petiole has several bipinnate (99) leaflets attached to it.

C. The CIRCUMSTANCES [of a leaf] (Plate III) attract attention because of something else (not its own structure); such as the *location, situation, insertion,* or *direction.*

k. The location in which they are attached to a part of the plant.

104. *Seminal*, fig. 88; it was initially a cotyledon (section 86; VI), and is the first [leaf] on the plant.

105. *Radical*; situated on the root (section 80).

106. *Cauline*, fig. 89; situated on the stem (section 82; A).

107. *Rameal*, fig. 90; situated on a branch (82).

108. *Axillary* (under the wing), inserted at the place from which a branch issues (82).

109. *Floral*, fig. 91; inserted at the place from which a flower springs (82; C).

l. The situation is the arrangement of the leaves on the stem of the plant.

110. *Star-shaped* (whorled), fig. 101; when more than two leaves surround the stem in a whorl.

111. *Sets of three, four, five, six*, etc., fig. 102; are forms to be counted as star-shaped (110): *Nerium, Brabejum*, and *Hippuris*.

112. *Opposite*, figg. 82–87 and 103; when two cauline leaves are placed directly opposite each other, cross-wise in pairs.

113. *Alternate*, fig. 104; when one [leaf] issues after another, as it were step by step.

114. *Scattered*; when they are very numerous in a plant, without any order.

115. *Crowded together*, fig. 105 [*sic*]; when they are so numerous that they occupy whole branches, with hardly any space left.

116. *Overlapping*, fig. 106; if they are crowded and upright (115), so that they partly cover each other.

117. *Bundled*, fig. 107; if several leaves issue from the same point: *Larix*.

118. *In two lines*; if all the leaves relate only to the two sides of the branch: *Abies, Diervilla*.

m. The insertion of a leaf is done from its base.

119. *Shield-shaped*, fig. 92; if the petiole is inserted into the disc of the leaf, (not at the edge or at the base): *Nymphaea, Hernandia*, and *Colocasia*.

120. *Petiolate*, fig. 93; if the petiole (section 82) is attached to the leaf at the edge of its base.

121. *Sessile*, fig. 94; if the leaf has no petiole (section 82.E) and is attached directly to the stem (82.A).

122. *Decurrent*, fig. 95; if the base of a sessile (121) leaf is further extended downwards along the stem, beyond the base: *Verbesina, Carduus, Sphaeranthus*.

123. *Embracing the stem*, fig. 96; if the base of the leaf encompasses the sides of the stem in every direction cross-wise.
Half-embracing the stem; differing from the preceding [by being so] to a lesser degree.

124. *Perfoliate*, fig. 97; if the base of the leaf encircles the stem cross-wise in every direction: *Bupleurum*.

125. *United*, fig. 98; if opposite leaves (112) are united to each other to make one.

126. *Sheathing*, fig. 99; if the base of the leaf forms a cylindrical tube, covering the branch (82): *Polygonum* and *Rumex*.

n. Direction.

127. *Turned to*; turning its side (not towards the sky, but) to the south: *Amomum*.

128. *Slanting*; when the base of the leaf faces the sky, and the tip faces the horizon: *Protea* and *Fritillaria*.

129. *Bent inwards* (curved inwards), fig. 82; when it is arched upwards towards the stem.
130. *Pressed close*; when the disc of the leaf is next to the stalk.
131. *Upright*, fig. 83; set on the stem at a very acute angle.
132. *Spread out*, fig. 84; set on the stem at an acute angle.
133. *Horizontal*, fig. 85; leaving the stem at a right angle.
134. *Leaning back*, fig. 86; curved downward, so that the tip becomes lower than the base; also called *bent back* by some people.
135. *Rolled back*, fig. 87; rolled back downwards.
136. *Hanging down*; directly facing the ground.
137 *Rooting*; if the leaf puts forth roots.
138. *Floating*; floating on the surface of the water: *Nymphaea* and *Potamogeton*.
139. *Submerged*; hidden below the surface of the water.

84. The SUPPORTS (81) are accessories to sustain the plant more aptly; nowadays seven of them are included in the reckoning: the *stipule*, the *bract*, the *spine*, the *prickle*, the *tendril*, the *gland*, and the *hair*.

1. The *STIPULE*, fig. 118b; a scale which is present on both sides at the bases of the petioles (82E) or of the peduncles (82D),[4] as they shoot out: the *Papilionaceae*, *Tamarindus*, *Cassia*, *Rosa*, *Melianthus*, *Liriodendrum*, *Armeniaca*, *Persica*, *Padus*, etc.
2. The *BRACT*, fig. 120; a name given to a floral leaf (83:100) when it differs from the other [leaves] in colour and shape: *Tilia*, *Fumaria bulbosa*, *Lavandulae Stoechas*, *Salviae Hormium*.
3. The *SPINE*, fig. 121; a sharp point thrust out from the woody part of the plant: *Prunus*, *Rhamnus*, *Hippophaë*, *Celastrus*, *Lycium*.
   In cultivation, it quite often disappears: as in *Pyrus*.
4. The *PRICKLE*, figg. 122 and 123; a sharp point of the plant, fixed only to the bark: *Rubus*, *Ribes*, and *Berberis*.
5. The *TENDRIL*, fig. 118; a spiral thread-like band by which the plant is tied to another body: *Vitis*, *Bannisteria*, *Cardiospermum*, *Pisum*, *Bignonia*.
6. The *GLAND*, fig. 119; a nipple that excretes liquid: *Urena*, *Ricinis*, *Jatropha*, *Passiflora*, *Cassia*, *Opulus*, *Turnera*, *Salix tetrandra*, *Heliocarpus*, *Bryonia zeyl[anica]*, *Acacia cornigera*, *Bauhinia aculeata*, *Armeniaca*, *Amygdalus*, *Morisona*.
   The location: the petioles, serrations of the leaves, and delicate stipules.
7. The *HAIR*, is a bristling excretory duct of the plant.
   Guettard made some important discoveries about these.

85. The *WINTER-BUD* (81) is the part of the plant that protects the embryo herb (81) from external injuries; and it is either a *bulb* or a *bud*.

1. The *BULB* is a winter-bud situated on the stock as it goes down (80). [It may be] *Scaly*, fig. 125; consisting of overlapping gills: *Lilium*.

Solid, fig. 126; consisting of solid tissue: *Tulipa.*

Tegumented, fig. 127; consisting of numerous teguments: *Cepa officinarum.*

Jointed, consisting of gills forming little chains: *Lathraea, Martynia,* and *Adoxa.*

2. The *BUD* is a winter-bud situated on the stock as it goes up (80). It consists either of stipules, or of petioles, or of the rudiments of leaves, or of scales made of bark.

Whereas most plants in *cold* countries possess buds, hardly any do in *hot* countries. Various trees *have no buds:*

Philadelphus, Frangula T., *Alaternus* T., *Paliurus* T., *Jatropha, Hibiscus, Bahobab, Justicia, Cassia, Mimosa, Gleditsia, Erythrina, Anagyris, Medicago, Nerium, Viburnum, Rhus, Tamarix, Hedera, Erica, Malpighia, Lavatera, Solanum, Asclepias, Ruta, Geranium, Petiveria, Pereskia* Pl., *Cupressus, Thuja,* and *Sabina.*

The *forms* of buds are various.

Deciduous in Dentaria, Ornithogalum, Lilium, and Saxifraga.

Bearing leaves, not flowers: Alnus.

Separate buds bearing leaves and bearing flowers: Populus, some species of Salix, and Fraxinus.

Bearing leaves and bearing female flowers: Corylus and Carpinus.

Bearing leaves and bearing male flowers: Pinus and Abies.

Bearing leaves and bearing hermaphrodite flowers: Daphne, Ulmus, Cornus, and Amygdalus.

Bearing leaves and flowers [together], as most trees do.

Loefling's dissertation *De Gemmis arborum* contains further information.

IV. FRUCTIFICATIO.

86. FRUCTIFICATIO (79 Vegetabilium pars temporaria, Generationi dicata, antiquum terminans, novum incipiens; hujus Partes VII. numerantur:

I. CALYX, Cortex plantæ in fructificatione præsens.

1. *Perianthium*, Calyx Plantæ (78) Fructificationi contiguus.
  a. *Fructificationis*, Stamina Germenque includens.
  b. *Floris*, Stamina absque Germine continens.
  c. *Fructus*, Germen absque Staminibus continens.
2. *Involucrum* f. 135. Calyx Umbellæ (29) a flore remotus.
  a. *Universale* umbellæ universali subjectum.
  b. *Partiale* umbellulæ partiali subjectum.
3. *Amentum* f. 137. Calyx ex Receptaculo communi paleaceo gemmaceo.
4. *Spatha* f. 132. 133. Calyx *Spadicis* (31) longitudinaliter ruptus.
5. *Gluma* f. 134. Calyx *Graminis*, valvis amplexantibus.
      Arista, mucro glumæ insidens.
6. *Calyptra* f. 136. Calyx *Musci* cucullatus, antheræ superimpositus.
7. *Valva* f. 139. Calyx *Fungi* membranaceus, undique lacerus.

II. COROLLA, Liber plantæ in Flore præsens.

8. *Petalum*, tegmen floris corollaceum.
    a. *Tubus* f. 142. a. corollæ monopetalæ pars inferior tubulosa.
    b. *Limbus* f. 142. b. corollæ monopetalæ pars superior dilatata.
        Campanulatus, ventricosus absque tubo.
        Infundibuliformis, conicus, tubo impositus.
        Hypocrateriformis f. 142 planus, tubo impositus.
        Rotatus planus, nulli tubo impositus. *+ brevissimo*
        Ringens, irregularis, in duo labia personatus.
    c. *Unguis* f. 144. a. Corollæ polypetalæ pars inferior basi affixa.
    d. *Lamina* f. 144. b. Corollæ polypetalæ pars superior patula.
        Cruciformis f. 144. petalis quatuor æqualibus patens.

Papilio-

❧ IV. THE FRUIT-BODY

86. The FRUIT-BODY (79), a temporary part of vegetables, dedicated to reproduction, ending the old and beginning the new; seven parts of it are included in the reckoning.

    I. The CALYX, the bark of the plant that is present in the fruit-body.
      1. The *perianth*, the calyx of the plant (78), adjoining the fruit-body.
        a. Of the *fruit-body*, enclosing the stamens and the ovary.
        b. Of the *flower*, containing the stamens, but not the ovary.
        c. Of the *fruit*, containing the ovary, but not the stamens.
      2. The *involucre*, fig. 135; the calyx of the umbel (29), distant from the flower.
        a. *Universal*, placed under the universal umbel.
        b. *Partial*, placed under the partial umbel.
      3. The *catkin*, fig. 137; a calyx consisting of a chaffy bud-like common receptacle.
      4. The *sheath*, figg. 132 and 133; a calyx on the stalk, torn length-wise.
      5. The *husk*, fig. 134; the calyx of a grass, with valves that clasp it.
          The *awn*, a sharp point situated on the husk.
      6. The *veil*, fig. 136; the calyx of a moss, placed over the anther.
      7. The *volva*, fig. 139; the membranous calyx of a fungus, rent all round.
    II. The COROLLA, the rind of the plant that is present in the flower.
      8. The *petal*, the corollaceous covering of the flower.
        a. The *tube*, fig. 142a; the lower, tubular part of a corolla consisting of a single petal.
        b. The *limb*, fig. 142b; the upper, expanded part of a corolla consisting of a single petal.
          Bell-shaped; bellying out, without any tube.
          Funnel-shaped; conical, placed on a tube.
          Salver-shaped, fig. 142; flat, placed on a tube.
          Wheel-shaped; flat, not placed on any tube.
          Ringent; irregular, mask-like, and [divided] into two lips.
        c. The *finger-nail*, fig. 144a; the lower part of a corolla consisting of several petals, fixed to the base.
        d. The *blade*, the upper, spreading part of a corolla consisting of several petals.

Cross-shaped, fig. 144; opening out with four petals, all alike.

Butterfly-shaped, irregular; with the lower petal boat-shaped, the upper going upwards, and those at the side solitary.

9. The *nectary*, the honey-producing part peculiar to the flower, figg. 145–8.

III. The STAMEN, the organ for the preparation of pollen.

10. The *filament*, the part that raises the anther, and has it attached to itself.

11. The *anther*, the part of the flower that is pregnant with pollen, which nature discharges.

12. The *pollen*, the powder of the flower, which will be burst by moisture and throw out elastic particles.

IV. The PISTIL, an organ adhering to the fruit, for the reception of the pollen.

13. The *ovary*, the rudiment of the immature fruit in the flower.

14. The *style*, the part of the pistil that raises the stigma from the ovary.

15. The *stigma*, the top of the pistil, which is damp with moisture from the bursting of the pollen.

V. The PERICARP, an organ pregnant with seeds, which it discharges when they are ripe.

16. The *capsule*, figg. 160, 159, and 161; a hollow pericarp, which dehisces in a definite way.

The *valvule*, fig. 159a; the wall by which the fruit is protected externally.

The *partition*, fig. 15b; the wall by which the fruit is internally divided into several chambers.

The *columella*, fig. 159c; the part that connects the internal wall with the seeds.

The *cell*, fig. 159d; an empty chamber to provide room for the seeds.

17. The *siliqua*, fig. 155; a pericarp with two valves, and seeds fixed to it along both seams.

18. The *pod*, fig. 154; a pericarp with two valves, and seeds fixed to it along one seam only.

19. The *conceptacle*, fig. 153; a pericarp with one valve, dehiscent length-wise at the side, separated from the seeds.

20. The *stone-fruit*, fig. 157; a pericarp filled [with tissue] and without any valve, containing a *nut*.

21. The *pome*, fig. 158; a pericarp filled [with tissue] and without any valve, containing a *capsule*.

22. The *berry*, fig. 158, a pericarp filled [with tissue] and without any valve, yet containing uncovered *seeds*.

23. The *cone*, fig. 138, a pericarp made out of a *catkin*.

VI. The SEED, a deciduous part of the vegetable, the rudiment of a new one, brought to life by the irrigation of the pollen.

24. The *seed* (properly), the fresh rudiment of a vegetable, irrigated with moisture and coated with a bag.

a. The *corcle*, the first beginning of a new plant within the seed.

> The *plumule*, the scaly part of the corcule, which goes up.
>
> The *rostellum*, the plain part of the corcule, which goes down.

b. The *cotyledon*, the lateral body of the seed, which is absorbent and drops off.

c. The *hilum*, the external scar on the seed, from its attachment to the fruit.

d. The *aril*, the peculiar outer coat of the seed, which comes off of itself.

e. The *coronule*, an epicalyx which adheres to it, and by which it becomes airborne.

> The *tuft*, fig. 162; a feathery or hairy corona which becomes air-borne.
>
> The *stipe*, a thread that supports [the tuft] and connects the tuft and the seed.

f. The *wing*, a membrane fixed to the seed, which becomes airborne and causes dissemination.

25. The *nut*, a seed covered with a bony outer shell.

26. The *layer*, the seed of a moss, stripped of its bark, discovered in 1750.

VII. The RECEPTACLE, the base by which the six [other] parts of the fruit-body are attached to each other.

27. The *peculiar receptacle*, which is concerned with the parts of only one fruit-body.

> Of the *fruit-body*, common to the flower and the fruit.
>
> Of the *flower*, the base to which the parts of the flower, except the ovary, are attached.
>
> Of the *fruit*, the base for the fruit, separate from the receptacle of the flower.
>
> Of the *seeds*, the base by which the seeds are fixed inside the pericarp.

28. The *common receptacle*, which has several florets attached to it, so that, if some are removed, this results in irregularity, fig. 140.

> The *palea*, a gill adhering to it, separating the florets, fig. 141.

29. The *umbel*, a receptacle extended from one and the same centre into proportionate thread-like peduncles.

a. *Simple*, not subdivided, for instance *Panax*.

b. *Compound*, with all its peduncles displaying smaller umbels at their tips.

> *Universal*, producing other, smaller umbels at its tips.
>
> *Partial*, the smaller umbel that issues from the universal one.

30. The *cyme*, a receptacle extended into erect bundled peduncles from one and the same universal centre, but with indeterminate partial [centres].

31. The *spadix*, fig. 133b; the receptacle of a *palm*, which grows out from within a sheath into small branches that bear fruit.

87. The parts of the FLOWER are the *calyx*, the *corolla*, the *stamen*, and the *pistil*.

The parts of the FRUIT are the *pericarp*, the *seed*, and the *receptacle*.

The parts of the FRUIT-BODY are therefore the *flower* and the *fruit*.

> The definition of the FLOWER is given variously by the authorities:
>
> *Jung's* [definition]:
>
> 'The flower is the more tender part of the plant, conspicuous in colour and shape or both, adhering to the rudiment of the fruit.'
>
> *Ray's*:
>
> 'The flower is the more delicate part of the plant, fleeting, conspicuous in colour and shape or both, coming before the fruit and usually adhering to it, and serving to protect and foster it when it is somewhat tender; and after it unfolds it soon either falls off or withers.'
>
> *Tournefort's*:
>
> 'The flower is a part of the plant that differs from the rest in form and substance; it usually adheres to the developing fruit, and apparently supplies it with the first nourishment, to deploy the very tender parts of the fruit.'
>
> *Pontedera's*:
>
> 'The flower is a part of the plant that differs from the rest in form and substance; and if the flower is furnished with a tube, it always either adheres to the embryo or is fixed as close to it as possible, and it serves for the benefit of the embryo; but if it has no tube, it does not adhere to an embryo at all.'
>
> *Ludwig's*:
>
> 'The flower is a part of the plant made of filaments and membranes, and it is distinguished from the rest of the lobes by its elegance and generally more subtle structure.'
>
> The definition of the FRUIT:
>
> *Jung's*: 'The fruit is a part of the plant that is annual, and it adheres to the flower and succeeds it; and when it has come to perfection, it falls off the plant of itself; and it is received by a suitable nursery, and makes the beginning of a new plant.'
>
> It is clear that the *calyx* is part of the flower, even thought it is often present on the fruit, for this reason, that the calyx never bursts into flower after the flowering.
>
> The calyx of the *Patagonula* is very much enlarged on the fruit.
>
> Many flowers are furnished with *deciduous calyces*, which fall off as soon as the flower unfolds, as in *Epimedium* and *Papaver*.

88. The essence of a FLOWER (87) consists in the *anther* (86) and the *stigma* (86).

The essence of a FRUIT (87), in the *seed* (86).

The essence of a FRUIT-BODY (87), in the *flower* and the *fruit*.

The essence of VEGETABLES (78), in the *fruit-body* (87).

It is difficult to elicit the character of the parts of plants, unless the two first premisses are assumed, those concerning the *pollen* and the *seed*.

1. The POLLEN is the powder of vegetables (Section 3), which will be burst when it is moistened with the appropriate liquid, and propulsively explode a substance which is not discernible by the naked senses.
2. The SEED is a deciduous part of the plant, pregnant with the rudiment of a new plant, and it is brought to life by the pollen.
3. The ANTHER is a vessel that produces and discharges *pollen* (1).[1]
4. The PERICARP is a small vessel that produces and discharges *seeds* (2).
5. The FILAMENT is the foot of the *anther* (3), by which it is attached to the vegetable.
6. The OVARY is the immature rudiment of the *pericarp* (4, or of the *seed* 2): it exists especially at the very time when the *anther* (3) discharges pollen (1).
7. The STIGMA is the dewy tip of the *ovary* (6).
8. The STYLE is a part of the *stigma* (7), and connects it with the *ovary* (6).
9. The COROLLA and the CALYX are the coverings of the *stamens* (1, 3, and 5) and the pistils (6, 7, and 8); and the latter springs from the bark-like *outer shell*, the former from the *rind*.
10. The RECEPTACLE is what attaches the aforementioned parts to each other (5, 6, and 9).
11. The FLOWER is produced from the *anther* (3) and the *stigma* (7), whether the coverings (9) are present or not.
12. The FRUIT is distinguished from the seed (2), whether the latter is covered by a pericarp (4) or not.
13. Every FRUIT-BODY possesses *anther* (3), *stigma* (7), and *seed* (2).
14. Every VEGETABLE is furnished with *flower* (11) and *fruit* (12); so that no species is without them.

The *essence* of a *seed* consists in the *corcule* (Section 86), which is attached to the *cotyledon*, and is enveloped by it, then closely clad with its own peculiar coating.

The *essence* of a *corcule* consists in the *plumule*, which is the point of the plant's actual life, a thing of very little mass; and like a *bud* it grows indefinitely; but the base of the plumule is the *rostellum*, and it goes down and puts out roots; it was formerly in contact with the parent.

The *layers of mosses* are seeds without coating or cotyledons, and so [they are] the plumule of an uncovered corcule, where the rostellum is fixed into the calyx of the plant.

89. The PERIANTH (86:I) differs from the *bract*, in that the former withers when the fruit is ripe, if not before; but the *floral leaflets* do not do so.

Examples of *bracts* are to be sought in *Melampyrum, Monarda, Salvia, Lavandula, Bartsia, Hebenstretia, Mussaenda, Tilia*, and *Fumaria*.

That a *bract* is often easily taken for a perianth is clearly shown in *Helleborus, Nigella, Passiflora, Hepatica*, and *Peganum*.

The *Perianth* is defined in the authorities as follows:

Malpighi: 'The calyx is the base and support of the flower; it fosters the stamens with its own physical structure of leaves, since it usually covers their growth too.'

Ray: 'The calyx is what holds the flower up, and is as it were its base and foundation; and so it is thicker and less conspicuous than the flower.'

Tournefort: 'The name calyx should be given to the hinder part of the flower, which is distinguished from the pedicel by a certain remarkable thickness.'

Ludwig: 'The calyx or perianth is the outer membrane of the flower.'

90. The COROLLA (86:IV [sic:II]) is distinguished from the PERIANTH (86:I) in that the *former* alternates in position with the stamens (86:III), but the *perianth* is placed opposite them.

Tournefort: 'The petals are the leaves that usually surpass the other parts in shape and colour, and which never become the involucre of a seed.'

Colonna was the first to call the petal a leaf of the flower.

That the stamens alternate with the petals, as the petals alternate with the perianth, so that now the stamens are placed opposite the lappets of the calyx, is clearly shown in perfect [flowers] *with four or five stamens*.

You should look for examples of this rule in *Chenopodium, Urtica*, and *Parietaria*, in which the corolla is lacking.

The doctrine that, where only one of the two, the perianth and the corolla, is present, it must be the corolla, is refuted by *Ammania, Isnarda, Peplis, Ruellia*, and *Campanula*, which often dispense with corollas, but not with calyces.

It is obvious to anyone that the calyx, since it springs from the back of the plant, is rougher and thicker than the corolla, which is produced from the tender, coloured, and soft rind. The boundaries between them are hardly ever defined, except by the colour; and this is not enough, for instance in *Bartsia*.

Most flowers bear coloured corollas, which are uncovered and apt to fall off during the time of flowering; then they become hardened, flourish, and persist, e.g. *Helleborus* and *Ornithogalum*.

It is obvious in *Daphnis* that nature has not set absolute boundaries between the calyx and the corolla; for there the two are joined together, and completely united at the edge, like the leaf of *Buxus*; see also [my] *Classes Plantarum* 5 n. 11.

Several [botanists] hold that *Euphorbia* has only one petal, since they take the calyx for the corolla; but the Indian annual species, with their very distinct white petals, show that the peltae of the flower are the petals.

91. The *number* of the PETALS (86:8) is to be taken at the base of the corolla; that of the *lappets* from the middle of the limb or blade.

> *Rivinus*: 'We count as many petals as the deciduous flower is divided into.'
>
> With only one petal: *Trientalis, Oxalis, Ledum, Anagallis, Veronica*.
>
> With five petals: *Alsines* [sic: *Alsine*]
>
> It is wrong to reckon *Oxycoccus* as four-petalled, even though it is divided into four parts; for they were originally one.
>
> *Rivinus*: 'Any flowers that adhere on the top of the fruit with their bottoms unbroken, have only one petal, even if they are not deciduous.
>
> On the other hand, any that nurse the fruit in their bosoms and consist of several petals sticking together, are to be reckoned according to the number of the petals, even though they do not disappear.'
>
> The number in those with several petals is determined by the arrangements of their insertions, as in *Hepatica*.
>
> The number of *petals* is commonly expressed by *Greek* words:
>
> *apetalus, petalodes, monopetalus, dipetalus, tripetalus, tetrapetalus, pentapetalus, hexapetalus*, and *polypetalus*.
>
> The *segments* of a corolla with only one petal are defined by various words ending in –oides: *pentapetaloides* and *tetrapetaloides*.
>
> It is rather difficult to divide *perianths* into those with one leaf and those with several leaves.
>
> *Vaillant*: 'If the *calyx* crowns the embryo, it has one leaf and constitutes a single body with the embryos.'
>
> 'The calyx has only one leaf, whenever the flower has only one petal': *wrong*.

92. The fruit-body's STRUCTURE of three kinds is observed by the botanist everywhere, in all parts of it (86): the *most natural*, that which *differs*, and the *peculiar*; and he must describe these with careful observation, according to four *measurements*: number, shape, proportion, and position.

> The four requirements for measurement, *number, shape, proportion*, and *position*, are so many pillars supporting the art.
>
> The *shape* is defined by the terms that describe the leaves: Section 83.
>
> The *proportion* is reckoned according to the relative heights of the parts.
>
> The *position* or the insertion or else the connexion of the parts, are the same.
>
> The following are quite often unsettled:
>
> *size, colour, smell*, and *taste*.

93. The MOST NATURAL structure (92) of the fruit-body is taken from a variety of existing [factors]: in α. the *number* (94), β. the *shape* (95), γ. the *proportion* (96), and δ. the *position* (97).

> The most natural [structure] occurs in very many plants:
> it usually possesses a calyx thicker and shorter than the corolla, which is tender and falls off;
> a pistil in the middle of the flower within the stamens; anthers resting on filaments; and stigmas resting on styles, etc.
> Fruit-bodies are all different, and yet in that they all agree.
> The most natural structure becomes familiar as a result of habit, and thereafter it is not noticed by a real botanist; but it is described at length by amateurs and the generality of travellers; and thereby they betray their ignorance, which is praised by the perverse.
> It seems to me that an eminent *writer on rustic matters*, though otherwise truly great, made little impression here.
> > As an *example*, I will set forth a character, crudely constructed: the CALYX green, erect, short, of five parts, with narrow segments, more narrow than long.[2] The COROLLA with five *petals*, flattened, narrower at the base, coloured, very thin, spread out, inserted into the receptacle within the calyx, descending to the base, alternating with the lappets of the calyx, falling before the fruit is ripe. Some *filaments* of the STAMENS, narrow, slightly rounded, narrower at the top, inserted into the receptacle within the corolla, alternate with the petals, but opposite the [lappets of the] calyx. A solitary *anther* in [each] individual filament, and it bursts and produces flour, and then withers. The PISTIL occupies the centre of the flower, within the stamens, and it has for its base the small *rudiment* of the fruit; and on the tip of that, the *styles* are situated, and they are narrower than the ovary; they are not flat but narrow, and separate from the corolla, and they display rather thick *stigmas* at their tips. After the plant has finished flowering, the ovary grows out and becomes the PERICARP, which fills the calyx, becomes hard, and displays a vestige of the style at its tip; eventually it bursts into several chambers and valvules, and throws out the taper-pointed seeds, which were fixed to the internal walls of the pericarp. *Who would say that it is* LINUM?

94. The most natural NUMBER (93) is that the *calyx* is divided into the same number of segments as the corolla, and that *filaments* correspond to them, each one furnished with a solitary anther. But the division of the *pistil* usually agrees with the chambers of the *pericarp* or the receptacles of the *seeds*.

> The number *five* is the most frequent in fruit-bodies, as is apparent in those with *five stamens*, those with *united anthers*, and others.

The *calyx* and the *corolla* turn out to be five-fold in most cases.
Lysimachia and *Linum* may serve as examples of the most natural number.

95. The most natural SHAPE (93) is that the *calyx* should be less open, and contain the gradually expanded *corolla*, which is furnished with erect *stamens* and *pistils*, gradually becoming thinner; and when all these, except the *calyx*, fall off, the *pericarp* swells and is enlarged, being packed with *seeds*.

> The *calyx* is a perianth that is somewhat erect, so as to support the corolla.
> The *corolla* quite often resembles the shape of a funnel, more or less.
> The filaments of the *stamens* usually become thin at the top; they are awl-shaped, erect, and slightly bent back at the tips.
> The styles of the *pistils* in most [plants] are erect and narrow, like the filaments.
> It is common knowledge that the *pericarp* swells and increases in size.

96. The most natural PROPORTION (93) exhibits a *calyx* smaller than the *corolla*, with *stamens* and *pistils* equal in length, if the flower is erect.

> A DROOPING flower displays a *pistil* longer than the *stamens*.
> A PROSTRATE flower display *stamens* and *pistils* bent down against the lower side.
> But an ASCENDING flower displays *stamens* and pistils hidden under the upper side.
> The *calyx* is shorter, when the bud of the fruit-body is present.
> The *corolla*, turns out to be large in most [flowers].
> In many, the *stamens* and pistils are hardly longer than the calyx.
> A *drooping* flower is so, to facilitate fertilization, Section 145.
> > Fritillaria, Campanula, Galanthus, and *Geranium*.
> > It follows that this is not caused by its own weight.
> It is *prostrate* in Cassia, and in every *flower with stamens in two sets*.
> It is *ascending* in didynamous gymnosperms.
> Where the *pistil* is shorter than the stamens, the anthers are shut in: *Saxifraga* and Parnassia.

97. The most natural POSITION (93) is that the *perianth* should envelop a *receptacle*, to which the *corolla* is joined alternately; on the inside the *filaments* correspond to the latter, and the *anthers* are situated on their tips; the *style* is situated on the tip of the ovary, and displays the *stigma* at its top. When these fall off, the ovary grows out into a *perianth*, which is supported by the calyx and encloses the *seeds* attached to the receptacle

of the fruit. The receptacle of the flower usually grows below, more rarely around or above.

> The arrangement of the fruit-body is as follows: the *calyx* will be outside the *corolla*, and the *stamens* inside the corolla, but they will be placed outside the *pistils*.
> The exceptions are comparatively rare, Section 111.
> > The receptacle connects the calyx, corolla, and stamens in four ways.
> > a. The receptacle is at the *base* of the ovary; as in most [plants]. The pistil goes off into fruit. *Tournefort*.
> > b. The receptacle is at the *tip* of the ovary; as in many [plants]. The calyx goes off into fruit. *Tournefort*.
> > c. The receptacle *encircles* the *ovary* or fruit: as in various *Saxifragas*.
> > d. The receptacle *goes about* the *ovary*, and it is extended above through the perianth, as in [plants] *with twenty stamens, Ribes, Mitella*, etc.
> The perianth and the corolla originate close to each other, so that if the corolla is situated on or under the ovary, the same is true of the perianth. Exceptions are rare. Section 100.
> > The perianth of *Hepatica* is distant from the corolla at its actual base; therefore it can hardly be a perianth, but rather an involucre; and this is confirmed by the related *Pulsatilla*.
> The ovary is small before fertilization; then it grows out large.
> > *Musa* is peculiar, since its ovaries are at their largest before flowering; and, if not fertilized, they quite often ripen when barren, though they hardly increase in mass.

98. The DIFFERING structure (93) of the fruit-body is derived from those parts that often differ in various plants.

> This will be the foundation of the genera and their characters.
> The more natural the class is, the less obvious is this structure.
> Every peculiar structure is a differing one, but not vice versa.

99. The CALYX differs as to α. number, composition, parts, and lappets. β. shape, evenness, edge and tip. γ. proportion. δ. location and duration.

> *Number:*   none: *Tulipa, Fritillaria*, and most of the *Lily family*.
>   single: *Primula* and *most* flowers.
>   double: *Malva, Hibiscus*, and *Bixa*.
> *Composition:*   overlapping, from the various scales' being placed one on top of another: *Hieracium, Sonchus*, and *Camellia*.
>   scalled, from the scales being spread out very extensively: *Carduus, Onopordum* and *Conyza*.

augmented (*with epicalyx*, Vaill[ant]), where a shorter, different series of leaves encircles the base of the calyx: *Corepsis, Bidens, Crepis*, and *Dianthus*.

with many flowers, shared by many florets: *Scabiosa* and [*flowers*] *with anthers united.*

Parts: with only one leaf: *Datura* and *Primula*.

with two leaves: *Papaver* and *Fumaria (including bulbosa)*.

with three leaves: *Tradescantia*.

with four leaves: *Sagina, Epimedium*, and *tetradynamia*.

with five leaves: *Citrus, Adonis, Gerbera*.

with six leaves: *Berberis*.

with ten leaves: *Hibiscus*.

Lappets: counted, especially in those with *only one leaf*.

entire: *Genipa*.

divided into two: *Utricularia*.

divided into three: *Alisma* and *Cliffortia*.

divided into four: *Rhinanthus*.

divided into five: *Nicotiana*.

divided into six: *Pavia*.

divided into eight: *Tormentilla*.

divided into ten: *Potentilla* and *Fragaria*.

divided into twelve: *Lythrum*.

Evenness: even: *Lychnis*.

uneven: *Helianthemum. Tournefort.*

with alternate [leaves] shorter: Tormentilla and *Potentilla*.

Shape: spherical: *Cucubalus*.

club-shaped: *Silene*.

bent back: *Asclepias*.

erect: *Primula* and *Nicotiana*.

Edge: perfectly entire: *in most [flowers]*.

serrated: species of *Hypericum*.

ciliated: species of *Centaurea*.

Tip: sharp: *Primula* and *Androsace*.

taper-pointed: *Hyoscyamus*.

blunt: *Nymphaea* and *Garcinia*.

with a single truncated denticle: *Verbena*.

Proportion: longer than the corolla: *Agrostemma, Sagina*, and species of *Antirrhinum*.

even with the corolla: species of *Cerastium*.

shorter than the corolla: *Silene*.

Location: of the flower: *Linnaea* and *Morina*.

of the fruit: *Linnaea* and *Morina*.

of the fruit-body: *Paeonia*.

Duration:	falling early, at the first opening of the flower: *Papaver* and *Epimedium.*
	falling late with the corolla: *tetradynamia* and *Berberis.*
	lasting until the fruit is ripe: *didynamia.*
THE INVOLUCRE:	with only one leaf: *Bupleurum.*
	with two leaves: *Euphorbia.*
	with three leaves: *Butomus* and *Alisma.*
	with four leaves: *Cornus.*
	with five leaves: *Daucus.*
	with six leaves: *Haemanthus.*
THE SHEATH:	with only one leaf: *Narcissus.*
	with two leaves: *Stratiotes*
	[with] overlapping [leaves]: *Musa.*

100. The COROLLA differs (98) as to α. petals, lappets, and nectaries (110); β. shape, evenness, and edge; γ. proportion; and δ. location and duration.

Petals:	the number of them is established by Rivinus' system: Section 61.
	With only one petal: *Convolvulus* and *Primula.*
	With two petals: *Circaea* and *Commelina.*
	With three petals: *Alisma* and *Sagittaria.*
	With four petals: *tetradynamia.*
	With five petals: *umbellate* [*plants*].
	With six petals: *Tulipa, Lilium,* and *Podophyllum.*
	With nine petals: *Thea, Magnolia,* and *Liriodendron.*
	With many petals: *Nymphaea.*
Lappets:	rather rare in those with many petals; frequent in those with only one petal; of the former:
	Two [lappets]: *Alsine* and *Circaea.*
	Three: *Holosteum* and *Hypecoum.*
	Four: *Lychnis.*
	Five: *Reseda.*
Nectaries:	about these, see Section 110.
Shape:	wavy: *Gloriosa.*
	folded: *Convolvulus.*
	rolled back: *Asparagus* and *Medeola.*
	twisted: *Nerium, Asclepias, Vinca,* and *natural order* 29.
Evenness:	even: *Primula.* uneven: *Butomus.*
	regular: *Aquilegia.* irregular: *Aconitum* and *Lamium.*
Edge:	scalloped: *Linum.*
	serrated: *Tilia* and *Alisma.*
	ciliated: *Ruta, Menyanthes,* and *Tropaeolum.*

with denticles placed in between: *Samolus, Sideroxylon* and *[plants] with rough leaves*. Ray.

with hairy surface: *Menyanthes* and *Hyperici Lasianthus*.

Proportion: very long: *Catesbaea, Siphonanthus, Brunfelsia*, and *Craniolaria*.

very short: *Sagina, Centunculus, Ribes*.

Location: the base of the corolla is near to the perianth, provided that it is present. There are rare examples of the corollas being distant from the calyx, with the ovary placed in between; as *Adoxa, Sanguisorba*, and *Mirabilis*.

Duration: lasting (till the fruit is ripe): *Nymphaea*.

falling early (when the flower opens): *Actaea* and *Thalictrum*.

falling late (when the flower falls off): *most [corollas]*.

decaying (it withers, but does not fall off): *Campanula, Orchis, Cucumis, Cucurbita, Bryonia*, and *those related to it* in natural order 45.

101.   The filaments of the STAMENS differ (98) as to α. number, β. shape, γ. proportion, and δ. position. The *anthers* differ as to α. number, chambers, and absence, β. shape and dehiscence, γ. attachment, and δ. position.

The FILAMENTS. The *number* differs, as in the sexual system.

Lappets: 2. *Salvia*; 3. *Fumaria*; 9. *Plants with two sets of stamens*.

Shape: like hair: *Plantago*.

flat: *Ornithogalum*.

wedge-shaped: *Thalictrum*

spiral: *Hirtella*.

awl-shaped: *Tulipa*.

emarginate: *Porrum*.

bent back: *Gloriosa*.

hairy: *Tradescantia* and *Anthericum*.

Proportion: uneven: *Daphne, Lychnis*, and *Saxifraga*.

irregular: *Lonicera* and *didynamia*.

very long: *Trichostema, Plantago*, and *Hirtella*.

very short: *Triglochin*.

Position: opposite the calyx: *Urtica*.

alternate with the calyx: *Elaeagnus*.

inserted into the corolla in those *with only one petal*, hardly in those *with many petals*.

sometimes inserted into the calyx in those *without petals*, as in *Elaeagnus*; and always in those with *twenty stamens*, and in *Oenothera* and those related to it in order 40.

They are commonly inserted into the receptacle, as are the calyx and the corolla.

The ANTHER: in *number* single,   in a solitary filament: *most* [*flowers*].

in [each of] three filaments: *Cucurbita*.

in [each of] five filaments: [*flowers*] *with united stamens*.

Two   in a solitary filament: *Mercurialis*.

Three   in a solitary filament: *Fumaria*.

Five   in [each of] three filaments: *Bryonia*.

in a solitary filament: *Theobroma*.

with *chamber*,   single: *Mercurialis*.

double: *Helleborus*.

triple: *Orchis*.

quadruple: *Fritillaria*.

with *absence*   of one: *Chelone* and *Martynia*.

of two: *Pinguicula* and *Verbena*.

of three: *Gratiola*, the *Bignonias*, and the *Geraniums*.

of four: *Curcuma*.

of five: *Pentapetes* and the *Geraniums*.

with *shape*   oblong: *Lilium*.

spherical: *Mercurialis*.

arrow-headed: *Crocus*.

angular: *Tulipa*.

horned: *Hamamelis*, *Erica*, the *Vacciniums*, and *Pyrola*.

with *dehiscence*   at the side: *most* [flowers], *Leucojum*.

at the tip: *Galanthus* and *Kiggelaria*.

going from the base to the tip: *Epimedium* and *Leontice*.

with *attachment*   by the tip: *Colchicum*.

by the base: *most* [flowers].

by the side: *Canna*.

by the nectary: *Costus*.

Position   in the filaments,   at the tip: *most* [flowers].

at the side: *Paris* and *Asarum*.

in the pistil: *Aristolochia*.

in the receptacle: *Arum*.

The POLLEN,   in *shape*,   prickly sphere: *Helianthus*.

perforated sphere: *Geranium*.

double sphere: *Symphytum*.

wheel-shaped and toothed sphere: *Malva*.

angular sphere: *Viola*.

knee-shaped sphere: *Narcissus*.

linden bark rolled into a ball: *Borrago*.

102. The PISTILS differ (98) as to α. number and lappets; β. Shape; γ. length and thickness; and δ; position; to wit, of the three parts.

> The three parts of the pistil: the *ovary*, the *style*, and the *stigma*.
>
> The ovary observes the laws of the pericarp; concerning this, see section 103.
>
> The style is always separate from the calyx and the corolla.
>
> > See *number* in the sexual system, where I have taken the number from the styles (if they are present, otherwise from the stigmas).

Lappets:	divided into two: *Persicaria* and *Cornutia.*
	divided into pairs: *Cordia.*
	divided into three: *Clethra* and *Frankenia.*
	divided into four: the *Rhamni.*
	divided into five: *Geranium.*
Shape:	cylindrical: *Monotropa.* angular: *Canna.*
	awl-shaped: *Geranium.* like hair: *Ceratocarpus.*
	thicker at the top: *Leucojum.*
Length:	very long: *Tamarindus, Cassia, Campanula, Scorzonera,* and *Zea.*
	very short: *Papaver.*
	with the same length as the stamens: *Nicotiana* and *most* [flowers].
Thickness:	thicker than the stamens: *Leucojum.*
	thinner than the stamens: *Ceratocarpus.*
	even with the stamens: *Lamium.*
Position:	at the tip of the ovary: most [flowers]: so there is no need to mention it.
	above and below the ovary: *Capparis* and *Euphorbia.* unless you take the lower part for an elongated receptacle.
	at the side of the ovary: [the order] *Icosandria polygynia,* and those related to it: natural order 35, *Hirtella* and *Suriana*
Duration:	lasting: *tetradynamia*
The stigma,	in *number,* single: in *most* [flowers].
	two: *Syringa.*
	three: *Campanula.*
	four: *Epilobium* and *Parnassia.*
	five: *Pyrola.*
Lappets:	rolled together: *Crocus.*
	like hair: *Rumex.*
	rolled back: *those with united anthers, Dianthus* and *Campanula.*
	bent left: *Silene.*
	[divided] into six parts: *Asarum.*
	divided into many parts: *Turnera.*

Shape: capitate: *Tribulus, Hugonia, Vinea,*
              Ipomoea, and *Clusia.*
         spherical: *Primula, Hottonia, Linnaea*
              and *Limosella.*
         egg-shaped: *Genipa.*
         obtuse: *Andromeda.*
         truncate: *Maranta.*
         pressed down sideways: *Actaea* and
              Daphne.
         emarginate: *Melia.*
         circular: *Lythrum.*
         peltate: *Sarracena, Nymphaea, Clusia,*
              and *Papaver.*
         crown-shaped: *Pyrola.*
         cross-shaped: *Penaea.*
         barbed: *Viola* and *Lantana.*
         gutter-shaped: *Colchicum.*
         concave: *Viola.*
         angular: *Muntingia.* streaked: *Papaver.*
         feathery: *Rheum, grasses, Triglochin*
              and *Tamarix.*
         downy: *Cucubalus* and *Lathyrus.*

Length: thread-shaped: *Zea.*
         with the same length as the style: *Genipa.*

Thickness: like a leaf: *Iris.*

Duration: lasting: *Sarracena, Hydrangea,*
              Nymphaea, and *Papaver.*
         withering, generally.

The stamens are completely separate from the style in every flower;

with the exception of *Canna, Alpinia,* and some of natural order 3, in which they adhere to the style.

Gynandras also adhere, but in a different way.

103. The PERICARP differs (98) as to α. number, compartments, valvules, and partitions; β. form, shape, and dehiscence; γ. enclosure; and δ. position.

The *number* divides the fruit externally into several parts, but not internally.
   No pericarp: *Gymnosperm,* Herm. *uncovered seeds*: Riv. *Thymus.*
   with one capsule: *Lychnis.*
   with two capsules: *Paeonia* and *Asclepias.*

with three capsules: *Veratrum* and *Delphinium*.

with four capsules: *Rhodiola*.

with five capsules: *Aquilegia*.

with many capsules: *Caltha, Trollius*, the *Hellebori*.

The *compartments* divide the fruit internally, not externally.

with one chamber: *Trientalis* and *Primula*.

with two chambers: *Hyoscyamus, Sinapis*, and *Nicotiana*.

with three chambers: *Lilium*.

with four chambers: *Euonymus*.

with five chambers: *Pyrola*.

with six chambers: *Asarum* and *Aristolochia*.

with eight chambers: *Radiola Lini*.

with ten chambers: *Linum*.

with many chambers: *Nymphaea*.

The *valvules* relate to the outside walls, just as the partition relates to the compartments.

with two valves: *Chelidonium* and *Brassica*.

with three valves: *Viola, Polemonium*, and *Helianthemum*.

with four valves: *Ludwigia* and *Oenothera*.

with five valves: *Hottonia*.

The *partition,*

parallel: *Lunaria* and *Draba*.

crossways: *Bisentella* and *Thlaspi*.

The *form,*

see Section 86, no. V.

The *shape,*

top-shaped: [blank].

inflated: *Cardiospermum* and *Staphylaea*.

membraneous: *Ulmus*.

three-edged, four-cornered, or five-cornered: *Averrhoa* and *Zygophyllum*.

jointed: *Ornithopus, Hedysarum*, and *Raphanus*.

The *dehiscence,* when the ripe fruit scatters its seeds.

at the tip, which has four teeth: *Dianthus*.

which has five teeth: *Alsine*.

which has ten teeth: *Cerastium*.

at the base, in three ways: *Triglochin* and *Campanula*.

in five ways: *Lectum*.

with corners, length-wise: *Oxalis* and *Orchis*.

by a pore: *Campanula*.

horizontally: *Anagallis, Plantago, Amaranthus, Portulaca*, and *Hyoscyamus*.

Every jointed fruit dehisces according to the joints, [each] containing a single seed: *Hypecoum, Hedysarum, Ornithopus, Scorpiurus*, and *Raphanus*.

The *enclosure*, with elasticity: *Oxalis, Elaterium, Momordica, Impatiens, Cardamine, Phyllanthus, Euphorbia, Justicia, Ruellia, Dictamnus, Hura, Ricinus, Tragia, Jatropha, Croton, Clusia, Acalypha.*

The *position* at the receptacle of the flower, and that either
    underneath: *Vaccinium* and *Ephilobium.*
    on top: *Arbutus* and *Tulipa.*
    on top and underneath: *Saxifraga* and *Lobelia.*

104. It is observed that the SEEDS differ as to α. number, the chambers; β. shape, substance, the coronule, and the aril; γ. size; δ. the corcule; and ε. the receptacle.

The *number* should be taken from Rivinus' system.
    with a single seed: *Polygonum* and *Collinsonia.*
    with two seeds: *umbellate* and *star-shaped [plants].*
    with three seeds: *Euphorbia.*
    with four seeds: *rough-leaved* and *whorled* [plants].
The *chamber* is single in most [plants].
    with two chambers: *Cornus, Xanthium, Locusta, Valeriana,* and *Cordia.*
The *shape*:
    girdled: *Arenaria* and *Bryonia.*
    heart-shaped, kidney-shaped, or egg-shaped.
    echinate: *Lappula Myosotidis.*
The *substance*:
    bony: *nuts, Corylus,* and *Lithospermum.*
    callous: *Citrus.*
The *coronule*:
    an epicalyx [issuing] from the perianth of the flower: *Scabiosa, Krantia, Ageratum* and *Arctotis.*
    a tuft, hairy (simple, thread-like): *Hieracium* and *Sonchus.*
        feathery (villous, compound): *Scorzonera* and *Tragopogon.*
        chaffy: *Bidens, Silphium, Tagetes,* and *Coreopsis.*
        none: (bare seed): *Tanacetum.*
The *aril*:
    which some call the veil: *Cossea, Jasminum, Cynoglossum, Cucumis, Dictamnus, Diosma, Celastrus,* and *Euonymus.*
The *size*:
    smallest: *Campanula, Lobelia, Trachelium* and *Ammannia.*
    greatest: *Coccus.*
The *position*:
    resting, scattered in pulp: *Nymphaea.*
    attached to the seam: *siliquous [plants].*

fixed to the columella: *Malva*.

situated on the receptacles: *Nicotiana* and *Datura*.

The *receptacle* of compounds is especially worthy of investigation.

in shape,   flat: *Achillea*.

convex: *Matricaria*

conical: *Anthemus* and *Melampodium*.

with surface bare: *Matricaria*.

dotted: *Tragopogon*.

villous: *Andryala*.

bristly: *Centaurea*.

chaffy: *Hypochaeris* and *Anthemis*.

The receptacles of the fruit of simple [flowers] are peculiar in *Magnolia*, *Uvaria*, and *Michelia*.

The HILUM is very conspicuous in *Cardiospermum* and *Staphylea*.

Next to the hilum is the *corcle*.

The situation of the corcle is either at the tip or at the base of the seed. *Cesalpino*.

The seedling is produced either on the part to which the peduncle is attached, or the exact opposite. If it is produced on the part to which the peduncle is attached, it puts out roots to the peduncle and leaves to the tip. If it is produced at the tip, it extends leaves towards the peduncle and roots to the tip, *Joseph ab Aromatariis, De seminibus* 4.5.6.

105. The PECULIAR (92) fruit-body is taken from the structure that is observed in very few genera.

It is contrasted with the natural structure, Section 93.

Examples: *Arum* has the stamens within the pistils.

Adoxa has the ovary between the calyx and the corolla.

Salvia has jointed filaments.

Eriocaulon has stamens situated on the ovary, and the corolla and calyx placed underneath the ovary.

Magnolia has a capitate receptacle of the fruit, with seeds like berries, hanging in a thread from the capsule.

106. The CALYX is usually less coloured than the corolla.

This means primarily the perianth, in the second place the involucre and the sheath.

The reason for this is apparent from the matter of which the calyx consists, derived from the bark of the plant.

Examples that indicate the contrary are comparatively rare.

Bartsia americana: the perianth is blood-red.

Cornus herbacea: the involucre is snow-white, the petals black.

Cornus americana: the involucre is red and heart-shaped.

Astrantia: the involucre is coloured.

Palms: the sheaths are blood-red.

Where there is no corolla, the perianth is usually more coloured, especially at the time of flowering, as in *Ornithogalum*, *Persicaria*, and *Polygonum*.

Where it is observed that the calyx or corolla are less coloured, there the leaves often assume colour, as in *Amaranthus tricolor*.

107. The RECEPTACLE of the flower (86) surrounds the perianth on the inside in plants with twenty stamens and others; and it adheres all round in Cucurbitaceae (77:45).

Study of the receptacle is a matter of great importance in the natural method.

Most plants have the receptacle of the flower inserted under the stamens and petals at the bottom of the flower.

Plants with twenty stamens (natural orders 45–48) have a calyx with a solitary leaf, and it is encircled all round on the inside by a line to which the stamens and the petals are attached. We observe calyces of that kind, which bear flowers, in other [plants] too: *Lythrum*, *Epitlobium*, *Oenothera*, *Ammannia*, *Isnarda*, *Peplis*, and *Elaeagnus*.

Cucurbitaceae (natural order 45). In these the receptacle covers the perianth all round on the inside, and the corolla is as it were glued to it [the perianth]. The same thing obtains in *Cactuses*.

The receptacle supporting the pericarps: *Passiflora*, *Capparis*, *Breynia*, *Arum*, *Calla*, *Dracontium*, *Pothos*, *Zostera*, *Nepenthes*, *Clutia*, *Helicteres*, and *Sisyrinchium*.

108. The FILAMENTS of the stamens are separate from a corolla with many petals, but are inserted into one with a solitary petal.

Vaillant observed this in those with a solitary petal.

Pontedera learnt, by dissecting 2,000 species, that flowers with a solitary petal display stamens inserted into the corolla, but those with many petals have them inserted into the receptacle of the flower.

Tournefort classifies the *Malvaceae* (natural order 44), which really have five petals, as is established from the base of the corolla, among those with a solitary petal.

Rivinus counts as the number of the petals, the number into which the flower is spontaneously resolved when it falls.

Flowers *with a solitary petal* have the stamens inserted into the petal:

So *Trientalis* has a solitary petal.

Oxalis, which just holds together at the bases of the petals, turns out to have a solitary petal.

Those *with many petals* display stamens that are separate from the petals.

But comparatively rarely an exception is allowed.

Statice, which has five petals, has filaments that are inserted into the narrow bases of the petals.

Melanthium, which has six petals, displays filaments that are inserted into the petals.

The *Lychnises* (natural order 42) quite often have alternate stamens attached to the narrow bases of the petals.

Granted that there are flowers that have a solitary petal, with stamens separate from the corolla, in [plants] with two horns (natural order 24): *Erica, Andromeda, Arbutus, Ledum, Azalea*, etc. as also *Cissus* and *Aloë*.

109. The ANTHERS are usually situated at the tips of the filaments.

There is a comparatively rare exception, where the anthers are glued to the side of the filament: *Paris* and *Asarum*.

The anthers are attached to the stigma, without filaments, in *Aristolochia*.

110. The NECTARY, if it is separate from the petals, is often sportive.

The honey fluid is secreted in most flowers.

The tubes of flowers with a solitary petal generally contain honey.

Pontedera maintained that this was a concoction of liquids or a liquid of the amnion, which entered fertilized seeds; but honey is to be found

even in male flowers: *Urtica* and *Salix*.

only in female flowers: *Phyllanthus* and *Tamus*.

in either sex: *Rusco, Clutia*, and *Kiggelaria*.

Vaillant maintained that the nectary was essential to the corolla; so he said that the nectaries of Nigella and Aquilegia were petals. He took the petals for the calyx.

It is established that the nectaries are *separate* from the corolla by these examples: *Aconitum* and *Aquilegia; Helleborus* and *Isopyrum; Nigella* and *Garidella; Epimedium, Parnassia*, and *Theobroma; Cherleria* and *Sauvagesia*.

[Plants] *with a spur*, and with a solitary petal: *Antirrhinum, Valeriana, Pinguicula*, and *Utricularia;*   with many petals: *Orchis, Delphinium, Viola, Impatiens*, and *Fumaria*.

Corollas with petals on the inside:	*Fritillaria, Lilium, Swertia, Iris, Hermamnia, Uvularia, Hydrophyllum, Myosurus, Ranunculus, Bromelia, Erythronium, Berberis*, and *Vallisneria.*
[Nectaries] crowning the corolla:	*Passiflora, Narcissus, Pancratinum, Olax, Lychnis, Silene, Coronaria,*

	Stapelia, Asclepias, Cynanchum, Nepenthes, Cherleria, Clusia, Hamamelis, and *Diosma.*
[Nectaries] constructed in a peculiar manner:	*Reseda, Cardiospermum, Amomum, Costus, Curcuma, Grewia, Urtica, Andrachne, Epidendrum, Helicteres,* and *Salix.*
[Nectaries] on the calyx:	*Tropaeolum, Monotropa, Biscutella,* and *Malpighia.*
[Nectaries] on the stamens, with anthers:	*Adenanthera.*
[Nectaries] on the stamens, with filaments:	*Laurus, Dictamnus, Zygophyllum, Commelina, Mirabilis, Plumbago, Campanula,* and *Roella.*
[Nectaries] on the pistils of the ovary:	*Hyacinthus, Iris, Butomus, Cheiranthus, Hesperis,* etc.
[Nectaries] on the receptacle:	*Lathraea, Helxine, Collinsonia, Sedum, Cotyledon, Sempervivum,* etc., *Mercurialis, Kiggelaria, Clutia, Phyllanthus, Melianthus,* and *Diosma.*

111. The PISTIL is usually placed inside the anthers.

> *Arum* is peculiar; for in it the receptacle is elongated into a club, and the pistils occupy the base and the stamens the upper part; and so the pistils are placed outside and around the stamens.
>
> *Calla aethiopica* is arranged in the same manner.
>
> *Rumex* is peculiar in the insertion of the stamens.

112. The STYLE is usually situated at the tip of the ovary, with few exceptions.

> *Jung* 39. 'The style is always attached to the tip of the fruit.'
>
> *Dillenius, respons.* 6. 'That there exists no style that does not arise from the middle of the flower, from the middle of the embryo, which occupies the middle of the flower, is well known to every beginner.'
>
> Various flowers are excepted:
> 1. *[Plants] with twenty stamens and many pistils* (natural order 35):
>   *Rosa, Rubus, Fragaria, Potentilla, Tormentilla, Dryas, Geum, Comarum, Sibbaldia, Agrimonia, Alchemilla,* and *Aphanes.*
> 2. *Suriana* and *Hirtella.*
> 3. *Passerina, Gnidia, Struthea,* and *Stelleria.*

113. The PERICARP is naturally closed, and is not filled with smaller pericarps, but is succulent and changes into a BERRY.

> It is completely closed in most [plants].
>> But *Reseda* and *Datisca* are always wide open.
>> *Parnassia* is wide open at the time of flowering; later it is closed.
> I do not accept that a pericarp may be naturally pregnant with smaller pericarps; but when pericarps appear to be concealed within another, then it is the common outer receptacle; as in *Magnolia*, *Uvaria*, and *Michelia*.
> A berry is a succulent fruit, properly from the pericarp, and improperly from any part. The purpose of the berry is that the seeds should be carried by animals: for example, *Viscum*.
> Peculiar and improper berries are frequent, where there is

a calyx:	*Blitum*, *Morus*, *Basella*, *Ephedra*, *Coix*, *Rosa*, and *Coriaria*.
a receptacle:	*Taxus*, *Rhizophora*, *Anacardium*, *Ochna*, *Laurus*, *Ficus*, *Dorstenia*, and *Fragaria*.
a seed:	*Rubus*, *Magnolia*, *Uvaria*, *Michelia*, *Prasium*, *Uvularia*, *Panax*, *Adonis*, *Crambe*, and *Osteospermum*.
an aril:	*Euonymus: Celastrus* demonstrated this.
a nectary:	*Mirabilis*.
a corolla:	*Poterium*, *Adoxa*, and *Coriaria*.
a capsule:	*Euonymus*, *Androsaemum* T., *Cucubalus*, and *Epidendrum*.
a dry berry:	*Linnaea*, *Galium*, etc., *Tetragonia*, *Myrica*, *Trientalis*, *Tropaeolum*, *Xanthium*, *Juglans*, *Ptelea*, *Ulmus*, *Comarum*, *Amygdalus*, and *Mirabilis*.
a capsule on the outside:	*Dillenia*, *Clusia*, *Nymphaea*, *Capparis*, *Breynia*, *Morisona*, *Stratiotes*, *Cyclamen*, and *Strychnus*.
a hollow [berry]:	*Staphylea*, *Cardiospermum*, and *Capsicum*
a conceptacle:	*Actaea*.
a pod:	*Hymenaea*, *Cassia*, *Inga* Pl., and *Ceratonia*.
a cone:	*Anona* and *Juniperus*.

> A berry does not dehisce naturally, because it is soft, and its purpose is different.
> Clustered berries are very conspicuous in *Adonis capensis*.

114. COMPLETE flowers are either single or clustered.

> This was *Vaillant's* classification of flowers.
>> A *complete* flower possesses a perianth and a corolla.
>> An incomplete flower lacks either a perianth or a corolla.
>>> A flower with no *petals* lacks a corolla, not a perianth.

A *bare* flower lacks a calyx, not a corolla.
A flower is more aptly called bare, when it lacks both corolla and calyx; this, however, is very rare.

The primary and most natural classification of plants has been taken from the cotyledons, [dividing them] into those with a solitary cotyledon and those with several cotyledons; where a division is made between simple and properly compound flowers, it is assumed that the plant is one with several cotyledons.

115. A flower is SIMPLE when no part of the fruit-body is common to several flowers.

A simple flower makes a chamber within a single perianth or corolla.
A compound or many-capsuled fruit cannot make a compound flower.

116. A flower is CLUSTERED when some part of the fruit-body is common to several florets; a clustered flower is called particularly either *compound* or *umbellate* or *sprouting*.

A *clustered* flower is formed, when several florets are united by means of some part of the fruit-body that is common to all of them, in such a way that, if one flower is removed, it destroys the shape of the whole of which it is part.
In these cases, what is common is the *receptacle* or the *calyx*.
A flower that is part of a clustered flower is called a *floret*.
There are seven principal forms of clustered flowers.

1. An *umbellate* flower has a receptacle divided into peduncles that all issue from the same centre.
2. A *sprouting* flower has a receptacle divided into peduncles that arise from the same universal centre, but with pedicels going forth hither and thither.
3. A *compound* flower has a receptacle that is expanded entire, with sessile flowers.
4. A clustered flower (properly so called) has an expanded receptacle with flowers situated on peduncles: for example *Scabiosa, Knautia, Dipsacus, Cephalanthus, Globularia, Leucadendron, Protea, Brunia, Barreria,* and *Statice T.*
5. A *catkin-shaped* flower has a thread-like receptacle with catkin scales:
   Xanthium, Ambrosia, Parthenium, and *Iva.*
   Alnus and *Betula.*
   Salix and *Populus.*
   Corylus and *Carpinus.*
   Juglans, Fagus, Quercus and *Liquidambar.*
   Cynomorion.
   Ficus, Dorstenia, Parietaria, Urtica,

Pinus, Abies, Cupressus, and *Thuja.*
Juniperus, Taxus, and *Ephedra.*

6. A clustered flower *with a husk* has a thread-like receptacle, whose base is furnished with a common husk.
   Bromus, Festuca, Avena, Arundo, Brisa, Poa, Aira, Uniola, Cynosurus, Melica, Elymus, Lolium, Tritieum, Secale, Hordeum, Scirpus, Cyperus and *Carex.*

7. A clustered flower *with a spadix* is one where the receptacle is within a sheath that is common to several florets.
   In *Palms* the spadix is subdivided.
   A simple one is covered all round by florets: *Calla, Dracontium,* and *Pothos.*
   Underneath: *Arum.*
   On one side: *Zostera.*

117. A COMPOUND flower is a clustered one containing several sessile florets that have a common entire receptacle and are contained in a perianth, but are furnished with anthers joined together to form a cylinder.

The properties of a compound flower are:
   a. a common receptacle that is enlarged and undivided.
   b. a common perianth encircling all the florets.
   c. 5 anthers joined together to form a cylinder.
   d. sessile florets with solitary petals.
   e. an ovary with its own solitary seed under [each] single floret.
It is essential for compound [flowers] to have anthers joined together to form a cylinder, and a solitary seed under [each] single floret.
Observe that there are compound flowers which have a calyx that is furnished with a solitary flower, for example, *Echinops, Stoebe, Corymbium,* and *Artemisia unica.*
Compound flowers are generally reckoned to be of three kinds:
   a. TONGUE-SHAPED (half-floreted, *Tournefort*), when all the little corollas of the florets are flat, and are spread out towards the outer side.
   b. TUBULAR (floreted, *Tournefort*), when the little corollas are all tubular and fairly even.
   c. RAYED, when the little corollas of the disc are tubular, and the flowers on the periphery are irregular:
      either with tongue-shaped florets on the periphery (rayed, *Tournefort*).
      or with tubular florets on the periphery: *Centaurea.*
      or with almost bare florets on the periphery: *Artemisia* and *Gnaphalium.*
A compound flower quite often consists of several florets, comparatively rarely fixed in number.

Tongue-shaped with 5 florets:	*Prenanthes.*
Tubular with 20 florets:	*Eupatorium scrophulariae folio* [with scrophularia leaf].

Tubular with 15 florets:	*Eupatorium perfoliatum.*
Tubular with 5 florets:	*Eupatorium digitatum.*
	Eupatorium Zeylanicum.
	Eupatorium secundum [my] Hortus Upsaliensis.
	Eupatorium quartum [my] Hortus Upsaliensis.
Tubular with 4 florets:	*Eupatorium volubile.*
Rayed with many ray florets:	*most [plants].*
Rayed with 20 ray florets:	*Arctotis.*
Rayed with 12 ray florets:	*Rudbeckia.*
Rayed with 10 ray florets:	*Tetragonotheca* and *Osteospermum.*
Rayed with 8 ray florets:	*Coreopsis* and *Othonna.*
Rayed with 5 ray florets:	*Achillea, Eriocephalus, Micropus, Seriphium, Sigesbeckia, Acmella, Melampodium, Chrysogonum,* and *Tagetes.*
Rayed with 3 ray florets:	*Sigesbeckia.*
With 1 floret:	*Milleria,* with 3 disc-florets.

118. An UMBELLATE flower (116) is a clustered one [formed] out of several florets situated on the receptacle into erect bundled peduncles all produced from the same point.

> But a CYME (116) is a clustered flower [formed] out of several florets situated on the receptacle into erect bundled peduncles, the first of which are produced from the same point, the later ones sporadically.
>
> An UMBEL exists, when all the peduncles issue from one centre, with an even periphery.
>
> It is a SIMPLE umbel, where the receptacle is thus divided into peduncles once: *Cornus* and *species of Spiraea.*
>
> An umbel is COMPOUND, when all the common peduncles are subdivided into little umbels.
>
> So a LITTLE UMBEL is a partial umbel. The peculiarities of umbellate flowers, properly so called, are:
>
> a. A common *receptacle* divided into peduncles originating in the same centre or point, even in the periphery, whether the umbel that issues from it [the centre] is flat or convex or concave.
>
> b. An *ovary* under the little corolla.
>
> c. 5 separate *stamens*, which eventually fall off.
>
> d. A *pistil* divided into two.
>
> e. Two *seeds*, attached to each other at the top.

It consists of an involucre and an umbel both with variations in the rays.

A universal involucre with 4 leaves:	*Hydrocotyle, Sison* and *Cuminum.*
A universal involucre with 5 leaves:	*Bupleurum, Scandix,* and *Bubon.*
A universal involucre with 7 leaves:	*Ligusticum.*
A universal involucre with 10 leaves:	*Artedia.*
A partial [involucre], halved:	*Ethusa, Coriandrum,* and *Sanicula.*
A partial [involucre] that falls off early:	*Ferula* and *Heracleum.*
An umbel with a *male* disc:	*Astrantia, Caucalis, Artedia, Oenanthe,* and *Scandix.*
A rayed [umbel], with relatively large petals at the edge:	*Tordylium, Caucalis, Coriandrum, Ammi,* and species of *Heracleum.*

A CYME, like an umbel, produces all its primary peduncles from the same centre, but scatters its partial ones sporadically: *Opulus,* the *Cornus* called *virga sanguinea* [bloody rod], and *Ophiorrhiza.*

A receptacle so extended indicates that there is an involucre; this is exhibited by the umbellate *Cornus,* of which *virga sanguinea* is a species that is branched with peduncles, like *Tinus* or *Viburnum.*

119. A LUXURIANT flower multiplies the coverings of the fruit-body so much that its essential parts are destroyed; it is either *multiple* or *full* or *prolific;* whereas *defective* is the term used for a flower that has no room for a corolla.

The covering of a flower is the perianth and the corolla.

It becomes luxuriant when some parts increase in number, so that there is no room for others.

A luxuriant flower generally results from luxuriant nourishment.

A DEFECTIVE flower, in our opinion, is one that does not produce a corolla, although it ought to produce it, and this generally happens as a result of lack of sufficient heat.

Ipomoea. [My] Flora Zeylanica 79.

Campanula pentagonia. [My] Hortus Upsaliensis n. 3.

Ruellia. [My] Hortus Upsaliensis 179.

Various Violas. [My] (Dissertatio de seminibus muscorum).

Tussilago Anandria.

Cucubalus. [My] Flora Suecica 363. [My] Flora Lapponica 181.

120. A MULTIPLE (119) flower is usually so called from a corolla that is multiplied, except for some stamens; and it is double or triple. The perianth and the involucre rarely make a multiple flower, the stamens hardly ever.

I distinguish a multiple flower from a full one, because in the *full* one the corolla is multiplied so much that there is hardly any place for stamens; the *multiple* one is so called from the augmented sequence of the corolla, double, triple, or quadruple.

So the DOUBLE flower is the first and smallest degree of fullness.

The TRIPLE one turns out to be augmented with a three-fold corolla.

Campanula folio urticae, flore duplici et triplici, [nettle-leaved, with double and triple flower]. Tournefort.

Stramonium flore altero alteri innato, [with one flower growing on another]. Vaillant

Stramonium flore violaceo duplici triplicive, [with violet-shaped, double or triple flower]. Tournefort.

Flowers with a solitary petal are quite often multiple; they turn out to be full comparatively rarely.

It is not rare for flowers with several petals to be multiple: for example *Hepatica* and *Anemone.*

The perianth rarely makes a multiple flower.

Dianthus Caryophyllus spicam frumenti referens [resembling an ear of corn]. E. N. C. cent. 3 p. 368 pl. 9; in this [flower] the scales of the calyx are indefinitely enlarged, so as to constitute the entire ear in a peculiar manner. See [my] *Hortus Cliffortianus* 164.

Alpine grasses turn out to be as it were full, when the husks grow out to form leaves: *Festuca spiculis viviparis* [with little ears that are viviparous][3]. [my] Flora Suecica 94.

Salix rosea, in which, after the destruction of the stamens or pistils by insects, the scales of the catkin sprout forth to form leaves.

Plantago rosea, when the bracts of the ear grow out to form leaves.

Be careful not to take a coloured perianth for a multiple one, even if there is a degree of abnormality: for example

Primula prolifera odorata [scented]. Tournefort.

Primula prolifera, flore majore [with a relatively large flower]. Tournefort.

Primula prolifera, flore purpureo [with a purple flower]. Tournefort.

121. A flower is FULL (119) when the corolla is multiplied so much that there is no room at all for any stamens.

It occurs when the stamens grow out to form petals; the latter fill up the flowers, and quite often suffocate the pistil, especially when there is no room for any stamens.

Those with several petals quite often become full.

> *Malus, Pyrus, Persica, Cerasus, Amygdalus, Myrtus, Rosa,* and *Fragaria.*
> *Ranunculus, Caltha, Hepatica, Anemone, Aquilegia,* and *Nigella.*
> *Papaver* and *Paeonia; Dianthus, Silene, Lychnis,* and *Coronaria.*
> *Lilium, Fritillaria, Tulipa, Narcissus, Colchicum,* and *Crocus.*
> *Cheiranthus* and *Hesperis; Malva, Alcea,* and *Hibiscus.*

Those with a solitary petal are comparatively rarely filled up.

> *Primula, Hyacinthus, Datura,* and *Polianthus.*

[For the fact] that full ones are barren, and the reason, [see] Section 150.

That full [flowers] do not constitute genera, [see] Sections 184–5.

Full flowers are the delight of *lovers of flowers* and *gardeners.*

122. Many natural orders of plants are unable to produce luxuriant flowers.

Of this sort especially are the following:

> *[Flowers] with no petals* (natural orders 48 and 53).
> *[Flowers] that are whorled* (natural order 58).
> *[Flowers] that are personate,* Tournefort, (natural order 59) except *Antirrhinum.*
> *[Flowers] with rough leaves,* Ray, (natural order 43).
> *[Flowers] that are star-shaped,* Ray, (natural order 44).
> *[Flowers] that are umbellate,* Ray, (natural order 22), except the prolific umbel.

The *Papilionaceae* (natural order 55) rarely produce full flowers: such as

> *Ternatea flore pleno caeruleo* [with a full dark blue flower] T.
> *Coronilla herbacea flore vario pleno* [with a variegated full flower] T.
> *Anthyllis vulgaris flore pleno* [with a full flower] observed by us.
> *Spartium.*

123. A flower becomes PROLIFIC (119) when some flowers spring forth within another flower (quite often a full one). A prolific [flower] is called *frondose,* when the off-shoot of the prolific one becomes leafy.

A flower is prolific, when one flower springs forth out of another.

Prolific flowers occur from an increased cause of fullness.

Prolific ones (not those of compound flowers) are made from the pistil, and so the offshoots spring forth from the centre of the flower.

A frondose prolific [flower] is a great rarity, seen in *Rosa, Anemone,* etc.

But a prolific [flower] with a flowery offshoot is frequent.

> *Ranunculus radice tuberosa, flore pleno et prolifero* [with a tuberous root and full prolific flower]. C. B.
> *Ranunculus, asphodeli radice* [with asphodel root], *prolifer miniatus* [scarlet]. C. B.
> *Anemone latifolia pavo dicta* [called the peacock], *major, prolifera.* C. B.

124. The PROLIFERATION (123) of simple [flowers] (115) comes from the pistil; but that of clustered ones (116) from the receptacle.

Proliferation comes in two sorts, either
a. Proliferation from the centre, or from a pistil that has sprung forth to form an offshoot, is carried out by a single pedicel, and it is formed in flowers that are not compound.
Dianthus: Caryophyllus altilis major, fore pleno prolifero [with a full prolific flower]. Flora March.
Ranunculus radice tuberosa, flore pleno et prolifero [with a tuberous root and full prolific flower]. Tournefort.
Ranunculus tuberosus anglicus polyanthos. Vaillant.
Anemone latifolia, pavo dicta [called the peacock], *prolifera.* Tournefort
Anemone pavota latifolia multiplex. [peacock-eyed, broad-leaved, multiple] Valent.
Geum flore uno alteri innato [with one flower growing on another]. Tournefort.
Rosa rubra prolifera.
b. Proliferation at the side, sending out numerous pedunculate offshoots from a common calyx, occurs in compound clustered flowers properly so called.
Bellis hortensis prolifera. C. B.
Calendula prolifera. C. B.
Hieracium falcatum proliferum. C. B. Prodromus 64.
Scabiosa, foliis gingidii [with *Gingidia* leaves], *prolifera.*
When umbellates proliferate, they reproduce a little umbel, so that a second umbel issues forth from the simple one.
Cornus: Periclymenum humile, flore flori innato [with flower growing on flower]. Acta Haffniensia IV p. 346.
A *supradecompound* umbel is formed similarly from a compound one.
Selinum: Thysselinum palustre lactescens, [occurs] by no means rarely.

125. The FILLING UP of simple flowers is done either with petals or with nectaries.

The filling of simple flowers is one thing, that of compound ones is another.
Aquilegia is filled up in three ways:
a. with petals multiplied: no room for nectaries, *Aquilegia flore roseo* [with rose-shaped flower]. C. B.
b. with nectaries multiplied: no room for petals. *Aquilegia flore multiplici* [with multiple flower]. C. B.
c. with nectaries multiplied and 5 petals remaining; so that the five petals persist and between them are placed the nectaries, three at every position; and they swallow each other mutually.

Nigella flore pleno [with full flower]: the 5 lower petals are ovate and entire: the rest, which fill it up, are much divided, three lobed and flat; therefore the latter have their origin in nectaries that are multiplied.

Narcissus is filled up either with petals and a nectary that have been multiplied, or with a full nectary and petals that have not been multiplied.

Delphinium is generally filled up with flat petals, and there is no room for a nectary.

The transformation of the English *Saponaria* is peculiar; it changes from [a plant] with five petals into one that actually has a solitary petal.

Peloria is extremely peculiar.

126. Flowers in a corolla with several petals are quite often multiplied (120); they are more frequently doubled in one with a solitary petal. But it is not contradictory to say that flowers may have a solitary petal and at the same time be full.

Kramer 6. maintained that flowers that have a solitary petal and are at the same time full, are contradictions; but this is refuted by *Colchicum, Crocus, Hyacinthus,* and *Polianthes.*

Flowers with a solitary petal are filled up by the lappets of the limb; those with several petals, by petals.

Opulus flore globoso [with spherical flower] C. B. provides a very rare example of fullness; for Opulus vulgaris possesses a sprout, which consists of numerous hermaphrodite bell-shaped florets on the disc; but on the periphery or ray, it consists of barren flowers [issuing] from a flat, wheel-shaped corolla: in Opulus flore globoso all the flowers of the disc turn out to be like those of the ray, to wit, with large round barren corollas, and so the Opulus is filled up in the manner of compound flowers, and only its size makes it barren. Hence even the nature of the *sprout* comes very close to being an umbel; and this is shown also by the *male Cornus,* which is umbellate, when it is compared with the *female Cornus* or *Ossea.*

127. COMPOUND flowers (117) are filled up (121) either by tubular petals or by flat ones.

It is to be understood that the above refers to compound flowers with united anthers.

These consist either of little corollas shaped like pipes (floreted T.), or of little corollas shaped like little tongues (demi-floreted T.) or else of little corollas, that are pipe-shaped on the disc and shaped like little tongues on the periphery (rayed T.).

If the above are correctly understood, filling up occurs in compound flowers in two ways.

a. [The flowers are made] full by the ray in rayed ones, while the ray, which is multiplied, leaves no room for the disc of the flower.

> *Helianthus, Calendula*, and *Chrysanthemum*.
>
> *Anthemis, Matricaria, Ptarmica*, and *Tagetes*.
>
> *Matricaria flore pleno* [with full flower]. C.B.
>
> *Centaureae Cyanus*.

b. [They are made] full by the disc, when the ray is not multiplied, but the little corollas of the disc are elongated, and less divided at the mouths; in some, even a flat ray becomes pipe-shaped.

> *Matricaria foliis florum fistulosis* [with pipe-shaped floral leaves]. H.A.P.
>
> *Bellis hortensis rubra flore multiplici fistuloso* [with multiple pipe-shaped flower] T.
>
> *Tagetes maximus rectus flore multiplicato* [with multiple flower]. Hermann, Leiden.
>
> *flore fistuloso duplicato* [with pipe-shaped double flower] Hermann, Leiden.
>
> *Serratulae Carduus in avena* occurs full of little corollas that are elongated and relatively large.

128. The full flowers (121) [made] of simple ones (115) differ from naturally compound [flowers] (118) inasmuch as the former, the full ones, have a common pistil in the centre of the flower; but the compound ones each have their own stamens and pistils.

> A rule may be useful to the beginner, for example:
>> A naturally compound flower, [made] out of demi-florets. *Hieracium*.
>> A simple flower, filled up and having several petals. *Lychnis*.
>>> Then *Hieracium* has the stamens and pistils that belong with the petals, but in *Lychnis* the rudiment of a common pistil will be present.
> Therefore compound [flowers] have fruit, stamens, and pistils peculiar to each single petal, but the pistils and fruit of the simple ones with full flowers are common.
> So *Nymphaea lutea* is not one with a compound flower, nor with a full one.

129. Compound flowers that are full of flat petals differ from their own [relatives] that are not full, inasmuch as in the latter the stigmas are elongated and the ovaries are enlarged and diverge.

> A rule separates half-floreted T. full [flowers] from those that are not full, and are shaped like a little tongue.
> *Scorzonera latifolia sinuata, floribus plenis* [with full flowers]. C.B. T.

Lapsana vulgaris, floribus plenis [with full flowers]. Frequent at Uppsala.

Tragopogon vulgare [*sic*: vulgaris] *flore pleno* [with full flower]. Seen at Uppsala in 1733.

Ovaries cylindrical, 12 times as long as the calyx, diverging.

The tufts of the seeds twice as large as in the natural [variety].

The petals, stamens, and styles as in the common [variety].

2 stigmas, thread-shaped, the same length as the petal, very long, not rolled back, but bent in various ways.

130. Full compound flowers [made] out of flat petals (127) differ from those that are naturally compound (117) with flat petals, inasmuch as the former, the full ones, are without anthers, whereas the natural ones possess them.

A rule serves to distinguish between the half-floreted T. ones and the rayed ones with full flowers, for example between *Hieracium* and *Chrysanthemum*.

a. Full compound flowers [made] out of flat petals are formed from rayed T. ones, when they are filled up, if the ray occupies the whole disc, as in *Chrysanthemum, Helianthus,* and *Calendula*.

b. Naturally compound [flowers] with flat petals are half-floreted, Tournefort, as *Hieracium, Leontodon,* and *Sonchus*.

These two, with their flowers a. and b., easily mislead beginners.

Half-floreted T. florets are never seen as other than hermaphrodite.

But full-rayed flowers are never seen furnished with anthers.

Thus the full flowers of Tagetes posses pistils without stamens, at [each of] the single florets; but Leontodon possesses both stamens and pistils.

131. In a naturally compound flower (117), if the ray is furnished with pistils, all the full flowers too are furnished with pistils; if it is without them, the full flowers also are without them.

The rayed T. [flowers] are quite often filled up so much that the ray leaves no room for the disc; then all the florets that fill it up turn out to be like a natural ray; for example, *Matricaria, Bellis, Chrysanthemum,* and *Tagetes* with full flower, are furnished with a particular style for [each] single floret or petal.

Helianthus, Calendula, and *Centaurea* with full flowers, when the disc is pushed out by a multiplied ray—you may observe that [each] single petal lacks a style, just like the ray.

So when no ray-floret, in a naturally rayed flower, possesses anthers, there results a very easy distinction between half-floreted T. [flowers] and full-rayed ones (130).

V. SEXUS.

132. Initio rerum, ex omni specie viventium (3) unicum sexus par creatum fuisse contendimus.

Oratio nostra de Telluris habitabilis incremento. Upf. & Lugdb.
1743. hanc Sententiam explicat.

Aqua quotannis subsidet; unde Tellus amplior evadit.

Plantæ diversæ indicant altitudinem perpendicularem terræ.

Fertilitas seminum in plantis sæpe insignis , ex una radice unica
æstate semina *Zeæ* 2000. *Inulæ* 3000, *Helianthi* 4000, *Pa-*
paveris 32000, *Nicotianæ* 40320.

Accedunt viviradices, perennitas, Gemmæ.

Gemmæ totidem Herbæ, ergo in una arbore, trunci vix spi-
thami latitudinem excedente , Herbæ sæpe 10000.

Disseminatio Naturæ stupenda est.

Aëris vis, præsertim vere & autumno procellæ.

Erigeron 3 Hort. cliff. 407. ex America disseminata per
Europam.

Fructus elevatur per caulem.

Scandentes itaque factæ, ut attollant fructum.

Capsulæ apice dehiscunt.

Volitantia Pappo plumoso: *Compositæ , Valeriana.*
piloso : *Compositæ, Stapelia, Xylon.*
calyce: *Compositæ, Scabiosa, Statice, La-*
gœcia, Brunia, Trifolium.
cauda : *Pulsatilla, Populus, Typha, Lagu-*
rus, Arundo, Saccharum.
Ala Seminis: *Abies , Liriodendrum , Betula,*
Plumeria, Bignonia, Conocar-
pus, Anethum, Artedia, Hespe-
ris, Corispermum, Thalictrum.
Pericarpii: *Acer , Fraxinus , Isatis, Begonia,*
Hæmatoxylon, Ulmus, Ptelea,
Dioscorea.
Calyce: *Humulus, Rajania, Rumex.*
Inflatione, ut volumen lævius evadat:
Calyce: *Physalis, Cucubalus, Trifolium.*
Pericarpio: *Colutea , Fumaria , Staphylæa ,*
Cardiospermum, Cicer.

Elasti-

🌺 V. SEX

132. We maintain that, in the beginning of things, a single sexual pair of every species of living [being] (3) was created.

> Our lecture, *De telluris habitabilis incremento*. Uppsala and Leiden, 1743, explains this opinion.
>
> Every year [the level of] the water sinks; as a result the ground turns out to be more extensive.
>
> Differing plants indicate the perpendicular altitude of the ground.
>
> In plants, the productivity of the seeds is often remarkable; from one root in a single summer [there may be] 2,000 seeds of *Zea*, 3,000 of *Inula*, 4,000 of *Helianthus*, 32,000 of *Papaver*, and 40,320 of *Nicotiana*.
>
> In addition, there are root-stocks, perennial duration, and buds.
>
> The shoots are equal in number to the buds; therefore on one tree, with the width of the trunk scarcely more than a span, [there may] often [be] 10,000 shoots.
>
> Nature's distribution of seeds is stupendous.
>
> [There is] the force of the wind, especially the storms in spring and autumn.
>
> Erigeron 3, [my] *Hortus Cliffortianus* 407, [the seeds are] distributed throughout Europe from America.
>
> The fruit [may be] supported by the stem.
>
> Climbing [plants] are so constructed that they hold the fruit up.
>
> Capsules dehisce at the tip.
>
> [There are] *those that float* by means of:-

a tuft that is feathery:	The *Compositae* and *Valeriana*.
a tuft that is hairy:	The *Compositae*, *Stapelia*, and *Xylon*.
the calyx:	The *Compositae*, *Scabiosa*, *Statice*, *Lagoecia*, *Brunia*, and *Trifolium*.
the tail:	*Pulsatilla*, *Populus*, *Typha*, *Lagurus*, *Arundo*, and *Saccharum*.
the wing of the seed:	*Abies*, *Liriodendron*, *Betula*, *Plumeria*, *Bignonia*, *Conocarpus*, *Anethum*, *Artedia*, *Hesperis*, *Corispermum*, and *Thalictrum*.
the wing of the pericarp:	*Acer*, *Fraxinus*, *Isatis*, *Begonia*, *Haematoxylum*, *Ulmus*, *Ptelea*, and *Dioscorea*.

the calyx	*Humulus, Rajania,* and *Rumex.*

inflation, so that the volume turns out to be lighter[1]:

the calyx:	*Physalis, Cucubalus* and *Trifolium.*
the pericarp:	*Colutea, Fumaria, Staphylea, Cardiospermum,* and *Cicer.*

Elastic [plants] propel their seeds a long way:

by the cartilage:	with three kernels (natural order 47), *Impatiens, Oxalis, Diosma,* and *Dictamnus.*
by the point:	*Justicia, Ruellia, Barleria,* and *Lathraea.*
by the fibres:	*Momordica, Cucumis,* and *Cardamine.*
by creeping:	*Crupina, Avena, Geranium, Sigesbeckia, Equisetum,* and *ferns.*

[Plants] *that stick* to animals by means of hooks, and which grow on rubbish heaps:-

by the calyx:	*Arctium, Agrimonia, Neurada, Rhexia, Asperugo, Rumex, Urtica, Parietaria, Plumbago, Linnaea,* and *Sigesbeckia.*
by the pericarp:	*Triumfetta, Bartramia, Heliocarpus, Petiveria, Triglochin, Martynia, Hedysarum, Glycyrrhiza, Scorpiurus, Vella, Circaea, Valantia,* and *Aparine.*
by the seeds:	*Cynoglossum, Myosotis, Verbena, Daucus, Caucalis, Sanicula, Bidens, Verbesina,* and *Arctopus.*

Animals that swallow whole seeds distribute them with advantage: *Viscus, Avena, Juniperus,* and *Epidendrum.*

Berries are produced to be distributed on account of their pulp.

In the process of chewing, squirrels, mice, and jackdaws scatter seeds around.

The mole, hedgehog, and earthworm dig [holes], so that the earth receives them.

Rivers, the sea, lakes, rain and *tides* all help.

Anastatica [demonstrates this] by a wonderful and stupendous example.

[There is] the natural conservation of seeds: *Cassia, Mimosa,* and *Cucumis.*

The bottom of the sea does not destroy seeds.

Resemblance deceives animals: *Salicornia* and *Medicago.*

The plants hide [seeds]: *Arachis, Trifolium, Lathyrus,* and *Valantia.*

They are protected from animals by weapons, the appropriate *spines, prickles,* and *stems.*

Fleshy [plants] are propagated by leaves.

Individual *trees* are each like a garden surrounded by a fence, in the wonderful plan of nature.

The ovary and the corcule in the seed [are made] out of marrow; therefore every reproduction is a continued multiplication.

133. That vegetables, though they are without sensation, are none the less just as much alive (3) as animals, is proved by their *origin, nourishment, age, movement, propulsion, disease, death, anatomy,* and *organism.*

Their *origin* is from a seed or bud.

Their *nutrition* is from the ground, very fine, *Kylbel*; together with water and the air, *Hales*.

Their *age[s]*: infancy, childhood, adolescence, manhood, old age: *trees* and *Hedera*.

Their *movement*: flowers *with demi-florets* and various others notice the hour of the day.

Calendula gives warning of rain on the morrow.

At night, these droop: *Draba, Parthenium foliis ovatis crenatis* [with egg-shaped scalloped leaves], and *Trientalis*;

    these wilt: *Impatiens* and *Amorpha*;

    these bend back: *Sigesbeckia* and *Triumfetta*;

    these close up: The *Mimosas, Papilionaceae*, and *Lomentaceae*;

    this lies down: *Tamarindus*.

But by day they are wide-awake with leaves opened out.

Reseda, Luteola, and flowers *with demi-florets* follow the sun.

[There is] a failure of movement due to shade or a wood; hence the differing heights in *Pinus* and others.

Propulsion [occurs]: for circulation does not exist in plants.

Disease[s]: heat, thirst, kibe, hunger, obesity, cancer, and insects.

Death is the opposite of life.

Anatomy: vessels, sacs, ducts, skin, and cuticle.

Organism: secretory vessels and glands.

134. Every living thing [is derived] from an egg: consequently even vegetables are; and that their seeds are *eggs* is shown clearly by their purpose, which is to produce offspring resembling their parents.

> *Harvey* declared that every living thing is produced from an egg.
>
> The purpose and essence of an egg consists in the point of life.
>
> Bobart discovered the seeds of *ferns*, I discovered those of *mosses*, Réaumur those of *lichens*, and Micheli those of *funguses*; about the larger [plants] there is no doubt.
>
> That the propagation of plants from seed and propagation from buds are of equal antiquity is shown by the examination of buds and by the early development of the flowering.

135. That vegetables are produced from an egg (134) is maintained by reason and experience; this is confirmed by the colytedons.

> No sane person has denied that *seeds* are present in every plant;
>
> and *polyps* are not without eggs.
>
> *Roots*, doubtless all of them, preserve the nature of a polyp, as is apparent in buds, root-stocks and runners.
>
> The presence of *cotyledons* in every plant that is produced proves that a seed was there.

> *Spontaneous generation* has long ago been exploded by experiments, and is completely disproved by the presence of cotyledons.

136. The *cotyledons* of animals are produced from the yolk of the egg, in which the point of life is inherent; therefore the seminal leaves of plants, which envelop the corcule (86:VI), are the same.

> Cotyledons and *seminal leaves* are synonymous in plants.
> *Lactiferous* cotyledons nourish the plumule up till the time when it puts forth roots, just as the placenta or cotyledons do in animals.
> Mosses and [plants] related to them lack cotyledons only. [My] Dissertation *De seminibus muscorum*.

137. That offspring is produced, not just from the egg, nor from the sperm alone, but from *both* together, is proved by *hybrid animals, reason*, and *anatomy*.

> Leeuwenhoek's *seminal worms* are not beings; yet they are corpuscles, but not in themselves alive; sometimes they are fertile.
> *Hybrid animals* [are derived] from different species; for example, the *mule* from a mare and an ass; it resembles neither parent exactly.
> *Anatomy:* study of the placenta and the umbilical cord.
> *Reason:* hereditary faults, dogs and hens.

138. All experience affirms that an egg, that has not been fertilized, does not germinate: so too the eggs of vegetables.

139. Every species of vegetable (157) is furnished with flower and fruit, even when sight does not reach them.

> The seeds of *mosses. Ourself.*
> The flowers of *Lemna* were drawn by *Valisnerius.*
> The flowers of *lichens* were observed by *Réaumur.*
> The flowers of *Pilularia* were investigated by *B[ernard de] Jussieu.*
> The stamens of *funguses* were described by *Micheli.*

140. Every (139) flower (88) is furnished with anthers (86) and stigmas (86).

> We discovered the anthers of *Isoëtes: Skånska resa.*

The stigma of *Parnassia* is lacking in the flower (it grows out later), but the ovary is wide open.

Mosses are perhaps the only [plants] without pistils, since their corcules are bare.

141. A flower (140) comes before every fruit, just as generation comes before birth.

Colchicum and *Hamamelis* flower in the autumn, and they produce their fruit the next year.

The fruit of *Musa* does not precede the flower, though the ovary is very large, and if it is not fertilized, it ceases to grow, but does not cease to ripen.

So the flower is always an antecedent of the fruit, and the fruit is always a consequence of the flower.

142. The FRUIT-BODY (88) consists of the *genitals* of plants (143–4); thus the FLOWERING (40) is the *generation*, the ripening of the FRUIT is the *birth*.

143. That the ANTHERS (140) are the *male genitals* of plants, and the POLLEN the true sperm, is clearly shown by their essence (88), precedence (141), position, time, and chambers, and by castrations and the structure of the pollen.

Position: The stamens of *Didynamistis* go up under the upper lip of the corolla, to which the pistil too inclines.

Most *monoecious* [plants] display flowers with stamens above those that bear pistils: *Zea* and *Ricinus*.

Time: *Monoecious* and *dioecious* male flowers perfect their anthers at the same time as the pistils perfect their stigmas.

Castration: On the contrary *Musa*, in which the flowers with stamens are later than those that bear pistils, displays barren fruit without seeds.

Those who diligently remove the staminal flowers of *Melo* do not obtain any fruit.

Chambers: The anthers have one, two, three, or four chambers, exactly like the pericarp, Section 101.

The structure of the pollen is peculiar and defined like the seeds, Section 101. All pollen is vesicular, and contains impalpable matter, which it blows out.

144. That the STIGMAS (140), which are attached to the ovary on all sides (97) are the *female genitals* is proved by their essence (88), precedence (141), position, time, fall, and cutting off.

    Position: *[Plants] with united anthers* are rarely barren, when the stigmas as it were perforate the anthers.

    Time: The stigma *flourishes* at the same time as the anthers blow the pollen out.

    Falling: In most [plants] the stigma *falls* and withers after the fall of the anthers; therefore the action [occurs] during the flowering.

    Cutting off: Thus *castration* [results] in every flower.

145. That the GENERATION (138) of vegetables is done by means of the falling of the pollen from the anthers onto the bare stigmas, so that the pollen bursts and blows out the *seminal breeze*, which is absorbed by the liquid of the stigma—this is confirmed by the eye, proportion, place, time, rain, palm-growers, drooping and submerged flowers, [plants] with united anthers, indeed by the genuine study of all flowers.

    The eye: *Moriland* believed that pollen entered the ovary; *Vaillant* maintained that its essence was extracted by means of a damp stigma; *B[ernard de] Jussieu* saw the pollen of *Acer* burst in liquid; and *Needham* confirmed that all pollen blows its seminal breeze out in liquid.

    Proportion: That the stigmas [first] incline to the anthers, and are then pulled out, is evident in *Dianthus*, *Passiflora*, and *Nigella*.
    When the pistil is very short, the anthers close up over the stigmas: *Saxifraga* and *Parnassia*.
    In *Celosia* the anthers close up while they blow the pollen out.
    In *Teucrium*, the corolla presses the anthers into the stigmas with its fingers.

    Place: [Plants] that bear pistils never arise naturally, without those that bear stamens in the same [piece of] ground; they are both produced from the same seed.

    Time: In *dicliny* [plants], the flowers are usually produced before the germination of the leaves, so that the leaves shall not cover the pistils: *Salix, Populus, Corylus*, etc.

    Rain soaks up the pollen, so that it cannot fall onto the stigmas. *Gardeners* notice this especially in [plants] that bear stone fruit and pomes.
    To *farmers* it [rain] is execrable in fields of cereals.

    Smoke does the same thing too, by absorbing the liquid of the stigma.

Palm-growers: The cultivation of *Pistacia* in the [Aegean] archipelago was well
 known to *Theophrastus, Pliny, Kaempfer*, and others: *Tournefort*.

Caprification was [the practice] of the ancients, and is still done by means of insects
 in the archipelago. See our dissertation *De ficu*.

Drooping flowers possess a pistil longer than the stamens, so that the pollen shall fall
 onto the stigma: *Campanula, Leucojum, Galanthus*, and *Fritillaria*.

Submerged flowers rise up for the time of flowering: *Nymphaea, Stratiotes,
 Myriophyllum, Potamogeton, Hydrocharis*, and *Valisneria*.

Disappointing [plants] with united anthers: Where there is no stigma, there is no
 fertilization: in the ray of *Centaurea, Helianthus, Rudbeckia*, and *Coreopsis*.

If the anthers of a solitary *Tulipa* are removed before the fall of the pollen, it turns out
 to be barren.

If *Brassicas* of several different varieties are sown in the same place for the time of
 flowering, they never produce distinct seeds.

The *Rhodiola* in the Uppsala garden was barren from the year 1702 until 1750, when
 a male [plant] was introduced; then it produced seeds.

Clutia was barren in most of the gardens in the Netherlands, but fertile seeds were
 observed at Leiden: I predicted that a male [plant] was present, and found it so.

Take care not to be put off by *Ficus, Humulus, Musa, Morus*, etc., which bear fruit in
 the absence of stamens; you must distinguish the parts of the fruit-body, the
 calyx, pericarps, and receptacles, from the seeds.

146. Therefore the C A L Y X is the *bedroom*, the C O R O L L A is the *curtain*,[2] the
 F I L A M E N T S are the *spermatic vessels*, the A N T H E R S are the *testicles*,
 the P O L L E N is the *sperm*, the S T I G M A is the *vulva*, the S T Y L E is the
 vagina, the [V E G E T A B L E] O V A R Y is the *[animal] ovary*, the P E R I C A R P
 is the *fertilized ovary*, and the S E E D is the *egg*.

The calyx could also be regarded as the *lips of the cunt* or the *foreskin*.

The corolla could be taken to stand for the *nymphae*.

The filaments, which deliver the juice to the anthers, are called the *spermatic vessels*.

The anthers are the *testicles*.

The stigma is the *vulva*, and corresponds to the part that secretes the *genital fluid* in
 the weaker sex.

The style corresponds to the *vagina* or to the *Fallopian tube*, though not very
 appropriately to the latter.

The [vegetable] ovary is the *[animal] ovary*, because it contains the rudiments of
 seeds.

The pericarp is the *fertilized ovary*, with the result that it produces fertile eggs.

That the seeds are *eggs* is evident from what has been said in Sections 134–5.

147. The *ground* is the STOMACH of plants, the *root* is their VESSELS THAT CARRY THE CHYLE, the *trunk* is their BONES, the *leaves* are their LUNGS, and *heat* is their HEART; for this reason, a plant was called, by the ancients, an *animal turned upside down*.

> Long ago, a plant was called an *animal* turned upside down; it draws juice, by means of its root, from very small particles of earth, as if by lacteal vessels; and that goes up through the rigid stem; in the branching of the latter, the genitals sprout forth.
>
> Plants have no *heart*, but heat does it all; and there is no need of a heart where an effect of perpetual motion is not necessary, and there is propulsion, not circulation, of liquids.
>
> The *leaves*, which are set in motion and keep breathing, correspond in this way to the lungs; but in themselves they are really analogous to a muscle, even though they are not fixed by the tail as in animals, since voluntary motion cannot occur in them.

148. A FLOWER (140) that contains *anthers* (143) is called MALE, one that contains *stigmas* (145) is called FEMALE, one that contains *both* (143–4) is called HERMAPHRODITE.

> Hermaphrodites are as frequent in vegetables as they are comparatively rare in animals; among the *worms* there are many that seem to be hermaphrodite; *snails* are certainly androgynous.
>
> The conjoining of the sexes is necessary in plants, since they cannot look for and go to a consort.

149. A PLANT with only male (148) flowers is called a MALE.
A PLANT with only female (148) flowers is called a FEMALE.
A PLANT with only hermaphrodite (148) flowers is called a HERMAPHRODITE.
A PLANT with male and female flowers together (a. b.) is called ANDROGYNOUS.
A PLANT with hermaphrodite and female or male flowers together is called POLYGAMOUS: but this generally consists of a *male* or *female hermaphrodite*.

> A HERMAPHRODITE plant displays, over the same root, flowers that are all furnished with *stamens* and pistils, as are *most genera*.
>
> An ANDROGYNOUS plant displays, over the same root, both *male* and *female* flowers together.

Ceratocarpus,	Zannichellia,	Callitriche,	Cynomorium.
Zea,	Coix,	Carex,	Axyris.
Typha,	Sparganium,	Phyllanthus,	Tragia.
Urtica,	Morus,	Plantago,	
Alnus,	Betula,	Buxus.	
Xanthium,	Ambrosia.	Iva,	Parthenium,

Amaranthus, some compound and umbellate [plants].

Zizania,	Rumex.		
Myriophyllum,	Ceratophyllum,	Sagittaria,	Poterium.
Quercus,	Corylus,	Carpinus,	Juglans,
Fagus,	Liquidambar.		
Pinus,	Abies,	Cupressus,	Thya,
Ricinus,	Jatropha,	Sterculia,	Croton,
Acalypha,	Theligonum,	Hura,	Hernandia.
Cucurbita,	Cucumis,	Trichosanthes,	Momordica,
Bryonia,	Sicyos,	Fevillea.	
Andrachne.			
Fucus,	Isoetes,	Pilularia.	
Bryum,	Hypnum,	Phascum,	Lycopodium.

A MALE or FEMALE displays, in the same plant, either solely *male* or only *female*
  flowers.

Najas.			
Valisneria,	Salix,	Rumex,	
Osyris,	Ficus,	Valeriana,	
Viscum,	Hippomane,	Myrica,	Hippophaë,
Morus,	Urtica.		
Antidesma,	Rhus,	Rhamnus.	
Pistacia,	Ceratonia,	Pisonia,	Zanonia,
Humulus,	Cannabis,	Acnida,	Spinacia.
Smilax,	Tamus,	Rajania,	Dioscorea,
Cissampelos,	Phoenix,	Borassus.	
Populus,	Diospyros,	Begonia.	
Laurus,	Mercurialis,	Hydrocharis.	
Carica,	Guilandina,	Kiggelaria,	Coriaria,
Datisca,	Cucubalus,	Silene.	
Aruncus.			
Cliffortia.			
Juniperus,	Taxus,	Ephedra.	
Napæa,	Croton.		
Ruscus.			
Clutia.			
Polytrichum,	Mnium,	Splachnum.	

A POLYGAMOUS (hybrid to others) [plant] necessarily consists of hermaphrodite flowers, but then it has others of either sex, and that in the following fashions;

male and female hermaphrodite flowers, [arranged] so that the flower on one side is hermaphrodite, and the other over against it is barren: as *Musa*.

female hermaphrodites and *male flowers* on the same plant: *Veratrum, Celtis, Aegilops*, and *Valantia*.

female hermaphrodites and *male flowers* on a separate plant: *Chamaerops, Panax, Nyssa*, and *Diospyros*.

male hermaphrodites, and *female flowers* on the same plant: *Parietaria* and *Atriplex*.

male hermaphrodites, and *female flowers* on a separate plant.

androgynous flowers, and *males* on a separate plant: *Arctopus*.

polygamous flowers from a female hermaphrodite, and a *male* on the same plant: a *female* on a separate one: *Gleditsia*.

hermaphrodite flowers, male and *female* on different plants: *Empetrum*.

150. NO LUXURIANT flowers (119) are natural, but all are abnormalities. For *full [flowers]* (121) have turned out to be eunuchs, and so always miscarry; the *multiple* ones (120), not to the same extent; the *prolific* (123) add to the deformity of the abnormal ones.

No completely full flower is propagated by seeds: therefore it is produced either from branches driven into the ground or from root-stocks: as *Dianthus, Lychnis, Hepatica, Cheiranthus, Tropaeolum, Rosa, Punica, Caltha, Ranunculus, Viola, Paeonia*, and *Narcissus*.

A few flowers, such as *Papaver* and *Nigella*, which should be called multiplied rather than completely full, produce complete seeds.

Luxuriant flowers multiply the corolla, to the harm of the stamens, which grow out into petals; and so *multiplied* flowers have lost most of their stamens, and full flowers have lost all of them. A multiple range of petals presents the appearance of a multiplied flower; but the flowers of the *Nymphaeas, Cactuses*, and *Mesembryanthemums* are not to be described as multiplied, because they turn out to be as they are without harm to the stamens.

Anyone who wishes to learn more about the *sex of plants* should go to [my] *Sponsalia plantarum*.

VI. CHARACTERES.

151. FUNDAMENTUM Botanices (4) duplex eſt: *Diſpoſitio & Denominatio.*

Syſt. nat. veget. 2. Fundamentum Botanices conſiſtit in Plantarum Diviſione & Denominatione Syſtematica; Generica & Specifica.

Claſſ. plant. 4. Nomina plantarum debent eſſe certa, adeoque impoſita certis Generibus.

Diſpoſitio eſt Denominationis fundamentum.

Scientia Botanices his cardinibus nititur; Sic plantæ omnes uno anno, primo intuitu, absque præceptore, ſine iconibus aut deſcriptionibus, conſtanti memoria addiſcuntur. Ergo, qui hoc novit, Botanicus eſt, alius non.

152. DISPOSITIO (151) Vegetabilium diviſiones ſ. conjunctiones docet; eſtque vel *Theoretica*, quæ Claſſe, Ordines, Genera; vel *Practica*, quæ Species & Varietates inſtituit.

Diſpoſitio plantarum, ex fundamento fructificationis, recentiorum inventum eſt.

Practica ab eo poteſt tractari, qui de Syſtemate nihil intelligit.

Theoretica curam Syſtematis gerit; hanc *Cæſalpinus*, *Moriſonus*, *Tournefortius* & alii excoluere.

153. Diſpoſitio Vegetabilium (152) vel *Synoptice* vel *Syſtematice* abſolvitur, & vulgo *Methodus* audit.

Synoptica diviſio ſeculo XVI & XVII maxime in uſu fuit.

Syſtematica vero ſeculo XVIII præcipue exculta fuit, incepta a Tournefortio & Rivino.

Methodici ſummi methodo *mathematica*, in ſcientia naturali, a ſimplicioribus ad compoſita adſcendunt, adeoque incepere ab Algis, Muſcis, Fungis, uti *Rajus*, *Boerhaavius &c.*

Naturalis inſtinctus docet noſſe primum proxima & ultimo minutiſſima, e. gr. *Homines*, *Quadrupedia*, *Aves*, *Piſces*, *Inſecta*, *Acaros*, vel primum *majores* plantas, ultimo *minimos* Muſcos.

Natura ipſa ſociat & conjungit *Lapides* & *Plantas*, *Plantas* & *Animalia*; hoc faciendo non connectit perfectiſſimas

G   Plan-

✿ VI. CHARACTERS

151. The FOUNDATION of botany (4) is two-fold, *arrangement* and *nomenclature*.

> [My] *Systema naturae, veget.* 2. 'The foundation of botany consists in the systematic classification and nomenclature of plants: generic and specific.'
>
> [My] *Classes plantarum* 4. 'The names of plants ought to be definite, and accordingly placed in definite genera.'
>
> Arrangement is the foundation of nomenclature.
>
> Knowledge of botany bears on these hinges; thus all plants become known in a single year, at first sight, with no instructor and without pictures or descriptions, by means of stable recollection. Therefore anyone who knows this is a botanist, and no one else is.

152. ARRANGEMENT (151) demonstrates the divisions and connections of the vegetables; and it is either *theoretical*, establishing classes, orders, and genera; or *practical*, establishing species and varieties.

> The arrangement of plants from the foundation of the fruit-body is a thing invented by comparatively modern [botanists].
>
> *Practical* [arrangement] can be managed by one who has no understanding of system.
>
> *Theoretical* [arrangement] carries with it [the need for] attention to system: such attention was cultivated by *Cesalpino, Morison, Tournefort*, and others.

153. The arrangement of vegetables (152) is effected either *synoptically* or *systematically*, and is commonly called *method*.

> *Synoptic* classification was mostly in use in the XVIth and XVIIth centuries.
>
> *Systematic* classification has been developed principally in the XVIIIth century, and was begun by Tournefort and Rivinus.

The most eminent methodical [botanists] ascend by a *mathematical* method in natural science, from relatively simple things to complex ones; so they have begun with algae, mosses, and funguses, as have *Ray, Boerhaave*, etc.

A *natural* instinct teaches [us] to learn about the nearest things first, and the most minute last, for example, *men, quadrupeds, birds, fishes, insects*, and *mites*; or the *larger* plants first, and the *smallest* mosses last.

Nature itself associates *minerals* and *plants*, *plants* and *animals*; in doing this, it does not connect the most perfect plants with animals that are described as the most imperfect, but combines imperfect animals with imperfect plants, for example,

the animalcule *Lernaeum* and the *alga Conferva*.

the alga *Spongia* and animal *corals;*

Taenia; the jointed *Conferva* and *Corallina*.

Lithoceratophyton B., a *vegetable* on the inside, a *mineral* on the outside, [derived] from an *animal*.

154. A SYNOPSIS (153) sets forth arbitrary divisions (152), larger or smaller, more or fewer; in general, it is not to be acknowledged by botanists.

A *synopsis* is an arbitrary dichotomy, which leads to botany like a road, but does not define its boundaries.

A synoptic *key to the classes* is technically obligatory, to prevent things that should be distinct from becoming confused.

Several methodical [botanists] have walked along this road: *Ray, Knaut*, and *others*.

155. A SYSTEM (153) separates the classes by 5 appropriate divisions: *classes, orders, genera, species*, and *varieties*.

These are illustrated by examples from other sciences.

Geography: *realm, province, district, parish, hamlet.*

Soldiering: *regiment, battalion, company, platoon, soldier.*

Philosophy: *g[enus] summum, intermedium and proximum, species, individual.*

Botany: *class, order, genus, species, variety.*

Botany owes these definitions of families to *Tournefort*.

The difference between a synopsis and a system is this:

for the synopsis: *a* 2; *b* 4; *c* 8; *d* 16; *e* 32.

for the system: *a* 10; *b* 100; *c* 1,000; *d* 10,000; *e* 100,000.

Therefore a system is superior to a synopsis.

156. The 'Ariadne's thread'[1] of botany is system (155), without which botany is chaos.

> Let an unknown Indian plant serve as an example; the amateur in botany may refer to all the descriptions, figures, and indexes, and he will not find the name unless by chance: but the systematic [botanist] will quickly determine whether it is an old or new genus.
>
> Honour lasting for all time will be given to the systematists who apply this thread, since all those that have been without it have lost their way and gone into the meanders of botany.
>
> True systematic authors or discoverers must be distinguished from compilers.
>
> A system by itself indicates even plants that are omitted; which enumeration in a catalogue never does.

157. We reckon the number of SPECIES (155) as the number of different forms that were created in the beginning.

> *The 5 classes of plants.* The number of species is the number of different forms produced by the Infinite Being from the beginning; and these forms have produced more forms, according to the laws laid down, but always ones that are similar to themselves. Therefore the number of species is the number of different forms or structures that occur today.
>
> [My] lecture *De telluris habitabilis incremento*, published at Uppsala and Leiden, explained many consequences that result from this argument.
>
> The root grows out into the shoot, and indefinitely, until the integuments burst into flower at the tip; and they form a seed contiguous with it, the ultimate end of vegetation. This seed falls, germinates, and as it were continues the plant in a different place; hence it produces an exactly similar offspring, as a tree produces a branch, a branch a bud, and a bud a shoot; so continuation is the generation of plants.
>
> That NEW SPECIES can come to exist in vegetables is disproved by continued generation, propagation, daily observations, and the cotyledons.
>
> Doubt has been removed by *Marchant* [in] *Acta Parisina*[2], 1719; *myself* in *Peloria*, 1744; *Gmelin* in his *inaugural lecture*, 1749. See [my] *Amoenitates academicae* 71.
>
> *Nymphoides T.* Shoot of *Nymphaea*, fruit body of *Menyanthes*.
>
> *Datisca*, male *Cannabis*; female *Reseda*.
>
> *Tragopogon*, [my] Hortus Upsaliensis 3, as if from a parent *Lapsana*.
>
> *Hyoscyamus*, Hortus Upsaliensis 2, as if from a parent *Physalis*.
>
> *Poterium*, Hortus Upsaliensis 2, as if from a parent *Agrimonia*.
>
> *Saxifraga*, [my] Flora Suecica 358, as if from a parent *Parnassia*.

Dracocephalum, Hortus Upsaliensis 6, as if from a parent *Nepeta*.
Species of Primula, alpine, as if from a parent *Cortusa*.
Carduus, Hortus Upsaliensis 1, degenerating into *Carduus Pyrenaicus*.
Numerous *Mesembryanthemums* at the *Cape of Good Hope*.
African *Geraniums*, similar in flower, *Cape of Good Hope*.
All the *Cactuses* in *America* alone.
Aloë, very numerous in Africa.
Numerous peculiar *varieties, Tournefort, Corollarium*.
A Virginian *Verbena*, observed by us.
A *Delphinium* observed by *Gmelin*.
Marchant's Mercurialis, with leaves much divided.

158. The number of VARIETIES (155) is the number of differing plants that are produced from the seed of the same species (157).

> A variety is a plant that is changed by an accidental cause: *climate, soil, heat, winds*, etc., and likewise it is restored by a change of soil.
> Kinds of variety are *size, fullness, curling, colour, savour*, and *smell*.
> Varieties could be excluded from botany, but
> > *Housekeepers* value large and curly ones.
> > *Gardeners* value full and coloured ones.
> > *Physicians* value savoury and smelly ones.
> A *Tropaeolum* in full flower [was brought] from America by order of Bewerning in 1684.

159. We say that there are as many GENERA (155) as there are similarly constructed fruit-bodies produced by different natural species (157).

> *Cesalpino*. 'If the genera are confused, it is inevitable that everything will be confused.'
> [My] *Classes Plantarum* 6. 'That all genera and species are natural is confirmed by things that are revealed, discovered, and observed.'
> [My] *Systema naturae* veg. 14. 'Every genus is natural, made in the first place such as it is; for this reason it is not to be capriciously split or stuck [to another], for pleasure, or according to each man's theory.'
> The *Ranunculi, Aconitums, Nigellas, Claytonias, Hibisci*, and *Passifloras* stand as examples of the general [terms], especially *a posteriori* [inductively].

160. A class (155) is the agreement of several genera (159) in the parts of the fruit-body (86) according to the principles of nature and art.

> *Tournefort*. 'A class is a collection of genera to which a certain common characteristic is peculiar, so that it is utterly different from all other genera of plants.'

That natural classes exist so created is clearly shown by many [kinds of plant]: *umbellate, whorled, siliquous, leguminous, compound, grasses*, etc.

Artificial classes are substitutes for natural ones, until the discovery is made of all the natural classes which more genera, that have not yet been discovered, will reveal; and then the most difficult distinctions between classes may become apparent.

We must take care not to lose 'Ariadne's thread'[1] (156) by imitating nature, as *Morison* and *Ray* did.

161. An ORDER (155) is a subdivision of the classes (160), so that it does not turn out to be necessary at one and the same time to distinguish more genera (159) than the mind can easily comprehend.

An order is a subdivision of the classes: for it is easier to distinguish 10 genera than 100.

162. *Species* (157) and *genera* (159) are always the work of NATURE; *variety* (158) is quite often the work of CULTIVATION; *class* (160) and *order* (161) are the work of NATURE and ART.

The *species* are very constant, since their generation is actual continuation.

That the *genera* are natural is proved by very many plants: the *Aconitums, Nigellas, Bignonias, Ranunculi, Mesembryanthemums, Zygophyllums, Geraniums*, and *Oxalises*.

That varieties are the work of cultivation is clearly shown by *horticulture*, which frequently produces and modifies them.

That most *classes* and *orders* are natural is clearly shown by the natural orders, Section 77.

163. HABIT is a certain conformity of vegetables that are allied and related in *placentation, radication, ramification, intorsion, gemmation, foliation, stipulation, pubescence, glandulation, lactescence, inflorescence*, and other things.

Formerly, *the EXTERNAL APPEARANCE* was commonly called habit.

C[aspar] Bauhin and the ancients have excellently divined the affinities of plants from their habit, so that the systematists themselves have quite often gone astray, where habit had drawn the right [conclusions].

The natural method is the ultimate purpose of botany (Section 77).

The *fruit-body*, a thing come upon by comparatively modern [botanists], first opened the way to the natural method, but even now this is not so well understood as to discover all the classes.

Just as *habit* in quadrupeds distinguishes wild beasts from farm animals, even if the teeth have not been inspected; so also in plants, it often shows forth the natural orders at first sight.

I. PLACENTATION is the arrangement of the cotyledons at the time of the germination of the seed.

1. *[PLANTS] WITHOUT COTYLEDONS*, where there exist no cotyledons at all: *mosses*.

2. *WITH A SOLITARY COTYLEDON* (though these are properly without cotyledons, when the cotyledons remain within the seed).

    perforated: *grasses*.
    one-sided: *palms*.
    reduced: *Cepa*.

3. *WITH TWO COTYLEDONS*

    unchanged: *pods, pomes, drupes*, and *Didynamia*.
    pleated: *Gossypium*.
    doubled: *Tetradynamia*, and *Malvas*.
    enveloped: *Helxine*.
    spiral: *Salsola, Salicornia, Ceratocarpus, Basella*, and *all the Holeraceae* 77:53.
    reduced: *umbellate [plants]*.

4. *WITH SEVERAL COTYLEDONS*

    Pinus 10.
    Cupressus 5.
    Linum 4?

II. RADICATION is the arrangement of the root, with stock descending [and] ascending, and with radicles. Examples are to be sought in Section 80.

    Bulbous and scaly: *Lilium*; coated: *Cepa*;
      doubled: *Fritillaria*; solid: *Tulipa*;
    Tuberous and palmate: *Orchis*; bundled: *Paeonia*;
      hanging down: *Filipendula* and *Elaeagnus*.
    Jointed (81): *Lathraea, Oxalis, Martynia*, and *Dentaria*;
    Thread-like: *Pastinaca, Daucus*, and *Raphanus*;
    Spherical: species of *Ranunculus*, species of *Chaerophyllum*, and *Bunium*.

III. RAMIFICATION is exhibited in the position of the branches, and the leaves follow this.

No branches, though the leaves are produced from the stem: *Dictamnus, Paeonia, Epimedium*, and *Podophyllum*.

Leaves that are opposite in plants and those that are alternate generally indicate entirely different vegetables, if you except a few, whereof there are some species with opposite leaves, and others with alternate ones: such as

    Euphorbia, Cistus, Lantara.
    Antirrhinum, Lilium, and *Epilobium*.

Leaves opposite below by the branches, alternate above by the flowers:
> *Antirrhinum, Jasminum.*
> *Veronica, Borago.*

Leaves alternate below, opposite above on the branches:
> *Potentilla* 1. [My]*Hortus Upsaliensis, Potamogeton.*

Leaves opposite below, in threes above: *Nerium.*

Leaves in threes below, alternate above (branches in threes below, and no others): *Ruscus.*

Leaves in fours below, alternate above: *Coreopsis* 2. *Hortus Upsaliensis, Antirrhinum* 3. *Hortus Upsaliensis.*

Anyone who desires to elicit the natural posture in plants that are different in ramification must go to the radical leaves.

IV. INTORSION is the bending of the parts towards one side.

The STEM turned round towards the left ☾ :
> *Taunus, Dioscorea, Rajania,*
>> *Menispermum,*
> *Cissampelos* and *Hippocratea.*
> *Lonicera.*
> *Humulus.*
> *Helxine.*

towards the right ☽ :
> *Phaseolus, Dolichos, Clitoria, Glycine,* and *Securidaca.*
> *Convolvulus, Ipomoea,*
> *Cynanchum, Periploca, Ceropegia,*
> *Euphorbia, Tragia,*
> *Basella,*
> *Eupatorium,*
> *Tournefortia.*

The *TENDRIL* turned round towards the right and back again.
> Most *Leguminosae* display tendrils of this kind.
> Smilax produces petioles that bear tendrils: *Piper* does just the same.

The *COROLLA* [turned round] towards the left*:
> *Asclepias, Nerium, Vinca, Rauwolfia, Periploca, Stapelia.*
> Towards the right: *Pedicularis* [my] *Flora Suecica* 505 and 507–8.
> *Trientalis* is peculiar, because all the petals lie to the right, overlapping on the other side.
> *Gentiana* is made to overlap against the sun, before it opens.

The *PISTILS*, towards the left: *Cucubalus* and *Silene.*

The *OVARIES* twisted towards the left: *Helicteres* and *Ulmaria.*

---

* [Turned] *towards the left*, that is what faces left [sic: corrected to *right*, see p. 369], if you suppose that you yourself are placed in the centre, and are looking south; and so [turned] *towards the right* is the opposite.

Inversion of the *FLOWERS*, when the upper lip of the corolla faces the ground, and the lower one faces the sky: *European Violas, Ocymum, Ajuga orientalis, species of Satyrium.*

Inclination [of the *FLOWERS*]: *Hyssopi Lophanthus, Nepeta* [my] *Hortus Upsaliensis* 3, *Pedicularis* [my] *Flora Suecica* 505, 507–8.

Spiral *SPIKES*: [Plants] with rough leaves, *Claytonia.*

Curving [of the spikes]: *Saururus, Mimosa, Petiveria, Papaver, Sedum rubrum, Lilium martagon.*

Various HYGROMETRICA, from the twisting of the fibres, occur in various plants.

The awn of *Avena* is twisted like a rope.

The aril of the seed of *Geranium*, which is [furnished] with a tail, is spiral.

Mnium, which is *Bryum* [my] *Flora Suecica* 903, [is furnished] with a peduncle, twisted below and oppositely twisted above.

V. GEMMATION is the construction of the bud out of leaves, stipules, petioles, or scales.

1. *FROM OPPOSITE PETIOLES*

a.	*Ligustrum*		*Laurus*
	Phillyrea		*Myrica.*
	Nyctanthes	c.	*Linnaea*
	Syringa		*Diervilla*
	Hypericum		*Lonicera.*
	Coriaria	d.	*Euonymus.*
	Buxus.	e.	*Fraxinus*
b.	*Jasminum*		*Acer*
	Vaccinium		*Aesculus*
	Arbutus		*Bignonia.*
	Andromeda	f.	*Opulus*
	Ledum		*Sambucus.*
	Daphne	g.	*Psidium.*

2. *FROM OPPOSITE STIPULES*

   Cephalanthus.   *Rhamnus catharticus.*

3. *FROM ALTERNATE PETIOLES*

a.	*Salix.*		*Ilex.*
b.	*Spiraea.*	e.	*Ribes.*
c.	*Genista*	f.	*Juglans*
	Solanum		*Pistacia.*
	Hippophaë.	g.	*Plumbago.*
d.	*Berberis*		

continued

4. *FROM ALTERNATE STIPULACEOUS PETIOLES*

a. *Sorbus*
Crataegus
Prunus
Mespilus germanica.

b. *Pyrus*
Malus.

c. *Cotoneaster*
Amygdalus
Cerasus

Padus.

d. *Melianthus*
Rosa
Rubus
Vitis
Robinia
Cytisus.

e. *Potentilla fruticosa*
Staphylea.

5. *FROM ALTERNATE STIPULES.*

a. *Populus.*

b. *Tilia*
Ulmus
Quercus
Fagus
Carpinus

Corylus.

c. *Betula.*
Alnus.

d. *Ficus*
Morus.

6. *ANOMALOUS*

a. *Abies.*

b. *Pinus.*

c. *Taxus.*

7. *NO [BUDS].*
[plants] without buds

Section 85. p. 65.

VI. FOLIATION is that folded arrangement which the leaves retain, so long as they are concealed within the bud or the shoots. It was overlooked by our ancestors, but admits the following [different] modes[3].

1. *[LEAVES] ROLLED IN,* when their lateral edges are rolled in spirally on either side inwards.

2. *ROLLED BACK,* when their lateral edges are rolled together spirally backwards.

3. *ROLLED OVER,* when the alternate edges grip the straight edge of the opposite leaf.

4. *ROLLED TOGETHER,* when the edge of one side encircles the other side of the same leaf, like a hood.

5. *OVERLAPPING,* when they lie one on another with flat surfaces in parallel formation.

6. *RIDING,* when the sides of the leaf close up in parallel formation, so that the inner ones are enclosed in the outer ones; which does not obtain in those that are doubled together (7).

7. *DOUBLED TOGETHER,* when the sides of the leaf are brought near to each other in parallel formation.

8. *FOLDED*, when they are folded together length-wise in folds, like the folded leaves in Section 83.

9. *SLOPED BACK*, when the leaves are bent back downwards, towards the petiole.

10. *ORBITED*, when the leaves are rolled in spirally downwards.

I. [LEAVES] ROLLED IN

Lonicera	*Alisma*
Diervilla	*Potamogeton natans*
Euonymus	*Nymphaea*
Rhamnus cathartica	*Saururus*
Pyrus	*Aster annuus*
Malus	*Humulus*
Populus	*Urtica*
Plumbago	*Hepatica*
Viola	*Sambucus*
Commelina annua	*Ebulus*
Plantago	*Staphylea.*

2. ROLLED BACK

Rosmarinus	*Polygonum*
Teucrium marum	*Parietaria*
Dracocephalum digit[atum]	*Primula*
Nerium	*Carduus*
Andromeda	*Cnicus*
Ledum	*Tussilago*
Epilobium irregulare	*Senecio*
Species of Salix!	*Othonna*
Rumex	*Potentilla fruticosa*
Persicaria	*Ptelea.*

3. ROLLED OVER

Dianthus	*Valeriana*
Lychnis	*Marrubium*
Saponaria	*Phlomis*
Epilobium oppositivum	*Salvia*
Dipsacus	*Prasium.*
	Scabiosa

4. ROLLED TOGETHER

Canna	*Commelina lutea*
Amomum	*Most grasses*
Calla	*Prunus*
Arum	*Armeniaca*
Piper	*Dodecatheon*
Hydrocharis	*Crepis*

continued

Lactuca

Hieracium

Sonchus sibiricus

　Tragopogon

Orobus

　Vicia

　Lathyrus

Solidago

　Aster

Pinguicula

Vaccinium

　Pyrola

Berberis

Brassica

　Armoracia

Symphytum

　Cynoglossum

Potamogeton perfoliatus

Eryngium

Menyanthes jointly

Saxifraga

Aralia

Dictamnus

Epimedium.

5. OVERLAPPING

Syringa

　Ligustrum

　Phillyrea

　Nyctanthes

Linnaea

Cephalanthus

Coriaria

Hypericum

Valantia

Justicia

Portulaca

Laurus

Daphne

Hippophaë

Ruscus

Cyanus perennis

Mespilus germanica

Campanula

　Polemonium

Sium.

6. RIDING

Hemerocallis

Some grasses

　Poa

Iris

Acorus

Carex.

7. DOUBLED TOGETHER

Quercus

　Fagus (right inside)

Corylus half (to one side)

　Carpinus (to one side)

　Tilia (to one side)

　Padus　　　spirally

　Cerasas　　spirally

　Amygdalus spirally

　Cotoneaster spirally

　Frangula

　　Alaternus

　　Paliurus

Juglans

Pistacia

Rhus

Fraxinus

Sorbus

Rosa

　Rubus

　Potentilla vulgaris

Comarum rolled over!

Bignonia

Cytisus

　Robinia

　Pisum

Most [plants] with two sets of stamens[4]

Melianthus

Pastinaca

　Heracleum

　Laserpitium

Poterium.

continued

8. FOLDED

Crataegus	*Ribes*
Betula	*Althaea*
Alnus	*Malva*
Fagus	*Humulus*
Vitis	*Urtica*
Acer	*Passiflora*
Opulus	*Alchemilla.*
Viburnum	

9. BENT BACK

Podophyllum	*Pulsatilla*
Aconitum	*Anemone*
Hepatica	*Adoxa.*

10. ORBITED

Ferns	*Some palms.*

VII. STIPULATION is the position and structure of the stipules at the bases of the leaves.

Stipules, no less than leaves, come forth different in different [plants].

a. None: *rough-leaved, didynamia, stellate* and *siliquous* [plants], *Liliaceae, Orchideae,* and *most compound* [plants].
Present: *Papilionaceae, Lomentaceae,* and [plants] *with twenty stamens*[5].

b. In pairs, or solitary on either side: in most [plants].
Solitary: *Melianthus* on the inside, *Ruscus* on the outside.

c. Falling late: *Padus, Cerasus, Amygdalus,* and order V [*stipules*] *of the buds.*
Persisting: *those with stamens in two sets* and *those with twenty stamens and several pistils.*[6]

d. Adhering: *Rosa, Rubus, Potentilla, Comarum,* and *Melianthus.*
Loose: in most [plants].

e. Within the leaves: *Ficus, Morus.*
Outside the leaves: *those with stamens in two sets, Alnus, Betula,* and *Tilia.*

VIII. PUBESCENCE is the plant's armour, which protects it from external injuries.
SCABROSITY is made up of particles that are scarcely visible to the naked eye, with which the surfaces of the plants are sprinkled.
The lynx-eyed GUETTARD was about the first to observe this.

Glandular [scabrosity],	millet-shaped [?] <miliaris>
	bladder-shaped: *Mesembryanthemum, Aizoa, Tetragonia.*
	lens-shaped.[7]
	spherical: *Atriplex, Chenopodium.*
	secretory.

continued

chain-shaped.

bag-shaped.

Bristly [scabrosity],

cylindrical.

conical.

with hooks.

bearing glands: *Ribes*.

forked: *Lavandula*.

axe-shaped: *Humulus*.

clustered, star-shaped: *Alyssum* and *Helicteres*.

clustered, simple: *Hippophaë*.

Jointed [scabrosity],

simple.

knotted.

with tails.

branchy: *Verbascum*.

feathery.

WOOL protects the plants from excessive heat.

Salvia canariensis, *Sideritis* canariensis, *Aethiops* Salviae, *Marrubium*, *Verbascum*, and *Stachys*.

Carduus eriocephalus and *Onopordum*.

MATTED HAIR protects the plants from the winds; quite often it possesses a *hoary* colour.

Tomex.

Medicago and *Halimus*.

THICK-SET HAIR wards off animalcules and tongues with its stiff bristles.

Cactus.

Malpighia.

Hibiscus and Rubus.

HOOKS stick onto passing animals.

Triglochides *Lappula*.

curved: *Arctium* and *Marrubium*.

Xanthium and *Petiveria*.

STINGS ward off unprotected animals with their poisonous prickling.

Urtica.

Jatropha, Acalypha, Tragia.

PRICKLES ward off particular animals.

Volkameria and *Pisonia*.

Hugonia (spiral or tendrilled [prickles]).

Caesalpinia, Mimosa, and *Parkinsonia*.

Capparis, Erythrina, and *Robinia*.

Solanum and *Cleome*.

Smilax, Convolvulus, and *Aralia*.

Duranta, Xylon, and *Drypis.*

Euphorbia, Tragacantha, and *Tragopogon.*

FORKS ward off animals.

Berberis, Ribes, and *Gleditsia.*

Mesembryanthemum and *Osteospermum.*

Ballota, Barleria, Fagonia, and *Poterium.*

SPINES ON BRANCHES ward off farm animals.

Pyrus, Prunus, and *Citrus.*

Hippophaë.

Gmelina, Rhamnus, Lycium, Catesbaea, and *Celastrus.*

Ulex and *Asparagus.*

Spartium, Achyronia, and *Ximenia.*

Ononis, Stachys, Alyssum, and *Cichorium.*

SPINES ON THE LEAVES.

Aloe, Agave, Yucca,

Ilex, Hippomane, Theophrasta, Carlina, Cynara,

Onopordum, Morina, Acanthus, Gundelia, Juniperus,

Salsola, Polygala, Ruscus, Borbonia, Statice, Ovieda,

and *Cliffortia.*

SPINES ON THE CALYX.

Carduus, Cnicus. and *Centaurea.*

Moluccella and *Galeopsis.*

SPINES ON THE FRUIT.

Trapa, Tribulus, Murex,

Spinacia, Agrimonia and *Datura.*

IX. GLANDULATION provides secretory vessels.

I. GLANDS OF THE PETIOLES:

Ricinus, Jatropha, Passiflora, Cassia, and *Mimosa.*

GLANDS OF THE LEAVES

on the serrations: *Salix.*

on the base: *Amygdalus, Cucurbita, Elaeagnus, Impatiens, Padus,* and *Opulus.*

on the back: *Urena, Tamarix* and *Croton.*

on the surface: *Pinguicola* and *Drosera.*

GLANDS OF THE STIPULES:

Bauhinia and *Armeniaca.*

CAPILLARY GLANDS:

Ribes, Antirrhinum quadrifolium, *Scrophularia, Cerastium,* and *Silene.*

PORES:

Tamarix, Silene viscaria.

2. FOLLICLES are vessels distended with air.

Utricularia produces at its root vessels that are almost round, inflated, and with two horns.

Aldrovanda puts forth pot-shaped semicircular follicles on its leaves.

3. UTRICLES are vessels filled with secreted liquid.

The extremity of the leaves of *Nepenthes* ends in a thread, and the thread ends in a cylinder of the size and shape of a penis, closed at the tip with a lid, which is opened on one side.

Sarracenia has hood-shaped leaves, almost like *Nepenthes*, but sessile at the root. *Marcgravia* puts forth vessels in the centre of the umbel, and they resemble the gaping corolla of *Galeopsis*, but are without the lower lip. [My] *Genera plantarum* 507.

X. LACTESCENCE is the supply of the liquid that flows out of a damaged plant.

White: *Euphorbia.*

              Papaver.

              Asclepias, Apocynum, Cynanchum, etc.

              [Flowers] with demi-florets T.

              Campanula, Lobelia, and *Jasione,*

              Cactus covered with tubercles.

              Acer, Selinum, and *Rhus.*

Yellow: *Chelidonium, Bocconia,* and *Sanguinaria.*

              Cambogia.

Red: *Rumex sanguinea.*

XI. INFLORESCENCE is the mode in which the flowers are attached to the peduncle of the plant; our predecessors called it the *mode* of flowering.

WHORLED Rj.: *Marrubium.*

BEARING CLUSTERS: *siliquous [plants]* (natural order 57).

SPIKED: (natural orders 1 and 2): the *Pipers* and *Mimosas.*

PANICULATE: *various grasses.*

Most flowers are AXILLARY, therefore the following are comparatively rare:

[FLOWERS] OPPOSITE THE LEAVES, where the flowers are directly opposed to the leaves, as *Piper, Saururus, Phytolacca, Dulcamasa, Vitis,* annual kinds of *Cistus, Cissus, Corchorus, Ranunculus aquatilis, Geranium.* ˙

[Flowers] BETWEEN THE LEAVES are placed between the leaves that are opposite, but alternately: *Asclepias.*

[Flowers] BESIDE THE LEAVES are placed at the sides of the bases of the leaves: *Claytonia, Solanum, [plants] with rough leaves.*

PETIOLAR flowers are on a peduncle that is inserted in the petiole: *Hibiscas* and *Turnera.*

Flowers that BEAR TENDRILS: *Cardiospermum* and *Vitis.*

[Flowers that grow] ABOVE THE AXIL: *rough-leaved* [plants], *Potentilla* 3. *[my] Hortus Upsaliensis.*

We pass over various things that are relevant to this: for example,

THE TIME *of germination*, before it emerges from the seed and the ground:

very short: the *tetradynamia*[8] and *Helxine.*

one year: *Hypecoum, Glaucium, Melampyrum* segetum, and *Ranunculus* falcatus.

two years: *Mespilus, Oxyacantha, Rosa,* and *Cornus.*
THE TIME *of gemmation,* or the opening of the buds.
THE TIME *of flowering,* annual; seasonal.
THE TIME *of thriving;* this is special and peculiar in various different [plants].

164. The primary (152) arrangement (155) of the vegetables is to be taken from the fruit-body alone.

Our predecessors argued that the fruit-body was insufficient, since few parts of it were known to them; but we have introduced them in a fully sufficient quantity.

CALYX.
    Perianthium *Riv.*   *Calix Tournef.*   [Perianth].
    Involucrum *Arted.*   [Involucre].
    Amentum.   *Julus T.*   [Catkin].
    Spatha *Linn.*   [Spathe].
    Gluma *Raj.*   *Locusta Raj.*   [Husk].
    Calyptra *Dill.*   [Veil].
    Volva *Michel.*   [Volva]
COROLLA *Linn.*   *Petalum T.*   [COROLLA]
    Petalum *Col.*   [Petal].
    Tubus or unguis.   *Unguis T.*   [Tube].
    Limbus or Lamina *L.*   [Limb].
    Nectarium *Linn.*   [Nectary].
STAMEN.   [STAMEN]
    Filamentum *Linn.*   *Stamen T.*   [Filament].
    Anthera *L.*   *Apex T.*   [Anther].
PISTILLUM   [PISTIL].
    Germen *L.*   *Ouarium B.*   [Ovary].
    Stylus *Boerh.*   *Pistillum T.*   *Tuba V.*   [Style].
    Stigma *Linn.*
PERICARPIUM *Riv.*   [PERICARP].
    Capsula *the ancients..*   *Capsula T.*   [Capsule].
    Siliqua *the ancients.*   *Siliqua T.*   [Silique].
    Legumen *Linn.*   [Pod].
    Conceptaculum *L.*   [Conceptacle].
    Drupa *Linn.*   [Stone-fruit].
    Pomum *the ancients.*   *Fructus carnosus.*   [Pome].
    Bacca *the ancients.*   [Berry].
    Strobilus *R.*   *Conus T.*   [Cone].
SEMEN.   [SEED].
    Coronula *L.*   [Coronule].
    Pappus *the ancients.*   [Tuft].

continued

Ala.		[Wing].
Arillus *L.*	*Calyptra T.*	[Aril].
Hilum.		[Hilum].
Nux *the ancients.*	*Nux T.*	[Nut].
Propago *Linn.*		[Layer].
RECEPTACULUM *Pont.*	*Placenta V.*	[RECEPTACLE]
Palea *Vaill.*		[Chaff].
Umbella *the ancients.*	*Umbella T.*	[Umbel].
Cyma *Linn.*		[Cyme].
Spadix *Linn.*		

Therefore all those [genera] that do not acknowledge their foundation in the fruit-body alone should be called CONTRIVED GENERA: for example

A Limodorum T. *with a fibrous root* would not be an Orchis.

A Bistorta T. *with a fleshy root* would not be a Polygonum.

A Rapa T. *with a humped root* would not be a Brassica.

A Sisarum T. *with a tuberous root* would not be a Sium.

A Hermodactylus T. *with a tuberous root* would not be an Iris.

A Sisyrinchium T. *with a bulb put onto it* would not be an Iris.

A Xiphium T. *with a tunicate bulb* would not be an Iris.

A Lilio-Fritillaria B. *with a scaly bulb* would not be a Fritillaria.

A Mesomora R. *with a herbaceous stem* would not be a Cornus.

An Anacampseros T. *with an upright stem* would not be a Sedum.

A Psyllium T. *with a branchy stem* would not be a Plantago.

A Bellis Leucanthemum M. *with a leafy stem* would not be a Bellis.

A Pilosella B. *with an uncovered scape* would not be a Hieracium.

A Suber T. *with fungous bark* would not be a Quercus.

A Larix T. *with bundled leaves* would not be an Abies.

A Genistella T. *with jointed leaves* would not be a Genista.

A Potamopithys B. *with leaves not stellate* would not be an Alsinastrum T.

A Quinquefolium T. *with digitate leaves* would not be a Pentaphylloides.

A Lupinaster B. *with digitate leaves* would not be a Trifolium.

A Dracunculus T. *with pedate leaves* would not be an Arum.

A Trichomanes T. *with pinnate leaves* would not be an Asplenium.

A Clymenum T. *with pinnate leaves* would not be a Lathyrus.

A Muscoides M. *with leaves frequently overlapping* would not be a Jungermannia.

A Lentiscus T. *with leaves* without an odd [leaflet] would not be a Terebinthus T.

A Faba T. *with leaves without tendrils* would not be a Vicia.

A Cytiso-Genista T. *with simple leaves in threes* would not be a Spartium.

A Colocasia B. *with leaves not auriculate* would not be an Arum.

A Cirsium T. *with leaves not spiny* would not be a C'arduus.

A Coronopus R. *with leaves pinnately cleft* would not be a Cochlearia.

A Coronopus T. *with indented leaves* would not be a Plantago.

An Ilex T. *with denticulate leaves* would not be a Quercus.

A Scorzoneroides *with indented leaves* would not be a Scorzonera.

An Anguria T. *with leaves divided into many parts* would not be a Cucurbita.

An Aleca T. *with leaves divided into many parts* would not be a Malva.

A Millefolium *with leaves minutely divided* would not be a Ptarmica T.

A Cicutaria T. *with cicuta leaves* would not be a Ligusticum.

A Cedrus T. *with cypress leaves* would not be a Juniperus.

A Ranunculoides V. *with capillary leaves* would not be a Ranunculus.

An Alhagi T. *with simple leaves* would not be a Hedysarum.

A Nissolia T. *with simple leaves* would not be a Lathyrus.

A Marsilea M. *with simple leaves* would not be a Jungermannia.

A Balsamita V. *with undivided leaves* would not be a Tanacetum.

A Cepa T. *with pipe-shaped leaves* would not be an Allium.

An Aphaca T. *with no leaves except stipules* would not be a Lathyrus.

A Mimosa T. *with sensitive leaves* would not be an Acacia T.

An Oxyoides G. *with pinnate, sensitive leaves* would not be an Oxalis.

An Aurantium T. *with heart-shaped petioles* would not be a Citrus.

A Calamintha T. *with branchy peduncles* would not be a Melissa.

A Colinus T. *with woolly peduncles* would not be a Rhus.

A Virga sanguinea D. *with a bare cyme* would not be a Cornus.

A Corona borealis *with a leafy tuft* would not be a Fritillaria.

A Stoechas T. *with a tufted spike* would not be a Lavandula.

A Carex D. *with small, androgynous spikes* would not be a Cyperoides T.

A Chamaepithys T. *with scattered flowers* would not be a Teucrium.

An Acinos D. *with scattered flowers* would not be a Thymus.

A Limonium T. *with scattered flowers* would not be a Statice.

A Chamaedrys T. *with whorled flowers* would not be a Teucrium.

A Thymbra T. *with whorled flowers* would not be a Satureja.

A Volubilis D. *with capitate flowers* would not be an Ipomoea.

A Polium T. *with cymose flowers* would not be a Teucrium.

A Castanea T. *with spiked flowers* would not be a Fagus.

A Fagopyrum T. *with spiked flowers and fibrous roots* would not be a Polygonum.

A Majorana T. *with spiked flowers, somewhat rounded* would not be an Origanum.

All GENERA CONTRIVED from their general shape should be rejected, unless
   they are also established from the foundation of the fruit-body: for example:

A Malus T. *with a peculiar shape* would not be a Pyrus.

A Cydonia T. *with a peculiar shape* would not be a Pyrus.

An Armeniaca T. *with a peculiar shape* would not be a Prunus.

A Cerasus T. *with a peculiar shape* would not be a Prunus.

A Lauro-Cerasus T. *with a peculiar shape* would not be a Prunus.

A Limon T. *with a peculiar shape* would not be a Citrus.

A Napus T. *with a peculiar shape* would not be a Brassica.

An Absinthium T. *with its outward shape* would not be an Artemisia.

An Abrotonum T. *with its outward shape* would not be an Artemisia.

A Bellidiastrum M. *with a peculiar habit* would not be a Doronicum.

A Euphorbia T. *with a leafless habit* would not be a Tithymalus T.

An Usnea D. *with a hairy habit* would not be a Lichen.

A Coralloides D. *with a caulescent habit* would not be a Lichen.

A Clavaria D. *with an unbranched habit* would not be a Coralloides T.

A Tuber T. *with a comparatively solid substance* would not be a Lycoperdon T.

A Fungoides M. *with smooth tissue on both sides* would not be an Elvela.

A Lycoperdoides M. *with cellular tissue* would not be a Lycoperdon.

An Amanita D. *with stiped cap* would not be an Agaricus.

A Phallus M. *with a stipe that has a volva at the base* would not be a Boletus M.

A Phalloboletus M. *with a cap that is free at the sides* would not be a Boletus M.

A Polyporus M. *with pores that cannot be distinguished* would not be a Boletus L.

An Erinaceus M. *with dense prickles* would not be a Ulex.

A Thysselinum *with milky juice* would not be a Selinum.

A Moly B. *with a sweet smell* would not be an Allium.

An Acetosa T. *with an acid taste* would not be a Lapathum T.

A Colocynthis T. *with bitter fruit* would not be an Angaria T.

165. Any vegetables that agree in the parts of the fruit-body (86) should not be separated in a theoretical arrangement (152), other things being equal.

> This pre-eminent discovery in the science of botany was discovered by *Gesner*, brought forward by *Cesalpino*, revived by *Morison*, and supported by *Tournefort*.
>
> *Rivinus*: 'Any that agree in flower and seed should be described by the same name, and the opposite is also true.'
>
> *Knaut*: 'Any plants, that flower and produce seed capsules in the same manner, belong to the same genus; and so also the contrary.'

166. Any vegetables that differ in the parts of the fruit-body (86), when notice has been taken of the things that ought to be noticed (162), should not be put together.

> This rule acts as the reverse of the preceding.

167. Every CHARACTERISTIC FEATURE ought to be elicited from the number, shape, relative size, and position of all the different parts of the fruit-body (98–104).

> It conveys the method by which every generic character is to be determined.
> The number of them does not exceed the 24 letters of the alphabet.

Let there be   VII parts of the *fruit-body*.

                    7 parts of the *calyx*.

                    2 parts of the *corolla*.

                    3 parts of the *stamens*.

                    3 parts of the *pistil*.

                    8 parts of the *pericarp*.

                    4 parts of the *seed*.

                    4 parts of the *receptacle*

                    ─────────────────

                        Total 38

                    The *number* of all the parts

                    The *shape*

                    The *position*

                    The *relative size*

                    ─────────────────

                        4

resulting in four times thirty-eight: 152.

Let the procedure be changed according to the 38 parts, and let it be 5736.

In this case the fruit-body suffices for at least 5736 genera: which could not ever exist.

Therefore it is a mistake to take [into consideration] the habit (section 163), colour, size, cotyledons, and other things, except those given [above].

Knaut forged many spurious genera from the manner of flowering.

So too, it is a mistake [to consider]

the habit of *Psyllus*.

the glabrosity of *Blattaria*.

the manner in which *Numularia* flowers.

the manner in which *Daphne Kn.* flowers.

the manner in which *Cracea* flowers, with many flowers.

the scattered flowers of *Limonium T.*

the genus *Inula* and the genus *Aster* separated on account of the colour *Vaillant*.

Cactus and *Melocactus* with one cotyledon, *Opuntia* with two.

Cupressus Europaea with two cotyledons, *Americana* with five.

168. The habit (163) should be carefully studied in private, to prevent the contriving of an erroneous genus for insufficient cause.

Experience, the mistress of affairs, quite often guesses the families of plants at first sight from their external appearance.

Examples confirm the rule:

Isopyrum, Nigella, Helleborus, and Caltha are different.

Sambucus and Ebulus; Trifolium and Triphylloides, should be linked.

An experienced botanist quite often distinguishes plants from Africa, Asia, America, and the Alps, but could not easily say by what feature. I do not know what is *grim, dry, and dark* about the AFRICAN plants' appearance, what is *proud and exalted* about the ASIATIC; *what is glad and smooth* about the AMERICAN, or *compressed and hardened* about the ALPINE.

The habit should be carefully studied in private, so that it shall not enter the ranks of the characteristic features and determine genera; this is proved by examples in *Boerhaave's Index, Leiden*; since no one would strive to recognize individual men from descriptions of their appearance.

Although *characters derived from habit* are unsatisfactory, yet at first sight they quite often indicate a plant clearly: they can be worked out as follows:

CARYOPHYLLEOUS [FLOWERS], natural order 42.

Placentation *with 2 cotyledons.*
Radication *fibrous.*
Ramification *opposite, jointed*, and *upright.*
Intorsion *of the pistil, to the left.*
Foliation *rolled over, lanceolate, and undivided.*
Stipulation *none.*
Pubescence *hardly noticeable.*
Inflorescence *with divisions in pairs.*

WHORLED [PLANTS], natural order 58.

Placentation *with two cotyledons.*
Radication *fibrous.*
Ramification *opposite* and *quadrangular.*
Foliation *rolled over* and *simple.*
Stipulation *none.*
Pubescence *slightly villous.*
Inflorescence *whorled, with bracts.*

ROUGH-LEAVED [PLANTS], natural order 43.

Placentation *with two cotyledons.*
Ramification *alternate.*
Foliation *rolled together, simple* and *undivided.*
Stipulation *none.*
Pubescence *scabrous.*
Intorsion *spiral with spike curved backwards.*
Inflorescence *with leaves at the side[s].*

UMBELLATE [PLANTS], natural order 22.

Placentation, *reduced, with two cotyledons.*
Radication *spindle-shaped.*
Ramification *alternate, somewhat rounded, upright.*
Foliation, *in pairs, supradecompound.*
Stipulation *none, sheathing the petiole.*
Pubescence *without prickles.*

Inflorescence *umbellate*.

LEGUMINOUS [PLANTS], natural order 55.

Placentation *unchanged and obliquely inserted with two cotyledons*.

Radication *fibrous, somewhat knotty*.

Ramification *alternate*.

Foliation *folded together or rolled together, feather-shaped*.

Stipulation *conspicuous*.

Pubescence *conspicuous*.

Intorsion *tendrilled*.

Inflorescence *somewhat spiked, one-sided*.

SILIQUOUS (PLANTS), natural order 57.

Radication *fleshily fibrous*.

Ramification *alternate*.

Foliation *rolled together, somewhat lyre-shaped*.

Stipulation *none*.

Pubescence *without prickles*.

Inflorescence *corymbose, terminal, without bracts*.

COLUMNIFEROUS [FLOWERS], natural order 34.

Placentation *with two cotyledons, heart-shaped, slightly pleated*.

Radication *fibrous*.

Ramification *alternate*.

Foliation *pleated, simple*.

Stipulation *narrow, outspread*.

Pubescence *somewhat tomentose*.

Inflorescence *axillary, pedunculate*.

FERNS, natural order 64.

Radication, *fibrous, bundled*.

Ramification, *in two rows*.

Intorsion *coiled*.

Foliation *coiled, feather-shaped*.

Stipulation *none*.

Pubescence *with bristling hairs*.

Inflorescence *at the back, sessile*.

169. Those [features] (167), that are effective to establish one genus, do not necessarily produce the same effect in another.

> You must understand that a character does not make a genus, but the genus makes the character.
>
> A character does not exist to form a genus, but to make it known.
>
> Examples prove this very clearly:
>
> The male of CLARICA has a *single petal*, the female has *five*.

The male flowers of JATROPHA have *single petals*, the female *five*.

One MYRICA has the seed *uncovered*, another has it *enclosed in a berry*.

One FRAXINUS has the flower *uncovered*, another has it *with a corolla*.

One GERANIUM has the corolla *regular*, another *irregular*.

One LINUM has *five petals*, another *four*.

One ACONITUM has *three capsules*, another *four*.

Hence Rivinus' controversy with Dillenius (Section 21).

Hence the greatest heresy in botany, which has generated countless spurious genera, causing very great harm to botany.

For example, one TRIFOLIUM has a *single petal*, another *four*; one has a *single seed*, another *many*. Some argue that one with a single petal and another with many petals, or one with a single seed and another with many, cannot serve in the same genus; therefore they establish spurious genera against nature.

170. Rarely is a genus observed, in which no part (167) of the fruit-body is erratic: for example:

An Arisarum T. *with a hooded sheath* would therefore not be an Arum.

An Astericus T. *with a star-shaped leafy calyx* would not be a Buphthalmum.

A Silybum V. *with spines on the calyx* would not be a Carduus.

A Moldavica T. *with a humped, two-lipped calyx* would not be a Dracocephalum.

A Tithymaloides T. *with a humped, irregular calyx* would not be a Euphorbia.

A Trionum L. *with an inflated calyx* would not be a Hibiscus.

A Ficaria D. *with a calyx that has 3 leaves and several petals* would not be a Ranunculus.

An Iva D. *with a humped calyx* would not be a Teucrium.

A Lunularia M. *with a common calyx divided into 4* would not be a Marchantia.

A Leucanthemum T. *with narrow scales on the calyx* would not be a Chrysanthemum.

A Cardiaca D. *with a 5-toothed calyx* would not be a Leonurus.

A Paronychia T. *with hooded leaves of the calyx* would not be a Herniaria.

A Pseudodictamnus *with funnel-shaped calyces* would not be a Marrubium.

An Anemone Ranunculus *with a corolla of five petals* would not be an Anemonoides.

A Linaria T. *with a tail-pointed corolla* would not be an Antirrhinum.

A Valerianoides V. *with a tail-pointed corolla* would not be a Valeriana.

A Bromelia P. *with a corolla of 3 petals* would not be an Ananas T.

An Opuntia T. *with a corolla of several petals* would not be a Melocactus T.

A Glaucia T. *with a rose-like corolla* would not be a Chelidonium.

A Polygonatum T. *with a tubular corolla* would not be a Lilium convallium.

A Centaurium minus T. *with a funnel-shaped corolla* would not be a Gentiana.

A Liliastrum T. *with a corolla of 6 petals* would not be a Hemerocallis.

A Borbonia P. *with a corolla like 3 leaves* would not be a Laurus.

A Benzoë B. *with a corolla divided into* 8 would not be a Laurus.

An Auricula ursi T. *with a salver-shaped corolla* would not be a Primula.

A Triphylloides T. *with a single-petalled corolla* would not be a Trifolium.

An Oxycoccus T. *with a 4-petalled corolla* would not be a Vaccinium.

A Bonarota M. *with a tubular corolla* would not be a Veronica.

A Zannonia P. *with a corolla of 3 petals* would not be a Commelina.

A Borraginoides *with a funnel-shaped corolla* would not be a Borrago.

A Horminum T. *with a helmet-shaped helmet and a concave beard on the corolla* would not be a Salvia.

A Sclarea T. *with a sickle-shaped helmet and a concave beard on the corolla* would not be a Salvia.

A Phelypaea P. *with a helmet on the corolla divided into two* would not be a Clandestina T.

A Murucuja T. *with an undivided nectary* would not be a Passiflora.

A Sherardia V. *[which has] two stamens* would not be a Verbena.

A Stellaris D. *[which has] stamens that are not flat* would not be an Ornithogalum.

A Porrum T. *[which has] stamens divided into three* would not be an Allium.

A Dodonaea P. *with a flower divided into three* would not be an Ilex.

A Hypocistis T. *with a flower divided into four* would not be an Asarum.

A Radiola D. *with a flower divided into* 4 would not be a Linum.

A Unifolium D. *with a flower divided into* 4 would not be a Convallaria.

A Bernhardia H. *with dioecious flowers* would not be a Croton.

A Petasites T. *with bundled flowers* would not be a Tussilago.

An Ananthocyclus V. *with floreted flowers* would not be a Cotula.

A Ceratocephalus V. *with rayed flowers* would not be a Bidens.

A Doria D. *with sparse radical flowers* would not be a Solidago.

I dismiss Ruppius' genera that are not defined, and those of Knaut and others, for which it is enough to bid them farewell, for example:

Medium Campanulae *with 5-chambered fruit.*

Speculum Veneris Campanula *with siliquous fruit.*

Cornucopioides Valerianae *with irregular flower.*

Limonioides Statices *with single-petalled flowers.*

Viscaria Silenes *with 5-chambered fruit.*

Tetragonolobus Loti *with angular fruit.*

Therefore, unless the rule is adopted, it turns out that there exist as many genera as species.

171. In most genera some peculiar (105) feature of the fruit-body is observed.

Denticles of the stamens in *Brunella, Torenia, Euphrasia, Alyssum,* and *Crambe.*

The stamen mutilated; *Curcuma, Chelone, Bignonia,* and *Martynia.*

The nectar-bearing pores in the narrow bases of the petals of *Ranunculus.*

The closed fissures inside the corolla of *Hydrophyllus*.
The tubular nectary of *Helleborus* and *Nigella*.
Hyoscyamus with its lidded capsule is distinct from *Physalis*.
The stamens of *Pancratium* inserted into the nectary, different from [those of]
    Narcissus.
The lateral nectary of *Reseda*, with varying corolla and pistil.
The 5-valved nectary of *Campanula*, with varying capsule and corolla.
The peculiar stigma of *Iris*, with the beard of the corolla varying.

172. If some feature of the fruit-body that is peculiar (105) or proper to its
genus (171) is not present in all the species, care must be taken to avoid
the accumulation of a greater number of genera.

    Erica and *Andromeda* were formerly in one genus, but the two-horned anthers are
        proper to *Erica*.
    Ranunculus formerly comprised *Adonis*, but *Adonis* has no nectar-bearing pores.
    Aloë and *Agave* constituted the same genus, but the stamens, which in the latter are
        not inserted into the corolla but into the receptacle, distinguish each of the two
        genera.

173. If a feature that is peculiar (105) to a genus is also found in a related
genus, care must be taken not to separate one and the same genus into
more genera than nature dictates.

    Sedum, Sempervivum, Rhodiola, Crassula, and *Tillea*; the *cotyledon* with nectaries
        adhering to the base of the pistil.
    Epilobium and *Oenothera* with tubular calyces.
    Mespilus, Crataegus, and *Sorbus*; the structure of the flower.
    Alnus and *Betula* with three florets above the leaflet of the catkin.

174. The more constant any part of the fruit-body (167) is in a relatively large
number of species, the more certainly indeed does it show a generic
feature.

    The nectary of *Hypericum* is constant, but not the siliqua.
    The berry of *Convallaria* is spotted, but not the corollas of *Lilium convallii,*
        Polygonatum, and *Unifolium*.
    The corolla of *Cassia*, but not the siliqua.
    The corolla of *Lobelia*, but not the fruits of *Lobelia, Cardinalis, Rapuntium,* and
        Laurentia.
    The calyx and corolla of *Verbena*, but note the stamens and seeds of *Sherardia. V.*

175. It is observed that in some genera one part of the fruit-body is relatively constant, in others another; but none is absolutely so.

> *Impatiens, Campanula, Primula, Papaver, Cistus, Fumaria,* and *Arbutus* show that the pericarp varies in species of the same genus.
> *Nymphaea* and *Cornus*: the calyx.
> *Vaccinium, Convallaria, Andromeda, Gentiana,* and *Linum*: the corolla.
> *Corispermum* and *Valeriana*: the stamens.
> *Ranunculus* and *Alisma*: the seeds.

176. If the flowers (87) agree but the fruits (87) differ, other things being equal, the genera should be united.

> *Cassia, Hedysarum, Sophora, Lavatera, Hibiscus,* and *Mimosa* show that similar flowers and different fruits exist within the same genus.

177. The shape (95) of the flower (87) is more definite than that of the fruit (87); yet the relative size (96) of the parts is very different, and not absolutely constant.

> It is obvious from many examples that the flower is more definite than the fruit:
> *Campanula, Medium,* and *Speculum veneris.*
> *Primula* and *Auricula ursi.*
> *Antirrhinum, Elatine,* and *Asarina.*
> *Alisma* and *Damasonium.*
> *Hibiscus* and *Malvaviscus.*
> *Cistus* and *Helianthemum.*
> *Fumaria* and *Corydalis.*
> *Arbutus* and *Uva ursi.*
> *Clematis* and *Viticella.*
> *Guilandina, Zygophyllum,* and *Papaver.*
> *Ranunculus, Hesperis,* and *Datura.*

178. The *number* (94) is more likely than the *shape* (95) to be erratic; yet it is very well explained by the proportion of the number. But flowers that differ in number in the same plant should be regarded according to the principal [number].

> Flowers that differ in number in the same plant.
> 5 flower[s] of *Ruta*, 4 flowers, confirmed by *Pseudo-Ruta Michel.*

5 of *Chrysosplenium*, 4 flowers, confirmed by *Saxifraga*.
5 of *Monotropa*, 4 flowers, confirmed by *uniflora*.
5 of *Tetragonia*, 4 flowers.
5 of *Euonymus*, 4 flowers, confirmed by *americana*.
5 of *Philadelphus*, 4 flowers.
4 of *Adoxa*, 5 flowers.
Therefore the principal number determines the natural one.

Numerical affinities proper to the fruit-body.

X and VIII of the flower.	*Stamens* in the same [fruit-body]:	of *Ruta* [and] *Monotropa*.
	Stamens in different [fruit-bodies]:	of *Vaccinium* and *Oxycoccus T*.
	Stamens in different [fruit-bodies]:	of *Stelleria* and *Passerina T*.
	Stamens in different [fruit-bodies]:	of *Sedum* and *Rhodiola* L.
	Stamens in different [fruit-bodies]:	of *Arenaria* and *Moerhingia*.
V and IV	*Corolla* in the same [fruit-body]:	*Euonymus* [and] *Ruta*.
	Corolla in the same [fruit-body]:	*Philadelphus* [and] *Monotropa*.
	Corolla in a different [fruit-body]:	*Linum* and *Radiola* D.
	Corolla in a different [fruit-body]:	*Anagallis* and *Centunculus*.
	Calyx in a different [fruit-body]:	*Nymphaea lutea* and *alba*.
	Stamens in the same [fruit-body]:	*Euonymus*.
	Stamens in a different [fruit-body]:	*Mimosa*.
IV and III	*Flower* in the same [fruit-body]:	*Asperula Rubeola T*.
	Flower in a different [fruit-body]:	*Paris europaea* and *americana*,
	Ilex and *Dodonea Plum*.	

V and III of the fruit very frequently proportioned
of *Ruta* with *Peganum*.
of *Euonymus* with *Celastrus*.
of *Nigella* with *Garidella*.
of *Cistus T*. with *Helianthemum T*.
of *Aconitum* with *Anthora*.
of *Viscaria* with *Silene*.

V and IV   of *Ruta*, *Euonymus*, *Philadelphus*, *Linum*, and *Rhodiola*.

179. The *position* (97) of the parts is absolutely constant. *Tournefort* (64) made much of the position of the receptacle (86) in the orders (161).

> *TOURNEFORT* understood more accurately than he stated, that the calyx goes off into the fruit, for it is the same as the ovary, whether it is raised above the perianth or pushed down below it, as in *Saxifraga* and *Geum T*.
> The position of the parts is very rarely changed in those of the same genus.

In flowers with twenty stamens[9] and some others, the receptacle goes round the
walls of the calyx on the inside; and so the corolla and filaments are inserted into
the calyx in *Fragaria, Pyrus*, and *Oenothera*.

180. *Rivinus* made too much of the *regularity of the petals* (61).

Some umbellate [plants] possess regular corollas, *others* irregular ones.
The *European Geraniums* possess a regular corolla, the *African* an irregular one.
Among the *plants with rough leaves*, which are regular, *Lycopsis* and *Echium* possess
an irregular corolla.

181. Nature has made the most of the nectary (110).

The nectary was not known even by name, until we defined it.

Orchis and *Satyrium*.	*Stapelia* and *Asclepias*.
Monotropa, Fumaria, and *Viola*.	*Diosma*.
Malpighia and *Bannisteria*,	*Campanula* and *Plumbago*.
Adenanthera and *Commelina*.	*Hyacinthus*.
Laurus and *Helxine*.	*Rhododendrum*.
Dictamnus.	*Cheiranthus* and *Sinapis*.
Zygophyllum.	*Kiggelaria* and *Clutia*.
Swertia.	*Aquilegia*.
Lilium and *Fritillaria*.	*Nigella*.
Hydrophyllum.	*Aconitum*.
Ranunculus.	*Parnassia*.
Hermannia.	*Epimedium*.
Berberis.	*Theobroma*.
Staphylea.	*Reseda*.
Passiflora.	*Grewia*.
Narcissus and *Pancratium*.	*Helleborus* and *Isopyrum*.
Mirabilis.	*Tropaeolum* and *Impatiens*.
Nerium.	

182. The stamens and the calyx, being much less liable to extravagancies, are
far more definite than the petals.

There are many corollas that differ in shape within the same genus.
The *Vacciniums, Pyrolas, Andromedas, Nicotianas,*
Menyanthes, Primulas, Veronicas, Gentianas,
Hyacinthi, Scabiosas, and *Narcissi*.
There are corollas that differ in number in a distinct species:
Ranunculus with five petals and with many petals.

Helleborus with five petals and with many.

Statice with five petals and with a single petal.

Fumaria with two petals and with four.

There are corollas that differ in number within the same species:

Carica and *Jatropha*.

183. It has been shown by countless examples that the *structure of the pericarp* (92), as described by our botanical predecessors, is less valid than they thought.

Very many contrived genera have been built up on the foundation of the fruit, and introduced without any regard for sound theory.

A Clandestina T. *with elastic fruit* would not be an Anblatum T.

A Trollius R. *with many-capsuled fruit* would not be a Helleborus.

A Sesamoides T. *with many-capsuled fruit* would not be a Reseda.

A Lycopersicon T. *with many-chambered fruit* would not be a Solanum.

An Ascyrum T. *with five-chambered fruit* would not be a Hypericum.

A Dortmanna R. *with two-chambered fruit* would not be a Rapuntium T.

A Helianthemum T. *with single-chambered fruit* would not be a Cistus.

An Androsaemum T. *with single-chambered fruit* would not be a Hypericum.

A Pavia B. *with single-chambered fruit* would not be an Aesculus.

An Asarina T. *with single-valved fruit* would not be an Antirrhinum.

An Elatine D. *with fruit dehiscent at the side* would not be an Antirrhinum.

A Nelumbo T. *with fruit pierced at the tip* would not be a Nymphaea.

A Raphanistrum T. *with jointed fruit* would not be a Raphanus.

A Cakile T. *with jointed fruit* would not be a Bunias.

An Ulmaria T. *with twisted fruit* would not be a Filipendula.

A Persica T. *with succulent fruit* would not be an Amygdalus.

A Cassia T. *with succulent fruit* would not be a Senna T.

An Inga P. *with succulent fruit* would not be an Acacia T.

A Malvaviscus D. *with succulent fruit* would not be a Hibiscus.

A Lobelia P. *with drupe-like fruit* would not be a Rapuntium T.

A Pereskia P. *with leafy fruit* would not be a Cactus.

A Sabina B. *with warty fruit* would not be a Juniperus.

A Bihai P. *with three-seeded fruit* would not be a Musa.

An Alaternus T. *with three-seeded fruit* would not be a Rhamnus.

A Frangula T. *with two-seeded fruit* would not be a Rhamnus.

A Dracunculus B. *with single-seeded fruit* would not be a Haemanthus.

An Onobrychis T. *with single-seeded fruit* would not be a Hedysarum.

A Malvinia D. *with uninflated fruit* would not be an Abutilon T.

A Cysticapnos B. *with inflated fruit* would not be a Fumaria.

An Impatiens R. *with attenuated fruit* would not be a Balsamina T.

A Guazuma P. *with net-like fruit* would not be a Cacao T.

A Paliurus T. *with shield-shaped fruit* would not be a Rhamnus.

An Alisma D. *with fruit without horns* would not be a Damasonium T.

A Securidaca T. *with sword-shaped fruit* would not be a Coronilla.

A Melo T. *with egg-shaped fruit* would not be a Cucumis.

A Melopepo T. *with grooved fruit* would not be a Cucurbita.

A Rapistrum T. *with fruit that is not dehiscent* would not be a Crambe.

A Radicula T. *with fruit like a silicle* would not be a Sisymbrium.

A Blattaria T. *with rounder fruit* would not be a Verbascum.

A Persea P. *with fruit pulpy all round* would not be a Laurus.

A Cururi P. *with fruit bearing seed at the tip* would not be a Seriana P.

A Bursa pastoris T. *with fruit that is not margined* would not be a Thlaspi.

A Nasturtium T. *with margined fruit* would not be a Lepidium.

A Valerianella T. *with fruit without a pappus* would not be a Valeriana.

An Anemonoides D. *with uncovered seeds* would not be an Anemone.

A Eupatoriophalacrum V. *with uncovered seeds* would not be a Verbesina.

A Leontodontoides V. *with seeds almost uncovered* would not be a Hyoseris.

An Atractylis V. *with seeds with a scarcely apparent corona* would not be a Carthamus.

A Carthamoides V. *with seeds that have a pappus* would not be a Carthamus.

A Zazintha T. *with seeds that have a pappus* would not be a Lapsana.

An Alypum N. *with seeds that have a pappus* would not be a Globularia.

A Xeranthemoides *with a feathery pappus* would not be a Xeranthemum.

An Asteropterus V. *with a feathery pappus* would not be an Aster.

An Acarna V. *with a feathery pappus* would not be a Cnicus.

An Achyrophorus V. *with a feathery pappus* would not be a Hypochaeris.

A Carlinoides V. *with a scarcely apparent pappus* would not be a Carlina.

A Viticella D. *with tail-pointed seeds* would not be a Clematis.

A Nymphoides T. *with seeds that have arils* would not be a Menyanthes.

A Karatas P. *with seeds without arils* would not be a Bromelia.

A Tragopogonoides V. *with seeds curved inwards* would not be a Tragopogon.

A Tinus T. *with pear-shaped seeds* would not be a Viburnum.

An Opulus T. *with heart-shaped seeds* would not be a Viburnum.

A Persicaria T. *with triangular seeds* would not be a Polygonum.

An Emerus T. *with cylindrical seeds* would not be a Coronilla.

A Foeniculum T. *with thick seeds* would not be an Anethum.

A Lens T,. *with lens-shaped[7] seeds* would not be a Cicer.

A Pepo T. *with leaves that are not emarginate* would not be a Cucurbita.

A Falcaria D. *with slender seeds* would not be a Sium.

A Cerinthoides B. *with 4 separate seeds* would not be a Cerinthe.

A Blaeria H. *with prickly seeds* would not be a Sherardia.

184. *Luxuriant flowers* (119), eunuchs (150), and mutilated flowers (119), being abnormalities, do not count in the establishment of the genera.

If *full flowers* were admitted as a character, then no number of petals could be assigned to a great number of plants, and in a great number of characters the stamens would be excluded.

Mutilated flowers (119) exclude the corolla, so that even, for instance, in the characters of *Campanula, Ipomoea, Ruellia*, etc., Section 119, the corolla would be excluded, contrary to the nature of all the rest of the species.

185. *Multiple* (120) and *full* (121) flowers are judged by the perianth and the lowest row of petals, as *prolific* ones (123) are judged by their offshoots.

A full flower cannot be referred to the genera, since it is abnormal; it will become natural, if it is sown in a thin soil.

The *perianth* does not change in a full flower; therefore a full flower is quite often referred to a genus according to the methods of the calycists; for example, *Hepatica, Ranunculus*, and *Alcea*.

The lowest row of a *corolla with many petals* remains constant in number even in full flowers; therefore the number of petals is quite often most easily elicited from that, as in *Papaver, Nigella*, and *Rosa*.

186. The CHARACTER is the definition[10] of the genus, and it exists in three forms; *the factitious* (188), *the essential* (187), and *the natural* (189).

The *generic character* is the same as the definition of the genus.

The *HABITUAL [Character]* from the habit (Section 163), which the ancients accepted, has now become intrinsically obsolete in the genera, since the discovery of the fruit-body (164).

187. The ESSENTIAL character (186) provides the genus to which it is applied with its most proper (171) and peculiar (105) feature.

By its unique pattern, the *ESSENTIAL* character distinguishes a genus from those of the same kind included in the same natural order.

The 6th edition of [my] Systema naturae contains many examples, which see:

Nyctanthes, Circaea, Gratiola, Salvia; Olax; Iris, Melica; Leucadendron.
Plantago, Epimedium; Hydrophyllum, Mirabilis, Hyoscyamus, Physalis.
Stapelia, Ceropegia, Aethusa, Parnassia, Statice.
Galanthus, Narcissus, Pancratium, Asphodelus, Aloë, Haemanthus.
Tropaeolum, Laurus; Anacardium; Dictamnus, Zygophyllum.
Melia, Dianthus, Phytolacca; Bixa, Reseda, Delphinium.
Aconitum, Nigella, Liriodendron, Uvularia, Ranunculus, Helleborus.
Hyssopus, Brunella, Scutellaria, Euphrasia, Torenia, Lathraea, Craniolaria.

Halleria, Acanthus; Crambe, Alyssum: Hermannia, Pentapetes.
Hibiscus; Polygala, Amorpha, Dalea, Psoralia; Theobroma;
Echinops, Inula, Centaurea, Corymbium, Jasione, Impatiens;
Sisyrinchium, Passiflora, Aristolochia, Helicteres, Arum, Zostera;
Zizania, Liquidambar, Clutia.

The shorter an essential character is, the more excellent it is.

188. The FACTITIOUS character (186) distinguishes the genus from other genera, but only from those of the same artificial order.

> The FACTITIOUS character distinguishes genera outside the natural order, whether it assumes fewer or more characteristic features.
>
> The diagnosis of the natural character is this, that it would never be effective for distinguishing genera in a natural order; therefore the characters, whether essential or natural, which cannot sufficiently distinguish the genera in a natural order, are to be called factitious.
>
> *Ray* defines a factitious character, stating that *the characteristic features of the genera should not be multiplied unnecessarily, and no more of them should be retained than is necessary for the certain determination of the genus.*

189. The NATURAL Character (186) adduces all (92–113) possible (167) generic features, and so it comprises the essential (187) and the factitious (188).

> All possible features are adduced by the natural character, except however those that are supplied by the most natural structure (93–97).
>
> I was the first to arrange these characters.
>
> It comprises in itself all the possible characters; it serves for every system, it lays a *foundation* for new systems, and it remains unchanged, even if unlimited new genera should be discovered every day.
>
> It is *corrected* only by the discovery of new species, with the exclusion, to be sure, of superfluous features: [my] *Genera plantarum, preface,* 20.
>
> The *speciality* and *use* of the natural character can be gathered from [my] *Genera plantarum, preface,* 18.

190. The *factitious* character (188) is secondary; the *essential* (187) is the best, but hardly possible in all cases; the natural one is worked out with great difficulty: but working it out is the basis (156) of all systems (53), and the infallible guardian of the genera; and it can be applied to every possible or actual system (26–37).

FACTITIOUS characters have been propounded by those who, in the artificial
method, have adduced features by which the genera that are arranged under the
same order could be distinguished; this has been done by Ray, Tournefort,
Rivinus, and many others, before our time.

ESSENTIAL characters are of the kind that are effective for distinguishing the most
closely related genera in a natural order, by one feature or another; so that they
cannot fail to be effective even where related genera are separated: very many of
these have been adduced by me in *Systema naturae*, 6th edition.

The *NATURAL* character exhibits all the possible characteristic features, so that it
serves all methods, however many may be discovered every day.

My Genera plantarum propounds natural characters, so that they serve everyone's
methods, and provide a foundation for old and new ones.

191. The *NATURAL* character (189) must be maintained by every botanist (7).

If the essential characters of all genera had been discovered, the recognition of plants
would turn out to be very easy, and many would undervalue the natural
characters, to their own loss. But they must understand that, without regard for the
natural character, no one will turn out to be a sound botanist; for when new genera
are discovered, the botanist will always be in doubt if he neglects the natural
character. Anyone who thinks that he understands botany from the essential
character and disregards the natural one is therefore deceiving and deceived; for
the essential character cannot fail to be deceptive in quite a number of cases.

The natural character is the foundation of the genera of plants, and no one has ever
made a proper judgement about a genus without its help; and so it is and always
will be the absolute foundation of the understanding of plants.

192. The natural character (189) will rehearse all the different (98) and peculiar
(105) features of the fruit-body that agree (165) throughout its individual
species (157); but it should be silent about those that disagree (166).

There is need of endless labour before the characters are defined in accordance with
all of their species.

All parts of the fruit-body should be examined, including those that elude vision,
even if a microscope has to be employed; (however, it is very rarely necessary to
employ it); for, if the fruit-body is unexplored, there is no certainty about the
genus. Anyone who wishes to investigate mites in flour or scurf must certainly
use microscopes.

And so Ray (wrongly) says: *The features should be obvious, palpable, and clearly visible to
anyone; for when the principal use of a method is to induce ignorant people and
beginners to investigate plants concisely without boredom or difficulty, one should not
propound features of the sort that need an attentive and careful observer, who would find
it necessary to bring a microscope with him.*

193. No character is infallible until it has been rectified in accordance with all its own (139) species (157).

> The most accomplished botanist, and he alone, achieves the best natural character; for it will be made by the agreement of the greatest number of species; for every species includes some superfluous feature.
>
> The natural character is made by the very accurate description of the fruit-body of the foremost species. All the other species of the genus should be compared with the foremost one, excluding all the features that disagree; eventually, it comes to pass that it [the character] has been worked out.

194. Inflorescence (163) will not provide a characteristic feature.

> The place that the fruit-body occupies in a herb is not a characteristic feature, although *Ray, Rivinus, Heucher, Knaut, Kramer,* and others thought otherwise; but it is the plague of botany, and is accordingly rejected by the principal botanists, Tournefort, Vaillant, and others.
>
> See the modes [in which it is joined] to the peduncle, Section 82 [D] and section 163 [XI].
>
> *Ray, Boerhaave, the followers of Rivinus,* and others allowed this a place in the character, so as to stick more closely to nature; but they lost touch with nature all the sooner.

195. The character must display the generic name on its façade.

> *Johrenius* in his *Hodegus* deduces the name from the character: for example, if the flower is of this kind, this will be its name. For we do not argue from the character to the name, but we determine the genus according to the generic name, and the essence of the name is comprised by the character.

196. Each form of fruit-body (86) in its natural character (189) should start a new line.

> This will be done, so that everything is propounded distinctly;
> hence, I ascertain the part in question at once;
> hence, I notice at once if anything is missing.

197. The name of a fruiting part (86) will begin a line (196) with different letters [see Section 199].

> Different letters, so that the character may be more easily examined and come out more clearly; but the opposite has been done by all our predecessors.

198. No character (192) should assume the feature (167) of an analogy, unless it is more familiar [to you] than [your] right hand.

> Every analogy is lame.
> An analogy [derived] from ourselves is always at hand.
>> One from animals is not agreed by everybody.
>> One from artificial [objects] is very variable.
>> *The tweezers* T. [in] the perianth of Urtica, [familiar] *to a surgeon.*
>> *The Polish mitre* T. in the corolla of Aconitum, *to historians.*
>> *The bishop's mitre* T. in the capsule of Mitella, *to divines.*
>> *The hyoid bones* T. in the filaments of Salvia, *to anatomists.*
>> *The adder's head* T. in the seeds of Anchusa, *to zoologists.*

199. The character should succinctly describe the features that agree (192).

> Therefore the first [duty] of the beginner will be to understand the technical terms.
> [My] *Genera plantarum, preface* 25. The pompous and their flourishes of eloquence are to be rejected.
> In a character, nothing is more abominable than a rhetorical style.
> The technical terms enable us to express our ideas in a few words.
> Let a *rhetorical character* of LINUM serve as an example.
>> The green outermost covering of the flower, which encloses the flower before it unfolds, is as it were cut into 5 equal parts at the base, but so that each part is longer than it is broad, and narrowed towards either end, the outermost lips ending in a point; besides, these 5 parts maintain a perpendicular posture, and are very short in proportion to the leaves of the flower; they do not fall off with the coloured leaves of the flower, but remain until the fruit is ripe. Within these leaves there are other leaves, also 5 in number, but delicate, coloured, and oblong also, but more and more enlarged towards the top, almost like an engineer's funnel; these are also far larger than the outer green leaves. Then next, within these five large coloured leaves of the flower, come some thread-like pieces, 5 in number, gradually attenuated into sharp points at the top. and they are almost perpendicular; and no longer than the outermost leaves of the flower; on the tops of the these rest the same number of minute particles which are undivided and thicker, and they scatter powder; and at the base they are split into 2 sharp pieces. When these pieces have been properly examined, a body is encountered in the centre of the flower, and this grows out to form the fruit; and at blossom-time it almost reproduces the shape of a sphere; and on top of it are placed five vegetable threads, which maintain the same thickness throughout, and have an almost perpendicular posture, and they have the same length as the 5 thread-like pieces just described; however, they are not headed or thickened at their tips, but bent a little outwards; when the blossoming is finished, the fruit becomes dry, roughly in the shape of a sphere, but faintly marked with 5 corners, and at its top it displays a sharp point; if you cut this fruit across, you will find

that it is divided on the inside into 10 chambers, and when it dehisces naturally, it opens out into 5 equal parts, within which 10 seeds have been concealed, and they are roughly egg-shaped, but longer, and pointed at one end, also a little compressed, with a surface as it were polished and smooth.

The same *character* of *Linum*, including all the same [features], in the language generally used by botanists.

CALYX: *Perianth* with 5 leaves; *leaflets* erect, lanceolate, pointed, small, and persistent.

COROLLA: funnel-shaped, 5 petals; with *petals* wedge-shaped, blunt, spread out and large.

STAMENS: 5 *filaments*, awl-shaped, erect, as long as the calyx; 5 other filaments, which are alternate, and wither; *anthers* arrow-shaped.

PISTIL: *Ovary* egg-shaped; 5 *styles* erect, thread-like, as long as the stamens; *stigmas* undivided, bent back.

PERICARP: *Capsule* somewhat spherical, somewhat pentagonal, with 5 valves and 10 chambers.

SEEDS: solitary, egg-shaped, rather flat, pointed and very smooth.

200. Plain TERMS (81–85) should be chosen; obscure and erroneous ones should not be allowed.

[My] *Genera plantarum, preface*, 26. Doubtful things should be omitted, rather than doubtfully defended.

Ray's Methodus. The features of the genera, both the principal and the subordinate, should be clear, distinct, and precisely defined; not obscure and indefinite, so that it is uncertain how far their meaning extends.

My own [terms].

MASCULUS Flos. *Sterilis* Tourn. *Paleaceus* Raj. *Abortiens* to others. [Male flower].

APETALUS Tourn. *Imperfectus* Riv. Knaut. Pont. *Stamineus*. Raj. *Incompletus* Vaill. [Without petals].

PETALOIDES Tourn. *Perfectus* Raj. Riv. Kram. Pont. [With petals].

CALYCULATUS. *Completus* Vaill. [With epicalyx].

IRREGULARIS Riv. *Difformis* Jung. Knaut. *Anomalus* Tourn. [Irregular].

RINGENS. *Labiatus* Tourn. *Barbatus* Riv. *Perfonatus* Tourn. [Gaping].

MULTIFIDUS. *Laciniatus* Tourn. *Monopetaloides* etc. to others. [Much divided].

COMPOSITUS Tourn. Riv. *Conglobatus* Pont. *Aggregatus* Kn. *Capitatus* Raj. [Compound].

PLANIPETALUS Raj. *Semiflosculofus* Tourn. *Lingulatus* Pont. *Cichoraceus* Vaill. [With flat petals].

RADIATUS Tournef. *Stellatus*. Moris. [Rayed].

DISCUS Tournef. *Umbo* Morif. [Disc].

ANTHERA. *Apex* Raj. Tourn. Riv. *Capsula. staminis* Malpigh. [Anther].

RECEPTACULUM Pont. *Sedes* Raj. *Placenta*, Boerh. *Thalamus* Vaill. [Receptacle].

AMENTUM Tourn. *Julus, Nucumentum, Catulus* to others. [Catkin].

STROBILUS. *Conus* to others. [Cone].

DRUPA kyber. lex. 150 [sic]. *Prunus* to others. *Fructus mollis ossiculo* [soft fruit with a stone]. Tournef. [Stone fruit].

GYMNOSPERMUS fructus. Herm. *Semina nuda.* Riv. [Fruit with uncovered seeds].

ANGIOSPERMUS fructus. Herm. *Semina Pericarpio* tecta Riv. [Fruit with seeds covered by the pericarp].

CLASSIS. *Ordo* Tournef. *Genus summum* Raj. Riv. [Class].

ORDO. *Sectio.* Tournef. *Genus subalternum* Raj. Riv. [Order].

201. Quite a lot of terms (199) must be excluded from the number of those that are requisite (200), and relatively few need to be enlarged [in meaning].

I have enriched botany with very many terms, Sections 82–86; for instance, involucre, spathe, corolla, anther, pollen, germen [ovary], stigma, legume [pod], drupe [stone-fruit], cyme, aril, stipule, scape, bract, peduncle, and gland.

[Technical] terms have saved anatomy, mathematics, and chemistry from amateurs; but the lack of them has damaged medicine.

The use of them is especially beneficial in speaking succinctly and thinking correctly, so long as they are provided with adequate definitions.[10]

Tournefort quite often adduced the *partition*, either athwart or parallel to the valves in siliquous [plants]; this should be taken with a grain of salt; it is called *parallel* when it agrees with the valves in breadth and cross-wise diameter; it is described as *athwart*, when the partition is narrower than the valves.

A *butterfly-shaped* corolla, so called from the insect, is commonly imagined as a boat, for its

Keel holds and encloses the stamens and pistils, and consists of two conjoined petals.

The *wings* are placed at the sides of the hull, only one on each side.

The *standard* is placed on top of the keel and the wings.

A *gaping* corolla is an irregular one with a solitary petal, and divided at its edge into two lips.

Rivinus' *helmet* is the upper lip.

The *beard* is the lower lip.

202. The character (192) should be kept without change in all systems (54–77), even the most diverse.

So long as the most eminent systematists were introducing new characters and new concepts of the genus, then the science of botany suffered grievously from exposure to barbarity, in the time of *Ray, Tournefort, Rivinus, Boerhaave, Knaut,* and others.

Now that most things are more settled, although new methods have been introduced, none of them is a disaster for botany, as is evident from the writings of *Gronovius, Royen, Wachendorff, Gmelin, Guettard, Dalibard,* and others.

203. A genus (159) can consist of a single (157) species, though it is more often composed of a large number of them.

Very many genera consist of a single species, for instance:

Parnassia,	*Epimedium,*	*Hydrophyllum,*	*Butomus,*
Tamarindus,	*Cornucopiae,*	*Diapensia,*	*Coris,*
Lagoecia,	*Gloriosa,*	*Petiveria,*	*Anacardium,*
Penthorum,	*Neurada,*	*Garcinia,*	*Mentzelia,*
Heliocarpus,	*Calligonum,*	*Hepatica,*	*Trichostema,*
Orvala,	*Halleria,*	*Dodartia,*	*Craniolaria,*
Obolaria,	*Limosella,*	*Anastatica,*	*Amorpha,*
Dalea,	*Corymbium,*	*Nepenthes,*	*Cynomorium,*
Hura,	*Valisneria,*	*Humulus,*	*Arctopus.*

Other genera consist of a large number of species, and they must be drawn to the attention of botanists species by species.

Mesembryanthemum,	*Euphorbia,*	*Allium,*	*Aloë,*
Sedum,	*Geranium,*	*Erica,*	*Statice,*
Convolvulus,	*Campanula,*	*Solanum,*	*Gentiana,*
Saxifraga,	*Silene,*	*Potentilla,*	*Ranunculus,*
Mimosa,	*Cassia,*	*Linum,*	*Chenopodium,*
Antirrhinum,	*Hypericum,*	*Hibiscus,*	*Eupatorium,*
Polygala,	*Phaseolus,*	*Astragalus,*	*Hedysarum,*
Aster,	*Gnaphalium,*	*Centaurea,*	*Buphthalmum.*
Carex,	*Salix,*	*Ficus.*	

204. What is valid for the generic character is also valid for the class character (160), though in the latter everything is applied more broadly.

The genus of the genera is the order, and the genus of the orders is the class. So here the rules given in Sections 164–202 are valid.

205. The class (160) is more arbitrary than the genus (159), and the order (161) more so than either.

In a description an erroneous designation of the class does less harm than an erroneous designation of the genus.

Be careful not to mislead, as a result of the affinity of the genera to the classes and the orders [and], not to reduce the natural orders to genera, nor indeed, eventually, the classes too.

a. *Malva, Althaea, Alcea, Lavatera, Urena, Hibiscus*, order 34.
b. *Sedum, Sempervivum, Cotyledon, Crassula, Tillea*, order 46.
c. *Cactus, Mesembryanthemum, Aizoon, Tetragonia*, order 46.
d. *Lychnis, Coronaria, Agrostemma, Silene, Dianthus, Saponaria, Cerastium, Spergula, Arenaria, Moerhingia, Sagina*, order 42.

In this way, from the *columniferous, stellate, umbellate, siliquous* and *whorled* orders, as many genera would be formed, but they would be so vast as to produce total confusion.

The *orchids* (order 4) would all be blended into the same genus, and the *delicacies* (order 3) into another, and those eventually into one; and *this would first cause the downfall of botany*, which would collapse because of the weight of particular genera.

206. The more natural the classes are, the more excellent, other things being equal.

Related [plants] agree in habit, manner of reproduction, properties, potencies, and use.

The daily labour of the most eminent botanists is exerted in these matters, and it is right that it should be exerted to the utmost.

Hence the natural method is the ultimate end of botany, and will continue to be so.

In the first place, three obstacles may present problems to the natural method.

a. Disregard for the *habit* of the plants, since the perfection of the teaching concerning the fruit-body, especially their fresh foliation, p. 119.
b. The absence of foreign genera not yet discovered.
c. The affinities of genera on either side. *Linnaea* is in the middle, between the stellate [plants], [order] 44; Valeriana [order] 18; and the Loniceras [order] 63.
Cornus connects the stellate [order] 44, the clustered [order] 18, and the bushy [plants], [order] 19.
Juncus links the Calamarias [order] 13, the grasses 14, and the Coronarias 9.
Dodecatheon brings Cortusa and Cyclamen together.
Hyoscyamus with inflated calyx, [my] Hortus Upsaliensis 2, is related to *Physalis*.
Hibiscus with flowering petioles, Hortus Upsaliensis 1, is associated with *Turnera* with similarly flowering petioles; nobody would trace the connexion of Turnera with Hibiscus, without taking notice of the fruiting petiole.

207. Classes and orders that are too long or too numerous cause great difficulties.

With the *sexual method*, the class *with five stamens* and [that] *with united anthers*[11] are absolutely enormous; and so with this system it is more difficult than with others to distinguish the genera in these [classes].

The order *with five stamens and one pistil*[12] is relatively difficult to distinguish, because of the large number of genera, and for this reason I leave the methods of others untouched.

Boerhaave has 33 classes, *Knaut* 8; and so the orders of the former will be more numerous and shorter, those of the latter fewer but absolutely enormous.

208. An order must place next to each other the genera that are more closely connected.

Ray's Methodus 5. 'It must be seen to that related plants are not separated, and that those that are dissimilar and unrelated are not linked together.'

Those that are more difficult to distinguish should be placed nearer[13] together.

Let the [order] with 4 stamens and one pistil serve as an example; the orders in it are the *stellate* 44, the *calycanthemi* 40, and the *clustered* 18; these natural orders should, if possible, be kept separate in the same order, and not confused, even if an easier division could be made in another way.

right	18.	*Leucadendron*	wrong	*Leucadendron* 18.
		Protea		*Asperula* 44.
		Cephalanthus		*Ludwigia* 40.
		Globularia		*Protea* 18.
		Dipsacus		*Sherardia* 44.
		Scabiosa		*Oldenlandia* 40.
		Knautia		*Cephalanthus* 18.
	44.	*Asperula*		*Spermacoce* 44.
		Sherardia		*Isnarda* 40.
		Spermacoce		*Globularia* 18.
		Hedyotis		*Hedyotis* 44.
		Knoxia		*Ammannia* 40.
		Dioidia		*Dipsacus* 18.
		Crucianella		*Knoxia* 44.
	40.	*Ludwigia*		*Scabiosa* 18.
		Oldenlandia		*Dioidia* 44.
		Isnarda		*Knautia* 18.
		Ammannia		*Crucianella* 44.

I would gladly have separated *Tournefortia*, in the [order] with five stamens and one pistil,[15] from the rough-leaved plants, on account of its berries, if nature had not resisted.

Those that serve in the same order ought not to be separated by the insertion of other genera, when they are so closely related that it is hardly possible to draw a line between them: for instance:

Alsine and *Arenaria*.

Primula and *Androsace*.

Lysimachia and *Anagallis*.
Lonicera and *Diervilla*.
Convolvulus and *Ipomoea*.
Chenopodium and *Beta*.
Solanum and *Capsicum*.

209. To cling to the habit of the plants (163), to such an extent that the usually accepted elements of the fruit-body (164) are set aside, is to seek folly instead of wisdom.

The flowering, or position of the fruit-body, in the axil, raceme, or cluster, (whence the characters derived from the habit, Section 168), has been the cause of an infinite number of genera that are false; especially with Knaut, Kramer, and such like.

The external form of the plants which was the Lydian stone[16] of the ancients and is the touchstone of the moderns, should be made much of by every botanist; but even in this there should be some moderation.

BIDENS (a) *with leaf not divided, Tournefort, [my] Flora Suecica* 664, displays precisely the fruit-body of the genus Bidens; but BIDENS (b) *with leaf not divided and with a flower rayed all round with golden petals, Morison*, displays perfectly the fruit-body of Coreopsis; whereas *a* has no ray, *b* is furnished with a ray of 8 barren petals; the former will be of the genus Bidens, but *b* of the genus Coreopsis. But it still remains doubtful whether the rayed plant *b* is a variety of the unrayed *a*; and if this were so, as seems likely, one may not separate these two plants; and I should not think it advisable to blend the genera Bidens and Coreopsis into one, lest the boundaries of the genera should eventually disappear. So here outward form and nature contend with principle and technical practice.

MORISON (Section 55), who has followed the thread of nature, ties his own thread of Ariadne into Gordian knots, which can be untied only with the sword:[17] for instance-
In class 7, *Anemone, Dryas* and *Hydrocotyle* are linked with the *Liliaceous [plants]*.
In class 5, *Oxalis, Fragaria* and *Epimedium* are linked with the *Leguminous [plants]*.
In class 9, *Eryngium, Bromelia* and *Cactus* are linked with the *Compound [plants]*.
In class 58, *All the Rough-leaved [plants]* are linked with the *Whorled [plants]*.
In class 11, *Plantago* is linked with the *Grasses*.

The division into trees and herbaceous plants seems to be so very natural, but it is essentially most misleading, and deceptive (Section 78.7).

BOTANY depends on fixed genera, and the following was its progression –
TOURNEFORT was the first to establish generic characters, from the principle of technical practice.
PLUMIER collected American plants into genera.
BOERHAAVE added some genera.
PETIT added on a very few more.

The *PARISIAN* Academicians, *Marchant, Isnard, Nissole,* and *Condamine* included several in the Transactions of the French [Academy].

VAILLANT began the reformation of botany.

The *JUSSIEU* brothers provided the Paris Transactions with some new [genera].

RUPPIUS and *DILLENIUS*, very close associates, worked to reform botany in Germany.

DILLENIUS, after becoming an Englishman, achieved several excellent [works].

PONTEDERA tried to improve botany in Italy.

MICHELI carefully tended Italian botany.

BUXBAUM gathered a few genera in the East.

AMMAN dealt with some in the Russian [empire].

HOUSTOUN investigated quite a number in his American journey, but died [there].

HALLER made a definitive examination of the plants of Switzerland.

GMELIN was the first to explore the plants of Siberia, in his arduous journey.

MONTI has recently become famous on account of a very rare genus.

IMYSELF have examined all the genera according to the principles of technical practice. I have reshaped their characters, and established them [the genera] as new.

Gronovius, Royen, and *Burman* have presented a very large number of very rare plants.

TOURNEFORT'S [GENERA]

Abies T.	Allium T. H.
Larix T.	Moly B.
Acanthus T.	Cepa T.
Acer T.	Porrum T.
Achillea V. *Ptarmica T.*	Scorodoprasum. *Mich.*
Millefolium T.	Aloë T.
Aconitum T.	Alnus T.
Actaea L. *Christophoriana T.*	Alsine T.
Adiantum T.	Alyssum T.
Adoxa L. *Moschatellina T.*	Alyssoides T.
Agrimonia T.	Vesicaria T.
Agrimonoides T.	Amaranthus T.
Ajuga L. *Bugula T.*	Amaryllis L. *Lilionarcissus T.*
Alchemilla L. *Alchimilla T.*	Ambrosia T.
Alisma D. R.	Ammi T.
Damasonium T. V.	Ammioides B.

continued

Amygdalus T.
   Persica T.
Anacardium L. *Acajou T.*
Anacyclus L. *Cotula T.*
   Santalmoides V.
Anagallis T.
Anagyris T.
Anchusa E. *Buglossum T.*
Andrachne L. *Telephioides T.*
Androsace T.
   Aretia *Hall.*
Anemone T.
   Anemonoides D.
   Anemone-Ranunculus. D.
Anethum T.
   Foeniculum T.
Anthemis L. *Chamaemelum* T. V.
Anthericum L. *Phalangium T.*
Antirrhinum T.
   Linaria T.
   Asarina T.
   Elatine D.
Anthyllis L. *Vulneraria T.*
Aphyllanthes T.
Apium T.
Apocynum T.
Aquilegia T.
Aralia T.
Arbutus T.
   Uva ursi T.
Arctium L. *Lappa T.V.*
Argemone T.
Aristolochia T.
Artemisia T. V.
   Abrotanum T.
   Absinthium T.V.
Arum T.
   Arisarum T.
   Dracunculus T.
   Colocasia B.
Aruncus L. *Barba caprae T.*
Arundo T.
Asarum T.
   Hypocistis T.
Asclepias T.

Asparagus T.
Asperugo T.
Asphodelus T.
Asplenium L. *Lingua cervina T.*
   Trichomanes T.
Aster T.V.
   Asteropterus V.
Astragalus T.
Astrantia T.
Athamanta L. *Meum T.*
Atriplex T.
Atropa L. *Belladonna T.*
Avena T.
Axyris L. *Ceratoides T.*

Ballotta L. *Ballotte T.*
Begonia T.
Bellis T. V.
   Bellis Leucanthemum *Mich.*
Berberis T.
Beta T.
Betonica T.
Betula T.
Bidens T.
   Ceratocephalus V.
Bignonia T.
Biscutella L. *Thlaspidium T.*
Biserrula L. *Pelecinus. T.*
Borrago T.
   Borraginoides. B. *Cynoglossoides 1.*
Brassica T.
   Rapa T.
   Napus T.
Bromelia Pl.
   Ananas T.
   Karatas Pl.
   Pinguin *Dill.*
Brunella T.
Bryonia T.
Bunias L. *Erucago T.*
   Kakile T.
Bunium L. *Bulbocastanum T.*
Buphthalmum T.
   Asteriscus T. V.
   Asteroides T. V.

continued

Anthemias *Mich.*
Bupleurum T.
Butomus T.
Buxus T.

Cachrys T.
Cactus L. *Melocactus T.*
    Opuntia T. *Tuna D.*
    Pereskia Pl.
Calendula Rp. *Caltha T.*
    Dimorphotheca V.
    Cardispermum [sic] *Trant.*
Calligonum L. *Polygonoides T.*
Caltha L. *Populago T.*
Campanula T.
Camphorosma L. *Camphorata T.*
Canna L. *Cannocorus T.*
Cannabis T.
Capparis T.
Capsicum T.
Cardamine T.
Cardiospermum L. *Corindum T.*
Carduus T.V.
    Cirsium T.V.
    Eriocephalus V.
    Polyacantha V.
    Silybum V.
Carex L *Cyperoides T.D.*
    Carex D.
Carica L. *Papaja P.*
Carlina T.V.
    Carlinoides V.
Carpesium L. *Conyzoides T.*
Carpinus T.
Carthamus T.V.
    Atractylis V.
    Carthamoides V.
Carum R. *Carvi T.*
Caryophyllus L. *Aromaticus T.*
    Caryophyllodendron V.
Cassia T.
    Senna T.
Catananche V. *Catanance T.*
Caucalis T.
    Celtis T.

Centaurea L. *Jacea T.V.*
    Cyanus T.V.
    Centaurium majus T.V.
    Calcitrupa V.
    Raponticum V.
    Raponticoides T.V.
    Amberboi V.
    Crocodilum V.
    Crupina D.
Cerastium D. *Myosotis T.*
Ceratonia L. *Siliqua T.*
Cerbera L.
    Thevetia L. *Ahouai T.*
Cercis L. *Siliquastrum T.*
Cerinthe T.
    Cerinthoides B.
Chaerophyllum T.
Cheiranthus L. *Leucojum T.*
Chelidonium T.
    Glaucium T.
Chelone T.
    Pentastemon *Mitch.*
Chenopodium T.
Chondrilla T.V.
Chrysanthemum T.
    Leucanthemum T. *Bellidioides V.*
Chrysosplenium T.
Cicer T.
    Lens T.
Cichorium T.V.
Circæa T.
Cistus T.
    Helianthemum T.
Citrus L. *Citrum T.*
    Aurantium T.
    Limon T.
Clematis L. *Clematitis T.*
    Viticella D.
Cleome L. *Sinapistrum T.*
Clinopodium T.
Clitoria D. *Ternatea T.*
Clypeola L. *Jon-Thlaspi T.*
Cneorum L. *Chamaelea T.*
Cnicus T.V.
    Acarna V.

continued

Cochlearia T.
   Coronopus Rp.
Coix L. *Lacryma Job.T.*
Colchicum T.
Colutea T.
Conium L. *Cicuta T.*
Convallaria L. *Lilium convallium T.*
   Polygonatum T.
   Unifolium D.
Convolvulus T.
Conyza T.
Corchorus T.
Coriandrum T.
Coris T.
Cornus T.
   Mesomora *Rudb.*
   Virga sanguinea D. *Ossea* Rp.
Coronilla T.
   Emerus T.
   Securidaca T.
Corylus T.
Cotyledon T.B.D.
Crambe T.
   Rapistrum T.
Crataegus T.
Crithmum T.
Crocus T.
Crotalaria T.
Croton L. *Ricinoides* T.
Crucianella L. *Rubeola* T.
Cucubalus T.
Cucumis T.
   Colocynthis T.
   Melo T.
   Anguria T.
Cucurbita T.
   Pepo T.
   Melopepo T.
Cuminum T.
Cunila L. *Marrubiastrum* T.
Cupressus T.
Cuscuta T.
Cyclamen T.
Cynara V. *Cinara* T.
Cynoglossum T.

Omphalodes T.
Cyperus T.
Cypripedium L. *Calceolus* T.
Cytisus. T.

Daphne L. *Thymelea* T.
Datisca L. *Cannabina* T.
Datura R. *Stramonium* T.
Daucus T.
Delphinium T.
Dentaria T.
Dianthus L. *Caryophyllus T.*
   Tunica Rp.
Dictamnus L. *Fraxinella* T.
Diervilla T.
Digitalis T.
Diospyros L.M. *Guajacana* T.
Dipsacus T.
Dodartia T.
Doronicum T.
   Bellidiastrum M.
Dorycnium T.
Dracocephalum T.
   Moldavica T.
Drosera L. *Ros solis* T.

Echinophora T.
Echinops L. *Echinopus T.V.*
Echium T.
Elaeagnus T.
Elatine T. *Alsinastrum T.V.*
   Potamopithys. Buxb.
Empetrum T.
Ephedra T.
Epilobium D. *Chamaenerium T.*
Epimedium T.
Equisetum T.
Erica T.
Eriophorum L. *Linagrostis* T.
Ervum T.
Eryngium T.
Erysimum T.
Erythrina L. *Corallodendron T.*
Erythronium L. *Denscanis T.*
Esculus L. *Hippocastanum T.*
   Pavia B.

continued

Euonymus T.
Eupatorium T.
Euphorbia L. *Tithymalus T.*
    Tithymaloides. T.
    Euphorbium *Isnard.*
Euphrasia T.
    Odontites D.

Fagonia T.
Fagus T.
    Castanea T.
Ferula T.
Ficus T.
Filipendula T.
    Ulmaria T.
Fragaria T.
Fraxinus T.
Fritillaria T.
    Corona imperialis T.
    Lilo Fritillaria B.
Fucus T.
Fumaria T.
    Capnoides T.
    Cysticapnos B. *Corydalis* D.
    Cucullaria J.

Galega T.
Galium L. *Gallium* T.
    Aparine T.L.
Garidella T.
Genipa T.
Genista L. *Spartium T.*
    Genistella T.
Gentiana T.
    Centaurium minus T.
Geranium T.
Geum L. *Caryophyllata T.*
Gladiolus T.
Glaux T.
Globularia T.
    Alypum Niss.
Gloriosa L. *Methonica T.D.*
Glycyrrhiza T.
Gnaphalium V.
    Helichrysum V. *Elichrysum T.*

Helichrysoides V.
Filago T.V.
Gomphraena L. *Amaranthoides*
    T. Caraxeron V.
Gossypium L. *Xylon T.*
Gundelia T. *Hacub V.*

Haemanthus T.
    Dracunculoides B.
Hedera T.
Hedysarum T.
    Onobrychis T.
    Alhagi T.
Helianthus L. *Corona solis* T.V.
Heliotropium T.
Helleborus T.
Helxine L. *Fagopyrum* T.
Hemerocallis L. *Lilio-Asphodelus T.*
    Liliastrum T.
Heracleum L. *Sphondylium T.*
Hermannia T.
Herniaria T.
    Paronychia T.
Hesperis T.
Hibiscus L. *Ketmia T.*
    Trionum L.
    Malvaviscus D.
Hieracium T.V.
    Pilosella T.V.
Hippocrepis L. *Ferrum equinum T.*
Hippophaes L. *Rhamnoides T.*
Hordeum T.
Humulus L. *Lupulus T.*
Hyacinthus T.
    Muscari T.
Hydrocharis L. *Morsus Ranae T. Stratiotes D.*
Microleuconymphaea B.
Hydrocotyle T.
Hydrophyllum T.
Hyoscyamus T.
Hypecoum T.
Hypericum T.
    Androsæmum T.
    Ascyrum T.
Hyssopus T.

continued

Ilex L. *Aquifolium.*
    Dodonea Pl.
Impatiens R. *Balsamina T.*
Imperatoria T.
Ipomœa L. *Quamoclit T.*
    Volubilis D.
Iris T.
    Xiphium T.
    Hermodactylus T.
    Sisyrinchium T.
Isatis T.

Jasminum T.
Jatropha L. *Manihot. T.*
    Bernhardia. *Houst.*
Juglans L. *Nux T.*
Juncus T.
Juniperus T.
    Cedrus T.
    Sabina B.

Kleinia L. *Cacalianthemum D.*
    Cacalia T.
    Porophyllum V.

Lactuca T.V.
Lagoecia L. *Cuminoides T.*
Lamium T.
Lapsana L. *Lampsana V.*
    Hedypnois T.
    Rhagadiolus T.V.
    Zazintha T.V.
    Rhagadiloides V.
Laserpitium T.
Lathraea L. *Clandestina T.*
    Anblatum T.
    Phelypaea T.
Lathyrus T.
    Clymenum T.
    Aphaca T.
    Nissolia T.
Lavandula D.
    Stoechas T.
Lavatera T.D.
Laurus T.

Borbonia P.
Persea P.
    Benzoë B.
Leontice L. *Leontopetalum T.*
Leontodon L. *Dens leonis T.V.*
    Taraxaconoides V.
Leonurus T.
    Cardiaca T.
Lepidium T.
    Nasturtium T.
Leucojum Rupp. *Narcisso-Leucojum T.*
Lichen T. *Lichenoides D.*
    Coralloides D.
    Usnea D.
Ligusticum T.
    Cicutaria T.
Ligustrum T.
Lilium T. *Lirium* Roy.
Linum T.
    Radiola D. *Chamaelinum V.*
            Linocarpon Mich.
Lithospermum T.
Lobelia P.
    Rapuntium T.
    Dortmanna *Rudb.*
    Laurentia *Mich.*
Lonicera R.
    Capriofolium T.
    Periclymenum T.
    Chamaecerasus T.
    Xylosteum T.
    Triosteospermum D.
    Symphoricarpos D.
Lotus T.
Lunaria T.
Lupinus T.
Lychnis T.
Lycoperdum T. *Bovista D.*
    Lycoperdiodes M.
    Lycoperdastrum M.
    Geaster M.
    Tubera T.
Lycopodium T.D.
    Lycopodioides D.

continued

Selago D.
Selaginoides D.
Lycopsis L. *Echioides* T.D.
Lycopus T.
Lysimachia T.
Lythrum L. *Salicaria T.*

Malope L. *Malacoides* T.
Malva T.
    Althaea T.
    Alcea T.
Mandragora T.
Marrubium T.
    Pseudodictamnus T.
Matricaria T.V.
Medicago T.
    Medica T.
Melampyrum T.
Melia L. *Azedarach.* T.
Melianthus T.
Melissa T.
    Calamintha T.
Menispermum T.D.
Mentha T.
    Pulegium R.
Menyanthes T.
    Nymphoides T.
Mercurialis T.
Mesembryanthemum D. *Ficoides* T.
Mespilus T.
Micropus L. *Gnaphaloides* T.
Mimosa T.
    Acacia T.
    Inga P.
Mirabilis R. *Jalapa* T.
Mitella T.
Molucella L. *Molucca* T.
Momordica T.
    Luffa T.D.
    Elaterium B.
Monotropa L. *Orobanchoides* T.
       Hypopithys D.
Morina T. *Diotheca D.*
Morus T.
Myagrum T.

Myrica D. *Gale T.D.*
Myrtus. T.

Narcissus T.
Neottia L. *Nidus avis* T.
    Corallorhiza Rupp.
Nicotiana T.
Nigella T.
Nymphaea T.
    Nelumbo T.

Ocymum T.
Oenanthe T.
Oenothera L. *Onagra* T.
Olea T.V.
Ononis T.
Ophioglossum T.
Orchis T.
    Limodorum T.
Origanum T.
    Majorana T.
Ornithogalum T.
    Stellaris D.
Ornithopus L. *Ornithopod.* T.
Orobanche T. *Aphyllon* Mitch.
Orobus T.
Oryza T.M.
Osmunda T.
Osteospermum L. *Chrysanthemoides* T.
    Monilifera V.
Osyris L. *Casia* T.
Oxalis L. *Oxys* T.
    Oxyoides *Garc.*

Paeonia T.
Papaver T.
Parietaria T.
Paris L. *Herba Paris* T.
Parnassia T.
Passiflora L. *Granadilla* T.
    Murucuja T.
Pastinaca.
Pedicularis T.
    Sceptrum Carolinum. *Rudb.*
Peganum L. *Harmala T.*
Periploca T.

continued

Peucedanum T.

Phaca L. *Astragaloides T.*

Phaseolus T.

Phellandrium T.

Philadelphus L. *Syringa T.*

Phillyrea T.V.

Phlomis T.

Physalis L. *Alkekengi T.*

Phyteuma L. *Rapunculus T.*

Phytolacca T.

Pimpinella R. *Tragoselinum T.*

Pinguicula T.

Pinus T.

Pistacia L. *Terebinthus T.*

    Lentiscus T.

Pisum T.

Plantago T.

    Coronopus T.

    Psyllium T.

Platanus T.

Plumbago T.

Plumeria T.

Podophyllum L. *Anapodophyllum T.*

Poinciana T.

Polemonium T.

Polygala T.

    Penaea P. *Chamaebuxus T.*

Polygonum T.

    Bistorta T.

    Persicaria T.

    Fagopyrum T. *Helxine L.*

Polypodium T.

Populus T.

Portulaca T.

Potamogeton T.

Potentilla L. *Pentaphylloides T.*

    Quinquefolium T.

Primula L. *veris T.*

    Auricula ursi T.

Prunus T.

    Armeniaca T.

    Cerasus T.L.

    Padus R. *Laurocerasus T.*

Psidium L. *Guajava T.*

Pulmonaria T.

Pulsatilla T.

Punica T.

Pyrola T.

Pyrus T.

    Malus T.

    Cydonia T.

Quercus T.

    Ilex T.

    Suber T.

Ranunculus T.

    Ranunculoides V.

    Ficaria D.

Raphanus T.

    Raphanistrum T.

Reseda T.

    Luteola T.

    Sesamoides T.

Rhamnus T.

    Frangula T.

    Alternus T.

    Paliurus T.

    Ziziphus T.

    Gervispina D.

Rheum L. *Rhabarbarum T.*

Rhinanthus L. *Elephas T.*

    Crista galli R.D.

Rhododendros L. *Chamaerrhododendros T.*

                 Memaecylon M.

Rhus T.

    Toxicodendron T.

    Cotinus T.

Ribes T.

    Grossularia T.

    Ribesium T.

Ricinus T.

Robinia L. *Pseudo-Acacia T.*

Rosa T.

Rosmarinus T.

Rubia T.

Rubus T.

Rumex L. *Lapathum T.*

    Acetosa T.

continued

Ruscus T.
Ruta T.
   Pseudo-Ruta *Mich.*

Salicornia T.
Salix T.
Salsola P. *Kali T.*
Salvia T.
   Horminum T.
   Sclarea T.
   Sambucus T.
   Samolus T.
Sanguisorba R. *Pimpinella T.*
Sanicula T.
Santolina T.V.
   Baccharis V.
   Sapindus T.
Sarracena T.
Satureja T.
   Thymbra T.
   Saxifraga T.
   Geum T.
Scabiosa T.
Scandix T.
   Myrrhis T.
Schinus L. *Molle T.*
Scilla L. *Lilio-hyacinthus T.*
     Hyacinthus stellaris R.
Scirpus T.
Scolymus T.V.
Scorpiurus L. *Scorpioides T.*
Scorzonera T.
   Scorzoneroides V.
Scrophularia T.
Scutellaria R. *Cassida T.*
Secale T.
Sedum T.
   Anacampseros T.
Selinum L. *Oreoselinum T.*
   Thysselinum T.
Senecio T.
   Jacobæa T.
Sicyos L. *Sicyoides T.*
   Bryonioides D.
Sida L. *Malvinda D.*

Abutilon T.
Sideritis T.
Sinapi T.
Sisymbrium T.
   Radicula D.
Sisyrinchium L. *Bermudiana T.*
Sium T.
   Sisarum T.
   Falcaria D.
Smilax T.
Smyrnium T.
Solanum T.
   Melongena T.
   Lycopersicon T.
Soldanella T.
Solidago V.
   Virga aurea T.V.
   Doria D.
Sonchus T.
   Crepis V.
Sorbus T.
Sparganium T.
Spartium L. *Genista T.*
   Cytiso-Genista T.
Spinacia T.
Spiraea T.
Spongia T.
Stachys R. *Galeopsis T.*
Staphylaea L. *Staphylodendron T.*
Statice T.
   Limonium T.
Styrax T.
Symphytum T.
Syringa R. *Lilac. T.*

Tagetes T.V.
Tamarindus T.
Tamarix L. *Tamariscus T.*
Tamus L. *Tamnus T.*
Tanacetum T.
   Balsamita V.
Taxus T.
Telephium T.
Teucrium T.
   Polium T.

continued

Chamaedrys T.
Chamaepithys T.
Iva D.
Thalictrum T.
Thapsia T.
Theligonum L. *Cynocrambe T.*
Theobroma L. *Cacao T.*
Guazuma P.
Thlapsi T.
Bursa pastoris T.
Thya T.
Thymus T.
Acinos D.
Tilia T.
Tordylium T.
Tormentilla T.
Trachelium T.
Tradescantia *Rup. Ephemerum T.*
Tragacantha T.
Tragopogon T.
Tragopogonoides T.
Trapa L. *Tribuloides T.*
Tribulus T.
Trifolium T.
Melilotus T.
Lupinaster *Buxb.*
Trifoliastrum *Mich.*
Triphylloides *Pont.*
Triglochin L. *Juncago T.*
Trigonella L. *Fœnum Graecum T.*
Triticum T.
Tropaeolum L. *Cardamindum T.*
Tulipa T.
Turitis T.
Tussilago T.V.
Petasites T.V.
Typha T.

Ulex L. *Genista Spartium T.*
Erinaceus T.?
Ulmus T.
Urtica T.

Vaccinium R. *Vitis idaea T.*
Oxycoccos T.

Valantia T. *Cruciata T.*
Valeriana T.
Valerianella T.
Valerianoides V.
Veratrum T.
Verbascum T.
Blattaria T.
Verbena T.
Sherardia V.
Blaeria H.
Kaempfera H.
Veronica T.
Bonarota M. *Paederota L.*
Viburnum T.
Tinus T.
Opulus T.
Vicia T.
Faba T.
Vinca L. *Pervinca T.*
Viola T.
Viscum T.
Vitex T.
Vitis T.

Xanthium T.
Xeranthemum T.V.
Xeranthemoides V.

Zea L. *Mays T.*
Zygophyllum L. *Fabago T.*

PLUMIER'S

Achras L. *Sapota P.*
Alpinia L. *Alpina P.*
Annona L. *Guanabanus P.*
Arachis L. *Arachidna P.*
Arachidnoides *Niss.*
Ascyrum L. *Hypericoides P.*

Barleria P.
Bauhinia P.
Bellonia P.
Besseria P.
Bocconia P.
Bontia P.

continued

Breynia P.
Brossæa? P.
Brunfelsia P.
Bucephalon P.

Caesalpinia P.
Calophyllum L. *Calaba P.*
          Calophyllodendron V.
Cameraria P.
Chrysobalanus L. *Icaco P.*
Chrysophyllum L. *Cainito P.*
Cissampelos L. *Caapeba P.*
Clusia P.
Columnea P.
Commelina P.
    Zannonia P.
Cordia P.
Cornutia P. *Agnanthus V.*
Crateva L. *Tapia P.*
Crescentia L. *Cujete P.*
Cupania P.

Dalechampia P.
Dioscorea P.
Dorstenia P.H.
Duranta? L. *Castorea P.*

Epidendrum L. *Vanilla P.*
Feuillea L. *Nhandiroba P.*

Fuchsia? P.

Gerardia P.
Gesneria P.
Guajacum P.
Guidonia P.
Guilandina L. *Bonduc. P.*

Helicteres L. *Isora P.*
Hernandia? P.
Hippocratea L. *Coa P.*
Hippomane L. *Mançanilla P.*
Hymenaea *Courbaril. P.*

Lantana L. *Camara P.*
          Myrobatindum V.
Loranthus V. *Lonicera P.*

Magnolia P.
Malpighia P.
Mammaea L. *Mammei P.*
Maranta P.
Marcgravia P.
Matthiola P.
Mentzelia P.
Morinda V. *Rojoc. P.*
Morisona P.
Muntingia P.
Musa P.
    Bihai P.

Ochna L. *Jabotapita P.*
Oldenlandia P.
Ovidea L. *Valdia P.*

Parkinsonia P.
Paullinia L. *Seriana P.*
    Cururu P.
Petiveria P.
Piper L. *Saururus P.*
Pisonia P. *Pentagonotheca V.*
Pistia L. *Koddapail P.*
Plinia P.
Pluknetia P.

Rajania L. *Jan-Raja P.*
Rauwolfia P.
Renealmia P.
Rheedea L. *van Rheedea P.*
Rhizophora L. *Mangles P.*
Rivina P *Solanoides* T.
Rondeletia P.
Ruellia P.

Sloanea P.
Spigelia L. *Arapabaca P.*
Spondias L. *Monbin. P.*
Suriana P.

Tabernaemontana P.
Thalia? L. *Cortusa P.*
Theophrasta L. *Eresia P.*
Tillandsia L. *Caraguata P.*
Tournefortia L. *Pittonia P.*
Tragia P.

continued

Triumfetta P.
Turnera P.

Ximenia? P.
Xylon L. *Ceiba P.*

BOERHAAVE'S

Acalypha L. *Ricinocarpus B.*
Blitum L. *Chenopodio-Morus B.*
       Morocarpus Rupp.
Chrysocoma D. *Coma aurea B.*
Clutia B.
Cortusa B.
Glechoma L. *Chamaecissus B.*
Glycine L. *Apios B.*
Hottonia B. *Stratiotes V.*
Knautia L. *Lychni-Scabiosa B.*
Leucadendron L. *Hypophylocarpodendrum B.*
    Lepidocarpodendrum B.
Phyllis L. *Bupleroides B.*
Protea L. *Conocarpodendron B.*
Scoparia L. *Samoloides B.*
Sesseli B.
Stratiotes L. *Aloides B.*
Tetragonia L. *Tetragonocarpus B.*

PETIT'S

Acorus L. *Calamus aromaticus P.M.*
Calla L. *Provenzalia P. Anguina* Trew.
Isnarda L. *Dantia* Petit.

THE PARISIANS'

Aizoon L. *Ficoidea* Niss.
Celastrus L. *Euonymoides* Isn.
Cinchona L. *Quinquina* Condam.
Coriaria *Niss.*
Lycium L. *Jasminoides* Niss. D.M.
Marchantia *March.*
    Lunularia *Mich.*
Parthenium L. *Partheniastrum* Niss. D.
Hysterophorus V.
Waltheria L. *Monospermalthaea* Isn.

VAILLANT'S

Andryala L. *Eriophorius V.*
Arctotis L. *Arctotheca V.*
Atractylis L. *Crocodyloides V.*
Boerhaavia V. *Antanisophyllum V.*
Cephalanthus L. *Plantanocephalus V.*
Celosia L. *Stachyarpagophora V.*
Chara V. *Hippuris D.P.*
Clavaria V. *Fungoides D.*
        Coralloides T.M.
Cotula V.
    Ananthocyclus V. Lancisia P.
Crepis L. *Hieracioides V.*
Elephantopus V.D.
Helenia L. *Heleniastrum V.*
Hyoseris L. *Taraxaconastrum V.*
    Leontodontoides V.
Hypochaeris V.
    Achyrophorus V.
Iva L. *Tarchonanthus V.*
Najas L. *Fluvialis V.M.*
Onopordum V.
Othona L. *Jacobaeastrum V.*
    Jacobæoides V.
Panax L.   *Araliastrum V.*
        Panacea Mitch.
Picris L. *Helmintotheca V.*
Prenanthes V.
Sagittaria L. *Sagitta V.D.*
Sphæranthus V.
Verbesina P. *Ceratocephaloides V.*
    Eupatoriophalacron V.D.
Utricularia L. *Lentibularia V.*
Zanichellia M. *Algoides V.*
        Aponogeton Pt.
        Graminifolia D.

JUSSIEU'S

Coffea L. *Coffe J.*
Corispermum J. *Rhagrostis* Buxb.
Neurada J.
Pilularia J.

continued

RUPPIUS' and DILLENIUS'

Adonis R.D.
Aegopodium L. *Podagraria D.*
Agaricus D.
 Amanita D.
Aphanes L. *Percepier D.*
Boletus D. *Suillus M.*
 Polyporus M.
Bromus L. *Aegilops D.*
Bryum D.
Callitriche L. *Stellaria D.*
Centunculus C. *Anagallidiastrum M.*
Conferva D.
Corrigiola L. *Poligonifolia D.*
Draba D.
Erigeron L. *Conyzella D.*
Fontinalis D.
Galeopsis L. *Tetrahit. D.*
Hepatica R.D.
Hippuris L. *Pinastella D.*
    Limnopeuce V.
Holosteum D.
Hydnum L. *Erinaceus D.*
Hypnum D.
Iberis R.D.
Illecebrum R. *Corrigiola D.*
Jungermannia R.M. *Lichenastrum D.*
 Muscoides M.
 Marsilea M.
Ledum R.
Lemma L. *Lenticula D.M.*
 Hydrophace *Buxb.*
Limosella L. *Plantaginella R.D.*
Mnium D.
Montia M. *Cameraria D.*
Myosotis R.D.
Myosurus R.D.
Peplis L. *Portula D.*
   Glaucoides M.
Peziza D. *Cyathoides M.*
Phallus M. *Morchella D.*
 Boletus M.
 Phallo-Boletus M.
Polytrichum D.

Sagina L. *Alsinella D. Spergula R.*
Scleranthus L. *Knawel R.D.*
Sempervivum R.
Serratula D.
Sherardia D.
Spergula D.
Sphagnum D.
Subularia *Raj.*
Trientalis R.
Ulva D.
Yucca D. *Cordyline* Roy.

DILLENIUS'

Achyranthes L. *Achyracantha D.*
Anacampseros L. *Telephiastrum D.*
Crassula D.
Eriocephalus D.
Isoetes L. *Calamaria D.*
Melochia D.
Pancratium D.
Patagonula L. *Patagonica D.*
Phlox L. *Lychnidea D.*
Porella L. *Poronia D.*
Sanguinaria D.
Sideroxylon D.
Silene L. *Viscago D.*
Spermacoce D.
Tetragonotheca D.
Urena D.

PONTEDERA'S

Chamaerops L. *Chamaeriphes P.*
Gelenia L. *Sherardia P.*
Ageratum L. *Carelia P.*
Anthospermum L. *Tournefortia P.*

MICHELI'S

Andromeda L. *Polifolia* Buxb.
 Ledum M. *Chamaedaphne* Bx.
Anthoceros M.
Blasia M.
 Byssus M.
 Aspergillus M.

continued

Botryitis M.
Cenchrus L. *Panicastrella M.*
Cynomorium M.
Drypis M.
Elvela L. *Fungoides D.*
    Fungoidaster M.
Eugenia M.
Frankenia L. *Franca M.*
Herminium L. *Monorchis M.*
Holcus L. *Sorghum M.*
Marsilea L. *Salvinia M.*
Mucor M.
    Mucilago M.
    Lycogala M.
Orvala L. *Papia M.*
Riccia M.
Ruppia L. *Bucca ferrea M.*
Tillæa M.
Tozzia M.
Trichosanthes L. *Anguina M.*
Valisneria M.

BUXBAUM'S

Ceratocarpus B.

AMMAN'S

Amethystea L. *Amethystina A.H.*
Cymbaria A.
Gmelina L. *Michelia A.*

Pentapetes L. *Pterospermadendron A.*
Siphonanthus L. *Siphonanthemum A.*

HOUSTON'S

Ammannia H.
Banisteria H.
Budleja H.
Conocarpus L. *Rudbeckia H.*
Gronovia *Mart.*
Heliocarpus L. *Montia H.*
Lippia H.
Lœselia L. *Royenia H.*
Martynia H.
Milleria H.
Mitreola L. *Mitra H.*
Petrea H.
Pontedera L. *Michelia H.*
Randia? H.
Volkamaria L. *Duglassia H.*

HALLER'S

Cherleria H.

GMELIN'S

Stelleria G.

MONTI'S

Aldrovanda M.

OURS

EUROPEAN.	ASIATIC.	AMERICAN.	AFRICAN
Acrostichum.	Adenanthera R.	Acnida *Mitch.*	Achyronia.
Aegilops.	Aeginetia.	Agave.	Anastatica.
Agrostemma.	Aeschynomene.	Amorpha.	Antholyza.
Agrostis.	Allophylus.	Arethusa G.	Arctopus L.B.
Aira.	Amomum.	Bartsia.	Barreria.
Alcea.	Anabasis.	*Staehelina H.*	Blaeria.
Alopecurus.	Antidesma B.	Bixa.	Bobartia.
Althæa.	Artedia.	Browallia.	Borassus.
Angelica.	Averrhoa.	Buchnera.	Borbonia.
Anisum.	Avicennia.	Callicarpa.	Bosea.

continued

OURS

EUROPEAN.	ASIATIC.	AMERICAN.	AFRICAN
Anthoxanthum.	Bartramia.	*Sphondylococcus M.*	Brabejum.
Arabis.	Basella.	Capraria.	Brunia.
Arenaria.	Cambogia.	Catesbæa G.	Burmannia.
Asperula.	Camellia.	Ceanothus.	Caryota.
Atragene.	Ceropegia.	Chionanthus.	Cassine.
Atraphaxis.	Cissus.	Chrysogonum.	*Maurocenia.*
Azalea.	Clerodendrum B.	Citharexylon J.	Chironia.
Baccharis.	Coccus.	Claytonia.	Cliffortia.
Briza.	Coldenia.	Clethra G.	Corymbium.
Bubon.	Connarus.	Collinsonia.	Corypha.
Bufonia S.	Cornucopiae.	Coreopsis.	Cynanchum.
Bulbocodium.	Costus.	Craniolaria.	Diosma.
Celsia.	Crinum.	Dalea.	Dracontium.
Cicuta.	Curcuma.	Dianthera G.	Eranthemum.
Comarum.	Cynometra.	Diodia G.	Exacum.
Coronaria.	Delima.	Eriocaulon G.	Gerbera.
Cressa.	Dillenia.	Gleditsia.	Gethyllis.
Cynosurus.	Dodonæa.	*Melilobus M.*	Glycine.
Dactylis R.	Elaeocarpus B.	Grissea.	Gnidia.
Diapensia.	Flagellaria.	Hæmatoxylum.	*Lachnaea R.*
Dryas.	Garcinia.	Hamamelis G.	Grewia.
Elymus.	Hedyotis.	*Trilopus M.*	Halleria.
Ethusa.	Hugonia.	Hemionitis.	Hebenstretia.
Festuca.	Jambolifera.	Heuchera.	Hirtella.
Galanthus.	Indigophora.	Houstonia G.	Ixia.
Gypsophila.	Jussiaea.	Hura.	Kiggelaria.
Horminum.	Ixora.	Hydrangea G.	Myrsine.
Jasione.	Kaempfera.	Itea G.	Passerina.
Inula L. *Helenium V.*	Lawsonia.	*Diconangia M.*	Paenaea.
	Melastoma B.	Limodorum R.G.	Pharnaceum.
Ischæmum.	Memecylon.	Liquidambar.	Philyca.
Isopyrum.	Mesua.	Liriodendrum.	Phœnix.
Lagurus.	Michelia.	Lonchitis.	Psoralea.
Andropogon R.	Microcus.	Ludwigia.	Roella.
Linnæa G.	Mimusops.	Medeola.	Royena.
Lolium.	Mussænda B.	Melampodium.	Selago.
Melica.	Myristica.	Melanthium.	Sesamum.
Melitis.	Nama.	Melothria L.M.	Stapelia.
Milium.	Nepenthes B.	Mimulus.	Stœbe.
Moehringia.	Nyctanthes.	*Cynorrhinchium M.*	Struthia
Mollugo.	Olax.		Tarchonanthus.
Nardus.	Pavetta.	Monarda.	Tetragonia.

continued

OURS

EUROPEAN.	ASIATIC.	AMERICAN.	AFRICAN
Panicum.	Phyllanthus.	Napæa A.C.	
Phalaris.	Polianthes.	Nyssa.	
Phleum.	Pontederia.	Obolaria.	
Poa.	Rumpfia.	Onochlea. *Angiopteris M.*	
Polycnemum.	Santalum.	Penthorum G.	
Poterium.	Saurus L.	Polypremum.	
Pteris.	Sigesbeckia.	*Symphoranthus M.*	
Rhodiola.	Sophora.	Prinos G.	
Saponaria.	Strychnus.	Proserpinaca	
Satyrium.	Thea.	*Trixis M.*	
Scheuchzeria.	Tomex.	Ptelea.	
Schoenus.	Torenia.	Rhexia.	
Sibbaldia.	Trewia.	Rudbeckia T.	
Sison.	Triopteris.	*Obeliscotheca V.*	
Splachnum.	Vateria.	Saccharum.	
Swertia.	Uvaria.	Samyda.	
Thesium.	Zannonia.	Sauvagesia.	
Vella.		Schwalbea G.	
Zostera.		Securidaca.	
		Silphium.	
		Stewartia.	
		Malacodendros M.	
		Tetracera.	
		Toluifera.	
		Trichomanes.	
		Trichostema. G.	
		Tridax.	
		Uniola.	
		Uvularia.	
		Xyris G.	
		Zizania.	
		Elymus M.	
		Zizophora.	
		Hedyosmos M.	

continued

VII. N O M I N A.

210. DENOMINATIO alterum (151) Botanices fundamentum, facta difpofitione (152), nomina primum imponat.

Nomina fi nefcis, perit & cognitio rerum.

Unicum ubi genus, unicum erit nomen. §. 215.

Veterum nomina plerumque præftantiffima; Recentiorum pejora fuere.

Critica Botanica Denominationem genericam, fpecificam & variantem rationibus & exemplis propofuit; hinc de ea in præfenti paucis.

211. Nomina vera plantis imponere *Botanicis (7) genuinis (26)* tantum in poteftate eft.

Botanicus novit genera diftincta, & nomina antea recepta.

Idiotæ impofuere nomina abfurda.

RELIGIOSA.

Pater Nofter Cyperus		*Chrifti oculus* After	
Bonus Henricus Chenopodium.		*Palma* Orchis	
Noli me tangere Impatiens		*Spina* Rhamnus	
Morfus Diaboli Scabiofa		*Lancea* Lycopus	
Filius ante Patrem Tuffilago		*Mariæ calceus* Cypripedium	
Herba Fumana Ciftus		*Chlamys* Alchemilla	
Mater herbarum Artemifia		*Stragula* Galium	
Surge & ambula Gentiana		*Veneris labrum* Dipfacus	
Fuga Dæmonum Hypericum		*Umbilicus* Cotyledon	
		Calceus Cypripedium	
		Pecten Scandix	
		Jovis *Barba* Sempervivum.	

NON SYSTEMATICI BOTANICI.

Bontiania Pt.
Breyniana Pt.
Ruyfchiana Pt.
Drakena Cluf. Dorftenia.

212. Nomina omnia funt in ipfa vegetabilis enunciatione vel *Muta*, ut Claffis (160,) & Ordinis (161); vel *Sonora*, ut Genericum (159), Specificum (157) & varians (158).

Nomen omne plantarum conftabit nomine Generico & Specifico;

Nomen

❧ VII. NAMES

210. NOMENCLATURE, the second (151) foundation of botany, after the arrangement has been done (152), should first of all apply the names.

> If you do not know the names of things, the knowledge of them is lost too.
> Where there is a single genus, there will be a single name. Section 215.
> The names given by the ancients are mostly excellent; those given by more recent [writers] have been inferior.
> [My] Critica Botanica proposed nomenclature for genera, species, and varieties, with reasons and examples; hence there is not much about that subject in the present work.

211. Only genuine (26) botanists (7) have the ability to apply names to plants.

> A botanist knows the genera that are distinct [from each other], and the names that have previously been accepted.
> Private individuals have applied absurd names.

> RELIGIOUS [NAMES]
> *Pater Noster* [*Our Father*] Cyperus
> *Bonus Henricus* [*Good King Henry*] Chenopodium
> *Noli me tangere* [*touch me not*][1] Impatiens
> *Morsus Diaboli* [*Devil's bit*] Scabiosa
> *Filius ante Patrem* [*the Son before the Father*][2] Tussilago
> *Herba Fumana* [*smoky herb*] Cistus
> *Mater herbarum* [*the mother of herbs*] Artemisia
> *Surge et ambula* [*arise and walk*][3] Gentiana
> *Fuga Daemonum* [*the flight of the demons*][4] Hypericum
> Christi *Oculus* [Christ's *eye*] Aster
> Christi *Palma* [Christ's *palm*] Orchis
> Christi *Spina* [Christ's *thorn*] Rhamnus
> Christi *Lancea* [Christ's *lance*][5] Lycopus

Mariae *Calceus* [St Mary's *slipper*][6] Cypripedium
Mariae *Chlamys* [St Mary's *mantle*] Alchemilla
Mariae *Stragula* [St Mary's *bed-straw*] Galium
Veneris *Labrum* [Venus' *lip*] Dipsacus
Veneris *Umbilicus* [Venus' *navel*] Cotyledon
Veneris *Calceus* [Venus' *slipper*] Cypripedium[7]
Veneris *Pecten* [Venus' *comb*] Scandix
Jovis *Barba* [Jove's *beard*] Sempervivum.

NON-SYSTEMATIC BOTANISTS
Bontiania Pt.
Breyniana Pt.
Ruyschiana Pt.
Drakena Clus. Dorstenia.[8]

212. In the verbal description of a vegetable, all the names are either *silent*, as those of the class (160) and the order (161); or *enunciated*, as those of the genus (159), species (157), and variety (158).

> Every plant-name must consist of a generic name and a specific one.
> The names of the class and the order must never be included in the name of the plant, but should be supplied by the understanding; hence Royen, who applied *Lilium* as a class name, rightly excluded it as a genus, and substituted the Greek term *Lirium*.

213. Any plants that agree (165) in genus should be designated by the same generic name (212)

Citrus T.	*Aurantium* T.	*Limon* T.
Pyrus T.	*Malus* T.	*Cydonia* T.

214. On the other hand (213), any plants that differ in genus (166) should be designated by different names (213).

Consolida.		Trifolium.	
major	Symphytum	*arborescens*	Cytisus
media	Ajuga	*acetosum*	Oxalis
minor	Brunella	*corniculatum*	Lotus
minima	Bellis	*falcatum*	Medicago
rubra	Tormentilla	*fragiferum*	Fragaria

continued

aurea	Cistus		*hepaticum*	Hepatica
regalis	Delphinium		*palustre*	Menyanthes
sarracenica	Solidago		*pratense*	Trifolium
palustris	Comarum		*spinosum*	Fagonia

215. The generic name must be *a single one* within the same genus (213).

Aconitum	caeruleum or *Napellus*.
	salutiferum or *Anthora*.
Aquifolium	or *Agrifolium*.
Jasminum	or *Geselminum* [sic: Gelsemium].

216. The generic name must be *the same* within the same genus.

Asclepias T.	*Vincetoxicum Hk.*	*Hirundinaria R.*
Limosella Ld.	*Plantaginella D.*	*Menyanthoides V.*
Hottonia B.	*Stratiotes V.*	*Myriophyllum R.*
Tetrahit D.	*Ladanum R.*	*Cannabina B.*
Radiola D.	*Linoides R.*	*Chamaelinum V.*

217. If one (215) and the same (216) generic name has been taken to designate different [genera], it will have to be excluded from one place or the other.

Aconitum	T.	Aconitum	*Cameraria*	Pl.	Cameraria.
"	R.	Helleborus	"	D.	Montia.
Asclepias	T.	Asclepias	Sherardia	D.	Sherardia.
"	Hk.	Stapelia.	"	V.	Verbena.
Caltha	T.	Calendula	"	Pn.	Galenia.
"	Rp.	Caltha.			

218. Anyone who establishes a new genus is obliged to apply a name to it.

Neither *Cacalia* nor *Cacaliastrum*, perhaps *Tithymaloides Kl.*
Plant with the leaf of a Methonica Plk.
Anonymous.

219. The generic name must be fixed unalterably, before any specific name is devised.

A specific name without a generic one is like a clapper[9] without a bell. Section 286.

220. No sane person introduces primitive generic names.

> All barbarous[10] names are regarded by us as primitive, since they are from languages not understood by the learned.
>
> [So are] doubtful appellations of plants, when it is hard to decide what language they are derived from.
>
> Osmunda,[11] Tanacetum.[12]

221. Generic names made from two entire and separate words are to be banished from the Commonwealth of Botany.

Bella donna T.	*Atropa.*
Centaurium majus T.	*Centaurea.*
Corona solis T.	*Helianthus.*
Crista galli D.	*Rhinanthus.*
Dens leonis T.	*Leontodon.*
Vitis Idaea T.	*Vaccinium.*

222. Generic names compounded of two entire and conjoined words are hardly to be tolerated.

> Appellations of this kind are very fine in Greek; Latin does not easily allow them.
>
> Com*aurea*. Chrysocoma. [Golden hair].
>
> We have allowed some [such] Latin words, but they are not therefore to be imitated in future; for example:
>
> Cornu*copiae* Sch.
>
> Ros*marinus* T.
>
> Semper*vivum* R.
>
> Sangui*sorba* R.

223. Hybrid generic names, from a Greek word and Latin one and such like, are not to be recognized:

Barbarous Latin:	Greek-Latin:
Tamar*indus* T.	Cardam*indum* T.
Mor*inda* V.	Chrysanthem*indum* V.
	Sap*indus* T.

224. Generic names, compounded of one partial generic appellation and another entire one, are unworthy of botanists.

*Ari*sarum T.	*Arum.*	*Allowable Greek names*
*Cann*Acorus T.	*Canna.*	*Elae*Agnus. Oleae Vitex[13]. [olive balm]
*Capn*Orchis B	*Fumaria.*	*Ciss*Ampelos. Hederae Vitis [ivy vine]
*Lilio*Narcissus T.	*Scilla.*	
*Lin*Agrostis T.	*Eriophorum.*	
*Lauro*Cerasus T.	*Padus.*	

225. A generic name, to which one or two syllables have been prefixed, to make it signify a genus widely different from that previously signified, should be excluded.

*Acri*Viola B.	*Tropaeolum.*
*Bulbo*Castanum T.	*Bunium.*
*Cyno*Crambe T.	*Theligonum.*
*Chamae*Nerium T.	*Epilobium.*
Jon-Thlaspi T.	*Clypeola.*
Leuco-Nymphaea B.	*Nymphaea.*
Micro-Nymphaea B.	Imaginary?
Micro-*Leuco*-Nymphaea B.	*Hydrocharis.*
*Chamae*Drys T.	*Teucrium.*
*Chamae*Pythis T.	*Teucrium.*
*Pseudo*Dictamnus T.	*Marrubium.*
*Pseudo*Orchis M.	*Orchis.*
*Pseudo*Ruta M.	*Ruta.*

226. Generic names ending in –*oides* should be banished from the market-place of botany.

Agrimonoides T.	*Agrimonia.*
Alyssoides T.	*Alyssum.*
Asteroides T.	*Buphthalmum.*
Astragaloides T.	*Phaca.*
Chrysanthemoides T.	*Osteospermum.*
Cuminoides T.	*Lagoecia.*
Cyperoides T.	*Carex.*
Cyperoides T.	*Carex.*
Nymphoides T.	*Menyanthes.*
Pentaphylloides T.	*Potentilla.*
Rhamnoides T.	*Hippophaës.*
Ricinoides T.	*Croton.*
Telephioides T.	*Andrachne.*
Tribuloides T.	*Trapa.*
Valerianell-Oides B.	*Valeriana.*
Alsin Astr-Oides Kr.	
Jon-Thlaspi-Oides Kr.	
CapnOides R.	*Fumaria.*
MalacOides T.	*Malope.*

227. Generic names, conflated out of other generic names with some syllable added at the end, are unacceptable.

Acetosella R.	Rumex.
Napellus R	Aconitum.
Myrtillus R.	Vaccinium.
Lappula R.	Myositis.
Pyrola T.	Pyrola.
Lupinaster B.	Trifolium.
Alsinastrum T.	Elatine.
Rapistrum T.	Myagrum.
Limonium T,	Statice.
Agrostarium M.	
Fabaria R.	Sedum.
Rosea R.	Rhodiola.
Adonia P.	Myosuros.
Balsamita V.	Tanacetum.
Camphorata T.	Camphorosma.
Lapathum T.	Rumex
Erucago T.	Bunias.
Arachidna V.	Arachis.
Saliunca B.	Valeriana.
Sediformis P.	Stratiotes.
Linophyllum P.	Thesium.
Corylifolia H.	Psoralea.
Corallodendron T.	Erythrina.
Alfinanthemos R.	Trientalis.
Morocarpus R.	Blitum.
Anemonospermos H.	Arctotis.
Fagopyrum T.	Helxine.
Fung-Oid-aster M.	Elvela.
Clathr-Oid-astrum M.	
Lent-icul-aria M.	Lemna.
Alsin-astri-formis P.	Montia.

228. Generic names that are similarly pronounced provide occasions for misunderstanding.

Alsine T.	Alsine.
Alsimoides Rj.	Bufonia.
Alsinella D.	Sagina.
Alsinastrum T.	Elatine.
Alsinastroides Kr.	
Alsinastriformis Pk.	Cameraria.

continued

Alsinanthemos Rj.	*Trientalis.*
Alsinanthemum Kr.	*Alsine.*
Lycogala M.	
Lycopersicon T.	*Solanum.*
Lycoperdon T.	*Lycoperdon.*
Lycoperdastrum M.	
Lycoperdoides M.	
Lycopodioides D.	*Lycopodium.*
Lycopodium D.	*Lycopodium.*
Lycopus T.	*Lycopus.*
Lycopsis R.	*Lycopsis.*
Juncus T.	*Juncus.*
Scirpus T.	*Scirpus.*
Cyperus T.	*Cyperus.*
Juncoides S.	*Juncus.*
Scirpoides S.	*Carex.*
Cyperoides S.	*Carex.*
Juncoidi affinis S.	*Scheuchzeria.*
Cyperoidi affinis S.	
Juncago T.	*Triglochin.*
Junco affinis S.	*Schoenus.*
Pseudo-Cyperus S.	*Schoenus.*
Scirpo-Cyperus S.	*Scirpus.*
Nymphaea T.	*Nymphaea.*
Nymphoides T.	*Menyanthes.*
Micro-Nymphaea B.	
Leuco-Nymphaea B.	*Nymphaea.*
Micro-Leuco-Nymphaea B.	*Hydrocharis.*

229. **Generic names that do not have a root derived from Greek or Latin are to be rejected.**

German	
Bovista D.	Lycoperdon
Beccabunga R.	Veronica
Brunella	Prunella
English	
Percepier D.	Aphanes
French	
Orvala L.	
Spanish	
Sarsaparilla	Smilax
Scorzonera T.	

continued

Italian
    Galega T.

Datura R.	Turkish	
Ketmia T.	Syrian	*Hibiscus*
Alhagi T.	Arabic	*Hedysarum*
Ribes R.	Arabic	
Doronicum T.	Arabic	
Tenga	Malayalam	
Alhatoda T.	Sinhalese	*Justicia*
Sesban	Egyptian	
Jabotapita P.	American	*Ochna*
Caapeba P.	Brazil	*Cissampelos*

We adopt barbarous names *as if they were new-born,* provided that we remake the
words that have to be excluded, forming them from Greek or Latin.
Thea [tea] (*Chinese*); θεα, goddess.
Coffea [coffee] (*Arabic*); κωφεω, to swell up.
Musa [banana] (*Arabic*); Antonius Musa.
Cassine (*American*); κασσυω, to plot,
Annona (Anona, *American*); from the harvest.
Mammea (*Mammei, American*); from the breast-like fruit.
Chara (*French*); χαρα, [joy], joy in water.
Pothos (*Sinhalese*); Ποθος, [longing], Theophrastus.
Basella (*Malayalam*); Basium [kiss], *Cosmet.*
Datura (*Turkish*); see [my] Hortus Cliffortianus.
Cheiranthus (Keiri, Arabic); χειρ, hand.
Jambolifera (Jambolines, *Indian*).
Toluifera (Tolu, *Indian*).
Indigofera (Indigo).
Ziziphora *Moris.*
Gratuitously accepted [names], that are barbarous in themselves, could perhaps be
reformed according to the method described above: for example:

Bixa O.	*Hura L.*
Genipa T.	*Urena D.*
Guajacum T.	*Santalum L.*
Tulipa T.	*Yucca T.*
Liquidambar.	*Curcuma. L.*

230. Generic names of plants that are identical with terms used by
zoologists and mineralogists should be reported to those parties, if they
were adopted later by the botanists.

Quadrupeds.

Taxus T.	*Meles.*
Elephas T.	Rhinanthus
Erinaceus D.	Hydnum.
Onagra T.	Oenothera.

Birds.

Acanthus	*Fringilla.*
Ampelis	Vitis.
Lagopus	Trifolium.
Meleagris	Fritillaria.
Oenanthe	*Motacilla.*
Phalaris	*Emberiza.*

Amphibians.

Natrix K.	Ononis.

Fish.

Buglossum T.	Anchusa.
Hippoglossum	Ruscus.
Dephinium T.	
Pastinaca T.	*Raja*

Insects.

Ephemerum T.	Commelina P.
Eruca T.	Brassica T.
Locusta R.	Valeriana T.
Phalangium	Anthericum.
Ricinus T.	*Acarus*
Scolopendrum T.	Asplenium.
Sphondylium T.	Heracleum.
Staphylinus R.	Daucus T.

Worms.

Balanus	Nepenthes.

Minerals.

Granatum. [Garnet]	Punica.
Hyacinthus T. [Jacinth].	
Heliotropium T. [Jasper].	
Molybdaena [*Lead ore*].	Plumbago T.

Heavenly [bodies]

Sol R. [*Sun*]	Helianthus.
Iris T. [Rainbow].	

Earthly [places]

China	Cinchona.
Molucca T.	Moluccella.
Stoechas T. [*Hyères*].	Lavandula.
Ternatea T.	Clitoria.

Moral [qualities]

Impatiens R.	
Patientia	Rumex.
Concordia	Agrimonia.

231. Generic names that are identical with the terms used by anatomists pathologists, healers and artisans, should be dropped.

Anatomical.

Auricula Hk.	[ear-lap]	Primula.
Clitoris Br.		Clitoria.
Epiglottis Kn.		Astragalus.
Priapus A.	[penis]	Nepenthes.
Umbilicus H.	[navel]	Cotyledon.

Pathological.

Paralysis.	[*or* palsy]	Primula.
Soda	[headache]	Salsola.
Sphacelus	[gangrene]	Salvia.
Verruca	[wart]	Lapsana.

Therapeutic.

Ptarmica T.	[causing sneeze]	Achillea.
Cardiaca T.	[for heart]	Leonurus.
Hepatica R.	[for liver]	Hepatica.
Vesicaria T.	[for bladder]	Alyssum.
Vulneraria T.	[for wound]	Anthyllis.

Commercial.

Candela	[candle]	Rhizophora.
Sagitta	[arrow]	Sagittaria.
Serra	[saw]	Bisserrula.
Muscipula	[mouse-trap]	Silene.
Corona	[crown]	Helianthus.
Camara	[arch]	Lantina.
Bursa	[purse]	Thlaspi.
Solea equina	[horse-shoe]	Hippocrepis.

232. Generic names that are inconsistent with any species of their genus are unsatisfactory.

Chrysanthemum [golden flower] *flore albo* [*with a white flower*].
Cyanus [blue] *luteus* [*yellow*].
Convolvulus [winding] *erectus* [*upright*].
Pilosella [hairy] *glabra* [*smooth*].
Holosteum [all bone] is *soft*.
Unifolium [single-leaved] is *two-leaved*.

233. Generic names, that are identical with the terms for classes or orders should be dropped.

Fungus	*Planta*
Alga	*Arbor* [tree]

Muscus [moss] Frutex [shrub]

Filix [fern] Suffrutex [sub-shrub]

Palma Herba

Lilium Vegetabile

234. Generic names that are diminutives wrung from Latin can be tolerated, though they are not very good.

1.	Alchemilla	from the alchemists: dew on the leaves.[14]
	Coronilla	from the cluster with a crown <corona>.
	Potentilla	from the potency of its effects.
	Pulsatilla	from the battering <pulsatio> of the flower by the wind.
	Tormentilla	from the torments of dysentery.
2.	Basella	from the painted kisses <basia>.
	Biscutella	from the double shield <scutum> of the fruit.
	Crucianella	from its crossed leaves.
	Hirtella	from the hairiness <hirtus> of the branches.
	Limosella	from its native slime <limus>.
	Mitella	from the mitre shape of the fruit.
	Moluccella	from the Molucca islands, its native place.
	Nigella	from the blackness of its seeds.
	Pimpinella	from its bipinnate leaves.
	Porella	from the pores in its seeds.
	Soldanella	from the coin solidus [shilling].
	Tremella	from trembling like a jelly.
	Trigonella	from the triangular corolla.
3.	Corrigiola	from the binding of the feet.
	Gratiola	from its gracious effect in medicine.
4.	Clypeola	from the shield-shaped <clypeus> fruit.
	Rhodiola	from the rosy smell of the root.
	Uniola	from the union of the husks.
	Medeola	from Medea [a sorceress], or from its Median potency.
	Mitreola	from the mitre shape of the fruit.
5.	Pyrola	from the pear shape of the leaves.
6.	Phaseolus	from the bean shape of the seed.
7.	Gladiolus	from the sword shape <gladius> of the leaves.
8.	Samolus	from the island of Samos.
9.	Tropaeolum	from the trophy shape of the marking.
10.	Asperula	from the roughness <asperitas> of the plant.
	Bisserrula	from the fruit shaped like a two-edged saw <serra>.
	Calendula	it flowers in all months.
	Campanula	the bell shape <campana> of the corolla.
	Crassula	from the thickness <crassities> of the leaves.
	Ferula	from its use for beating <ferire> [children].
	Filipendula	the roots hang by a thread <filum>.

continued

	Lavandula	useful for baths <*lavare*, wash>.
	Patagonula	from Patagonia, its native place.
	Primula	from the precocity of its flowers.
	Pinguicula	from the fatness <*pinguedo*> of the leaves.
	Sanicula	from the healing <*sanare*> of wounds.
	Serratula	from the saw-like <*serra*> leaves.
	Spergula	from the scattering <*spargere*> of the seed.
11.	*Ranunculus*	from the frogs <*rana*>, its fellow-denisens.
	Convolvulus	from the convolutions of the stem.
	Humulus	from the ground <*humus*>, soil.
12.	*Asparagus*	from the roughness <*asperitas*> of the plant.
13.	*Asperugo*	from the roughness of the plant.
	Mollugo	from the softness <*mollities*> of the plant.
14.	*Borrago*	from the heart in action, to the ancients.
	Medicago	introduced by the Medes.
	Plantago	a plant to be to touched <*tangere*>.
	Plumbago	from a leaden <*plumbum*> mark placed on it.
	Selago	Silego?
	Solidago	from making wounds whole.
	Tussilago	from overcoming a cough <*tussis*>.
15.	*Pastinaca*	from a place to be trenched <*pastinare*>.
	Proserpinaca[15]	from creeping forwards <*proserpere*>.
	Securidaca	from the axe shape <*securis*> of the fruit.
	Portulaca	from a little door <*portula*>.
16.	*Reseda*	from the assuaging <*resedare*> of pain.
	Lactuca	from the milk <*lac*> of the plant.
	Urtica	from burning <*urere*> when touched.
	Lantana	from the flexibility <*lentor*> of the branches.
	Spinacia	from the thorns <*spina*> of the fruit.
	Salsola	from the saltiness of the plant.
	Salix	from its growing out by leaps <*salire*>.
	Sedum	from its sitting <*sedere*> on rocks.
	Ledum	from its hurting <*laedere*> by smell.
	Lamium	from an enchanted witch <*lamia*>.
	Juncus	from a slip to be grafted <*jungere*>.
	Cornus	from the horn-hardness <*cornu*> of the fruit.
	Juglans	from Jove's glans [penis].

235. Adjectives are inferior to substantives as generic names, and are therefore not very good.

1.	*Arenaria*	from its native soil [*arena*, sand].
	Convallaria	from its location [*convallis*, enclosed valley].
	Clavaria	from its appearance [*clavus*, nail].
	Capraria	from its use for food [for a goat, *caper*].
	Cochlearia	from the shape of the leaves [*cochlea*, shell].

continued

	Coriaria	from its commercial use [in tanning leather, *corium*].
	Coronaria	from its commercial use [*corona*, garland].
	Crotalaria	from the shape of the pods [κροταλον, castanet].
	Craniolaria	from the shape of the fruit [κρανιον, skull].
	Cymbaria	from the shape of the fruit [κυμβη, boat].
	Dentaria	from the shape of the root [*dens*, tooth].
	Flagellaria	from the shape of the leaves [*flagellum*, whip].
	Fragaria	its fragrance.
	Fritillaria	from the appearance of the petals [*fritillus*, dice-box].
	Fumaria	from smoke (*fumus*) as of the earth.
	Globularia	from the shape of the flower.
	Herniaria	from its use in medicine.
	Lunaria	from the shape of the fruit [*luna*, moon].
	Matricaria	from its use in medicine [*matrix*, womb].
	Obolaria	from the shape of the leaves [οβολος or οβελος coin or spike].
	Pulmonaria	from its use in medicine [*pulmo*, lung].
	Parietaria	from its location [*paries*, wall].
	Persicaria	from its peach-tree (περσικη) leaves.
	Sagittaria	from the shape of the leaves [*sagitta*, arrow].
	Sanguinaria	from the bloody juice [*sanguis*, blood].
	Saponaria	from the potency of the leaves [*sapo*, soap].
	Scoparia	from its commercial use [*scopa*, broom].
	Scrophularia	from scrofulae in medicine.
	Scutellaria	from the shape of the calyx [*scutellum*, small shield].
	Stellaria	from the shape of the flowers [*stellula*, little star].
	Subularia	from the shape of the leaves [*subula*, awl]
	Utricularia	from the shape of the appendages of the root [*utriculus*, bladder or womb].
	Uvaria	from the grape shape [*uva*] of the fruit.
	Uvularia	from the cluster shape of the inflorescence [*uvula*, small cluster].
2.	*Eriophorum*	from the wool [εριον] of the fruit.
	Echinophora T.	from its prickly fruit [εχινος, hedgehog or sea urchin].
	Rhizophora	from branches that take root [ριζα].
	Ziziphora Moris.	from the zizi of the Indians.
	Jambolifera	from Jambolus [or –um].
	Toluifera	from the tolu balsam.
	Indigofera	from the colour indigo.
3.	*Clitoria*	from the shape of the corolla [clitoris].
	Imperatoria	from the potency of the root [*imperator*, commander].
4.	*Hepatica*	from the shape of the leaves [ηπαρ, liver].
5.	*Scabiosa*	from its use in medicine [scabies, mange].
	Passerina	from the shape of the flower [*passer*, sparrow or turbot].
	Angelica	from the potency of the root and seed.
	Impatiens	from the elasticity of the root [so not suffering harm].
	Gloriosa	from the splendid flower.
	Mirabilis	from the variously coloured flowers.
	Amethystea	from the colour of the flowers.

continued

Pedicularis	from its potency in medicine [*pediculus*, louse].
Trientalis	from the size of the plant [*triens*, one-third of a foot].
Digitalis	from the finger-shape [*digitus*] of the corolla.
Fontinalis	from its location in springs [*fontes*].
Turritis	tall and narrow [*turris*, tower].
Sempervivum	from its evergreen leaves [*viror*].
Momordica	from its seeds, which seem to be chewed [*mordere*, bite].
Bistorta	from the shape of the root [twice twisted].
Saxifraga	from its location in cracks in rocks [*saxum*, stone; *frangere*, break].
Sanguisorba	from its potency in medicine [*sanguis*, blood; *sorbere*, suck].
Passiflora	from the instruments of the Passion, which the flower is said to resemble.

From a place.

Moluccella	from the Molucca islands.
Athamanta	from a city in Thessaly.
Parnassia	from Mount Parnassus.
Marrubium	from a town in Italy [Marruvium, *S. Benedetto*].
Smyrnium	from the city of Smyrna [*Izmir* in Turkey].
Nepeta	from a town in Italy [Nepete, *Nepi*].
Arethusa	from a city in Syria [*Er Rustan*].
Arabis	from the region of Arabia.
Punica	from the city of Carthage [in Tunisia].
Santolina	from a region of Aquitaine [*Saintonge*].
Thapsia	from a town in Africa [Thapsus (Baltah) in Tunisia].
Colchicum	from Colchis a city of Armenia; [sic: the western part of *Georgia*].
Cerasus	from the city of Cerasus [near *Khersoun*, Turkey].
Samolus	from the island of Samos [in Greece].
Agaricus	from Agaria, a city of Sarmatia [Eastern Europe].
Iberis	from the region of Iberia.
Patagonula	from Patagonia, a region of South America.
Carica	from Caria, a region of Asia [Minor, Turkey].
Ligusticum	from Liguria, a region of Italy.

236. Generic names should not be misused to gain the favour, or preserve the memory, of saints, or of men famous in some other art.

It is the only prize available to botanists; therefore it should not be misused.

[FROM THE NAMES] OF SAINTS[16]

Albert	Arabis
Antony	Epilobium
Benedict	Geum [Herb Bennet]
Christopher	Actaea [Herb Christopher]
Gerard	Aegopodium [Herb Gerard]
George	Valeriana
William	Agrimonia

James	Senecio
John	Hypericum [St. John's Wort]
Cunegund	Eupatorium
Ladislas	Gentiana
Laurence	Sanicula
Paul	Primula
Peter	Panetaria
Philip	Isatis
Quirinus	Tussilago
Rupert	Geranium [Herb Robert]
Simeon	Malva
Stephen	Circaea
Valentine	Paeonia
Zachary	Centaurea
Barbara	Erysinum
Catherine	Impatiens
Clare	Valeriana
[Holy] Cross	Nicotiana
Mary	Tanacetum
Odile	Delphinium
Rosa	Paeonia

OF DIVINES

Seriana Pl.	Paullinia
Uvedalia Pet	Osteospermum. [Robert Uvedale].
Levisanus Pet.	

OF FAMOUS MEN

Phelypaea T.	Lathraea [Raimond Balthasar, Marquis de Phélypeaux]
Buccaferrea M.	Ruppia
Bonarota M.	Veronica [Michelangelo Buonarroti]

237. I retain generic names derived from poetry, imagined names of gods, names dedicated to kings, and names earned by those who have promoted botany.

Names from poetry that are commonplace in the lore of the ancients.

Ambrosia	*Adonis*	*Amaryllis*	*Canna*
Nepenthes	*Crocus*	*Phyllis*	*Syringa*
Cornucopiae	*Centaurea*	*Circaea*	*Smilax*
Protea	*Chironia*	*Medeola*	*Mentha*
Actaea	*Achillea*	*Andromeda*	*Myrsine*
Narcissus	*Paeonia*	*Daphne*	
Hyacinthus	*Cerbera*		

[NAMES] OF GODS

Asclepias	[Asclepius (Aesculapius), the god] of physicians.
Mercurialis	[Mercury, the messenger] of the gods.
Hymenaea	[Hymen, the god] of marriage.
Serapias	[Serapis, a god] of Egypt.
Satryrium	[the satyrs], demons of desire.
Satureja	a satyr.
Sterculea	[Sterculius, the god] of the dung-hill, [alluding to the smell of the flowers.]
Ixora	[a god] of Malabar, [to whom the flowers were offered.]
Tagetes	[Tages], a grandson of Jove.
Musa	[the Muse, goddess] of the sciences.
Nymphaea	[the nymphs] of the waters.
Najas	[the Naiads, nymphs] of the springs.
Nyssa	[Nysa], a nymph.
Melissa	[a nymph, patroness] of honey.
Dryas	[the dryads, nymphs] of oak-trees.
Atropa	[Atropos], the last of the Furies [*sic*: Fates].
Napaea	[a nymph] of the groves.
Herminium	[Hermes]

OF KINGS

Eupatorium	(Mithridates)[Eupator] of Pontus.
Gentiana	[Gentius] of Illyria.
Lysimachia	[Lysimachus] of Sicily.
Telephium	[Telephus] of Mysia.
Teucrium	[Teucer] of Troy.
Valeriana	[Valerian, Roman Emperor].
Carlina	[Charles, i.e. Charlemagne]
Philadelphus	[Ptolemy II] of Egypt.
Pharnaceum	[Pharnaces] of Pontus.
Artemisia	[wife] of Mausolus.
Althaea	[wife] of Oeneus.
Helenia	[Helen, wife] of Menelaus.

OF PATRONS

Borbonia	(Gaston) Sr. [de Bourbon, Duke of Orléans].
Eugenia	Prince [Eugène of Savoy.]
Bignonia	[Jean-Paul Bignon], abbé.
Petrea	Lord Petre.
Sherardia	[William Sherard], consul at Smyrna [Izmir].
Cliffortia	[George Clifford] Dr. of Civil and Canon Law.
Stewartia	[John] Stewart [sic: *Stuart*], Earl [of Bute]
Maurocennia	[Andrea Morosini], Counsellor of Venice.
Bosea	[Bose] Counsellor of Leipzig.
Begonia	[Michel Bégon], governor [of Canada].
Poinciana	[M. de Poinci], governor [of the Antilles].

OF FALSE CLAIMANTS

Nicotiana	J[ean]. Nicot.[17]
Cinchona	[Countess of Chinchón.]
Euphorbia	[Euphorbus, physician.]

238. Generic names that have been formed to perpetuate the memory of a botanist who has done excellent service should be religiously preserved.

> This, the only and pre-eminent reward for such labour, should be religiously preserved and fairly awarded.
>
> The reasons are to be sought in [my] *Critica botanica*.

Aeginetia L.	*Clusia* P.
Aldrovanda M.	*Coldenia* L.
Alpinia P.	*Columnea* P.
Ammannia H.	*Commelina* P.
Artedia L.	*Cordia* P.
Averrhoa L.	*Cornutia* P.
Avicennia L.	*Cortusa* B.
Barleria P.	*Crateva* L.
Bauhinia P.	*Crescentia* L.
Bellonia P.	*Cupania* P.
Besleria P.	*Dalea* L.
Blaeria H.	*Dalechampia* P.
Bobartia L.	*Dillenia* L.
Bocconia P.	*Dioscorea* P.
Boerhaavia V.	*Dodartia* T.
Bontia P.	*Dodonaea* P.
Breynia P.	*Dorstenia* P.
Bromelia P.	*Duranta* P.
Brossaea P.	*Fagonia* T.
Browallia L.	*Fevillea* L.
Brunfelsia P.	*Frankenia* L.
Burmannia L.	*Fuchsia* P.
Caesalpinia P.	*Galenia* L.
Camellia L.	*Garidella* T.
Cameraria P.	*Gerberia* L.
Catesbaea L.	*Gerardia* P.
Celsia L.	*Gesneria* P.
Cherleria H.	*Gleditsia* L.

continued

Gmelina L.	Ovieda L.
Grewia L.	Parkinsonia P.
Grislea L.	Paullinia L.
Gronovia H.	Penaea P.
Guilandina L.	Petiveria P.
Halleria L.	Pisonia P.
Hebenstretia L.	Plinia P.
Hermannia T.	Plukenetia P.
Hernandia P.	Plumeria T.
Heucheria L.	Pontederia H.
Hippocratea P.	Rajania P.
Houstonia L.	Randia H.
Hugonia L.	Rauwolfia P.
Isnarda L.	Renealmia P.
Jungermannia R.	Rheedia P.
Jussiaea L.	Rivina P.
Kaempferia L.	Robinia L.
Kiggelaria L.	Royena L.
Kleinia L.	Rudbeckia L.
Knautia L.	Ruellia P.
Lawsonia L.	Ruppia L.
Linnaea G.	Sauvagesia L.
Lobelia P.	Scheuchzera L.
Loeselia L.	Sherardia D.
Lonicera P.	Sibbaldia L.
Ludwigia L.	Sigesbeckia L.
Magnolia P.	Sloanea P.
Malpighia P.	Spigelia L.
Maranta P.	Staehelina L.
Marchantia M.	Stapelia L.
Marcgravia P.	Suriana P.
Marsilea M.	Swertia L.
Martynia H.	Thalia P.
Matthiola P.	Theophrasta P.
Mentzelia P.	Tillaea M.
Mesua L.	Tillandsia L.
Michelia L.	Tournefortia P.
Milleria H.	Tabernaemontana P.
Moerhingia L.	Tradescantia R.
Monarda L.	Tragia P.
Montia M.	Trewia L.
Morisona P.	Triumfetta P.
Muntingia P.	Turnera P.
Musa L.	Valantia T.

continued

Vateria L.
Vallisneria M.
Volkameria L.
Waltheria L.

Ximenia P.
Zanichellia L.
Zannonia L.

[NAMES] OF DISCOVERERS

Collinsonia L.
Claytonia L.
Diervilla T.
Nicotiana T.
Torenia L.

Sarracenia T.
Bartramia L.
Stellera G.

OF TRAVELLERS

Bannisteria H.	Virginia
Bartsia [from Bartsch]	Surinam
Brunia	Orient
Clutia	Barbary
Garcinia	India
Gundelia	Orient
Knoxia	Ceylon
Lippia	Abyssinia
Mitchella	Virginia
Oldenlandia	Africa
Sarracenia	Canada
Torenia	China
Barreria	Gall. æquinoct. [sic][18]

APPROVED [NAMES]

Budleja H.
Justicia H.
Richardia H. Richardson
Schwalbea G. physician.
Lavatera T.

Morina T.
Hottonia B.
Tozzia M. assiduous abbé.
Blasia [Blaes] M. monk, botanist.
Riccia M. senator, knight.

EXCLUDED

Chomelia L. a species of Rondeletia.
Pavia B. a species of Aesculus.
Bonarota M. a species of Veronica.
Buccaferrea M. Count.
Franka physician.

Laurentia professor of medicine.
Puccinia professor of anatomy.
Salvinia professor of Greek
Targionia physician.

239. Generic names that have been given without doing any harm to botany ought to be tolerated.

Those that we have given in Sections 214–17, 220–33, and 236 are harmful. We describe as harmful those that are

1. Incompatible with the genus: Sections 215–17.
2. Wrongly formed: Sections 220–9.
3. Wrongly given: Sections 231–3, 236.

Obscure LATIN names, of which we do not know the sources, or which turn out to be of doubtful origin, are acceptable but not to be imitated: for example

Abies	Iris
Acer	Juniperus
Allium	Laurus
Alnus	Ligustrum
Apium	Lilium
Aralia	Linum
Arbutus	Lolium
Arundo	Lupinus
Atriplex	Malva
Avena	Milium
Bellis	Opulus
Berberis	Panicum
Betula	Papaver
Carduus	Paris
Carex	Pinus
Carpinus	Pisum
Centunculus	Populus
Cicer	Porrum
Cicuta	Prunus
Cotula	Quercus
Cucumis	Rosa
Cucurbita	Rosmarinus
Cunila	Rubia
Equisetum	Rubus
Ervum	Rumex
Aesculus	Ruscus
Ficus	Salicornia
Genista	Sambucus
Hedera	Scirpus
Illecebrum	Secale
Ilex	Solanum
Inula	Sorbus

continued

Tamarix	Vinca
Tilia	Viola
Triticum	Viscum
Verbena	Vitex
Veronica	Vitis
Viburnum	Ulmus
Vicia	Ulva

Obscure GREEK names remain in use, however great the difficulty in eliciting them, even if, once elicited, they are still doubtful:

Achras	Dorycnium
Aloë	Elatine
Amomum	Elvela
Anagyris	Epimedium
Aparine	Erinus
Atraphaxis	Eringium
Blitum	Exacum
Boletus	Fucus
Borassus	Geum
Byssus	Glaux
Cactus	Gossypium
Cassia	Hibiscus
Carum	Isatis
Celtis	Itea
Cenchrus	Lathyrus
Cerasus	Lemma
Cissus	Lichen
Cistus	Lotus
Citrus	Lycium
Cneorum	Lythrum
Coccus	Malope
Coix	Melia
Colutea	Melica
Comarum	Melochia
Corylus	Memecylon
Costus	Mespilus
Crataegus	Morus
Croton	Myrica
Cuminum	Myrtus
Cycas	Nardus
Cytisus	Nerium
Daucus	Ochna

continued

Oryza	*Seseli*
Penthorum	*Sicyos*
Pentapetes	*Sida*
Peplis	*Sinapis*
Peziza	*Sisymbrium*
Phaca	*Sium*
Phillyrea	*Spartium*
Phleum	*Sphagnum*
Phlomis	*Spiraea*
Phoenix	*Spongia*
Piper	*Statice*
Pistacia	*Strychnus*
Platanus	*Styrax*
Polemonium	*Tamus*
Pothos	*Taxus*
Prasium	*Thalictrum*
Prinus	*Thesium*
Ptelea	*Tridax*
Rhamnus	*Thya*
Rhus	*Vella*
Saccharum	*Ulex*
Samyda	*Xyris*
Scandix	*Zea*
Scilla	*Zizania*
Sesamum	

Several names that have been corrupted by false readings of ancient authors have undergone remarkable transformations.

Agrimonia	for	Argemonia
Ajuga	for	Abiga
Aquilegia	for	Aquilina
Betonia	for	Vettonica
Brassica	for	πρασικη
Buxus	for	πυξος
Coriandrum	for	Coriannum
Diapensia	for	διαπενθης
Euphrasia	for	ευφροσυνη
Gomphrena	for	Gromphena P.
Lupulus	for	Upulus
Malope	for	μαλαχη
Melochia	for	μολοχη
Pimpinella	for	Bipennula
Santolina	for	Sanctolina

continued

Spinachia	for	Spanachia
Verbascum	for	Barbascum
Verbesina	for	Forbesina
Veronica	for	Vetonica
Betula	for	Betulla
Equisetum	for	Equiselis
Myrsine	for	Myrsinum P.
Melothria	for	Melothron P.
Phleum	for	Phleos P.
Spiraea	for	Spiraeon P.

Thya, wrongly Thuja or Thuya.

240. Generic names that show the essential character or habit of the plant are the best.

[1.] The GREEK DERIVATIONS of plant names are very difficult to explain, and so guesses are often accepted as satisfactory.

The essential character is rarely captured in plants, though it is the best.

Adenanthera	anther of the gland	αδενος	ανθηρα
Triopteris	triple wing	τρεις	πτερον
Epilobium	violet on a silique	επι λοβου	ιον
Helicteres	spiral	ελιξ	
Tetracera	quadruple horn	τετρας	κερας
Trichosanthes	hair-like flower	τριχος	ανθος

The habit indicates a likeness from which an image is produced, and from the image a name.

1.	Glycyrrhiza	ριζα	root	γλυκυς	sweet
	Ophiorrhiza	ριζα	serpent's	οφις	
2.	Clerodendrum	δενδρον	tree	κληρος	lucky
	Epidendrum	above	επι		
	Leucadendrum	"	white	λευκος	
	Liriodendrum	"	lily-like	λειριον	
	Rhododendrum	"	rose-like	ροδον	
3.	Haematoxylum	wood	bloody	αιμα	
	Ophioxylum	ξυλον	serpent-like	οφις	
	Sideroxylum	"	like iron	σιδηρος	
4.	Eriocaulon	stem	woolly	εριον	
	Caucalis	καυλος	lying down	κεω Ambr.	
5.	Calophyllum	leaf	beautiful	καλος	
	Caryophyllum	φυλλον	of a nut	καρυα	
	Ceratophyllum	"	horn-like	κερας	

continued

	Chaerophyllum	"	glad	χαιρω
	Chrysophyllum	"	golden	χρυσος
	Hydrophyllum	"	aquatic	υδωρ
	Myriophyllum	"	countless	μυριος
	Podophyllum	"	foot-shaped	πους, ποδος
	Triphyllum	"	in threes	τρεις
	Zygophyllum	"	conjoined	ζυγος
6.	Chrysocoma	top κομη	golden	χρυσος
7. a.	Amaranthus[19]	flower ανθος	unfading	μαραινω
	Cephalanthus		capitate	κεφαλη
	Cheiranthus	"	hand-like	χειρ
	Chionanthus	"	snowy	χιων
	Dianthus	"	Jove's	Διος
	Galanthus	"	milky	γαλα
	Haemanthus	"	bloody	αιμα
	Helianthus	"	of the sun	ηλιος
	Loranthus	"	of thongs	λωρος
	Melianthus	"	honeyed	μελι
	Phyllanthus	"	of leaves	φυλλον
	Rhinanthus	"	long-nosed	ριν
	Scleranthus	"	without juice	σκληρος
	siphonanthus	"	tubular	σιφων
	Sphaeranthus	"	spherical	σφαιρα
	Tarchonanthus	"	Arabs' tarchon = [tarragon]	
b.	Achyranthes	"	chaff-like	αχυρον
	Aphyllanthes	"	without leaves	αφυλλος
	Menianthes	"	monthly	μην
	Nyctanthes	"	of night	νυξ
	Prenanthes	"	prone	πρηνης
	Polianthes	"	urban	πολις
	Trichosanthes	"	hair-like	θριξ
c.	Cerinthe	"	waxy	κηρος
	Oenanthe	"	vine	οινη
d.	Anthoxanthum	"	of flowers	ανθος
e.	Melanthium	"	black	μελας
f.	Chrysanthemum	"	golden	χρυσος
	Eranthemum	"	of the earth	ερα
	Mesembryanthemum	"	in the middle	μεσος
		"	of the ovary	εμβρυων
	Xeranthemum	"	dry	ξηρος
8.	Trichostema	stamen στημων	hair-like	θριξ
9.	Adenanthera	anther	glandular	αδενος
	Dianthera	ανθηρα	double	δις

continued

10.	Ceratocarpus	fruit	horny	κερας
	Conocarpus	καρπος	conical	κωνος
	Elaeocarpus	"	olive-shaped	ελαια
	Heliocarpus	"	of the sun	ηλιος
	Callicarpa	"	beautiful	καλος
11.	Tetragonotheca	capsule θηκη	four-cornered	τετραγωνος
12.	Anthospermum	seed	of the flower	ανθος
	Cardiospermum	σπερμα	of the heart	καρδια
	Corispermum	"	of a bug	κορις
	Lithospermum	"	stony	λιθος
	Menispermum	"	moon-shaped	μηνη
	Osteospermum	"	bony	οστεον
13.	Diospyros	grain	Jove's	Διος
	Isopyrum	πυρος	similar	ισος
	Melampyrum	"	black	μελας
14.	Chrysobalanus	stone-fruit βαλανος	golden	χρυσος
15.	Aegilops	face	nanny-goat's	αιξ
	Echinops	οψις	hedgehog's	εχεινος
	Mimusops	"	ape's	μιμω
	Coreopsis	"	bug's	κορις
	Galeopsis	"	cat's	γαλη
	Lycopsis	"	wolf's	λυκος
16.	Bucephalum	head	ox's	βους
	Dracocephalum	κεφαλη	snake's	δρακων
	Eriocephalus	"	woolly	εριον
17.	Leontodon	tooth οδους	lion's	λεων
18.	Cynoglossum	tongue	dog's	κυων
	Ophioglossum	γλωσσα	serpent's	οφις
19.	Melastoma	mouth στομα	black	μελας
20.	Buphthalmum	eye οφθαλμος	ox's	βους
21.	Antirrhinum	nose ριν	equivalent	αντι
22.	Arctotis	ear	bear's	αρκτος
	Hedyotis	ους, ωτος	pleasant	ηδυς
	Myosotis	"	mouse's	μυς
23.	Tragopogon	beard πωγων	goat's	τραγος
	Callitriche	hair	beautiful	καλος
	Polytrichum	θριξ	much	πολυς

continued

24.	Anthoceros	horn	}	of a flower	ανθος
	Tetracera	κερας	}	quadruple	τετρας
25.	Cynomorium	penis [μοριον]	}	dog's	κυων
26.	Cynometra	womb μητρα	}	bitch's	κυων
27.	Alopecurus	tail	}	fox's	αλωπηξ
	Cynosurus	ουρα	}	dog's	κυων
	Lagurus	"		hare's	λαγως
	Leonurus	"		lion's	λεων
	Myosurus	"		mouse's	μυς
	Saururus	"		lizard's	σαυρος
	Scorpiurus	"		scorpion's	σκορπιος
	Hippuris	"		horse's	ιππος
28.	Calligonum	knee	}	beautiful	καλος
	Chrysogonum	γονυ	}	golden	χρυτος
	Polygonum	"		multiple	πολυ
	Theligonum	"		female	θηλυ
29.	Polycnemum	shin κνημη	}	multiple	πολυ
30.	Arctopus	foot	}	bear's	αρκτος
	Elephantopus	πους	}	elephant's	ελεφας
	Micropus	"		small	μικρος
	Ornithopus	"		bird's	ορνις
	Lycopus	"		wolf's	λυκος
	Lycopodium	"		wolf's	λυκος
	Aegopodium	"		nanny-goat's	αιξ
	Chenopodium	"		goose's	χην
	Clinopodium	"		of a bed	κλινη
	Melampodium	"		black	μελας
	Polypodium	"		multiple	πολυ
31.	Asplenium	spleen	}	none	α privative
	Chrysosplenium	σπλην	}	golden	χρυσος
32.	Bupleurum	side	}	ox's	βους
33.	Camphorosma	smell	}	of camphor	
	Diosma	οσμη	}	of Jove	Διος

2. ANIMALS that provide names of plants.

Leontice	lion	λεων		
Arctium	bear	αρκτος		
Cynara	dog	κυων		
Cynanchum	dog	κυων	strangle	αγχω

continued

Apocynum	dog	κυνος	away from	απο
Lycoperdon	wolf	λυκος	fart	περδω
Onopordon	donkey	ονος	fart	περδω
Ononis	donkey	ονος		
Hippophaë	of a horse	ιππος	splendour	φαω
Hippomane	of a horse	ιππος	madness	μανια
Sisyrinchium	pig's	υς	snout	ρυγχος
Hyoscyamus	pig's	υς, υος	bean	κυαμος
Hyoseris	pig's	υος	lettuce	σερις
Orobus	ox	βους	excite	ορω [sic]
Tragacantha	goat's	τραγος	thorn	ακανθα
Hieracium	hawk	ιεραξ		
Geranium	crane	γερανος		
Chelidonium	swallow	χελιδων		
Struthia	sparrow	στρουθιον		
Dracontium	snake	δρακων		
Echium	adder	εχις		
Chelone	tortoise	χελωνη		
Delphinium	dolphin	δελφιν		
Coris	bug	κορις		
Melittis	bee	μελιττα		
Myagrum	flies	μυια	catch	αγρευω
Astragalus	vertebra	αστραγαλος		
Acorus	pupil	κορη	privative	α
Ophrys	eye-brow	οφρυς		
Orchis	testicle	ορχις		
Phallus	penis	φαλλος		
Cotyledon	concavity	κοτυλη		
Splachnum	internal organ	σπλαγχνον		
Bubon	groin	βουβων		
Pteris	wing	πτερον		
Lythrum	gore	λυθρον		

3. IMPLEMENTS adopted as names on account of the plants resemblance to them.

Gomphrena	peg	γομφρος [sic: γομφος]
Brabeium	sceptre	βραβειον [prize]
Atractylis	spindle	ατρακτος
Lonchitis	lance	λογχη
Caltha	basket	καλαθος
Cercis	shuttle	κερκις
Prinos	saw	πριων
Ceropegia	candlestick	κηροπηγιον
Lychnis	lamp	λυχνος

continued

Phlox	flame	φλοξ		
Selinum	moon	σεληνη		
Cestrum	hammer	κεστρα		
Sideritis	iron	σιδηρος		
Stratiotes	army	στρατος		
Oenothera	wines	οινος	os [*sic*]–	θηρα
Othona	linen cloth	οθονη		
Delima	file	[λιμα in Modern Greek]		
Lagoecia	hare's	λαγως	form	οικος

4. COMPOSITION used in [the formation of] a name.

Agave	admirable	αγαυος		
Adoxa	glory	δοξα	privative	α
Aphanes	invisible	αφανης		
Adiantum	moisten	διαινω	privative	α
Cleome	closed	κλειομαι		
Clathrus	railings	κλαθρον		
Aeschynomene	restrain	ισχω	be ashamed	αισχυνομαι
Mimosa	versatile actor	[μιμος]		
Mimulus	masked actor	[μιμος]		
Silene	frothy	σιελιζω		
Ascyrum	roughened	σκιρος [*sic*]	privative	α
Holosteum	whole	ολος	bony	οστεον
Asarum	bound	σειρον [*sic*]	privative	α
Erythrina	red	ερυθρος		
Erythronium	red	ερυθρος		
Aizoon	ever	αει	living	ζωον
Ageratum	old age	γηρας		
Bulbocodium	bulb	βολβος	wool	κωδιον
Asphodelus	trip	σφαλλω	privative	α
Bryonia	germinate	βρυω		
Bryum	germinate	βρυω		
Acanthus	thorn	ακανθα		
Dolichus	long	δολιχος		
Schinus	split	σχιζω		
Xylon	wood	ξυλον		
Clematis	vine cutting	κλημα		
Periploca	around	περι	entwining	πλοκη
Schœnus	rope	σχοινος		
Osyris	branchy	οζυρις [*sic*]		
Tomex	stuffing	[θωμιξ]		
Gnaphalium	stuffing	γναφαλον		
Sonchus	hollow	σομφος		

continued

Acalypha	beautiful	καλος	by touch, αφη, privative. α	
Drosera	dew	δροσιον		
Neurada	sinew	νευρα		
Drypis	tear	δρυπτω		
Cnicus	scratch	κνεω		
Cotyledon	concavity	κοτυλη		
Sparganium	bandage	σπαργανον		
Zostera	belt	ζωστηρ		
Platanus	broad	πλατυς		
Corypha	summit	κορυφη		
Cyclamen	circle	κυκλος		
Corymbium	cluster	κορυμβος		
Stachys	ear of corn	σταχυς		
Acrostichum	top	ακρος	line	στιχος
Staphylea	bunch	σταφυλη		
Andrachne	manly	ανδρειος		
Achyronia	chaff	αχυρον	husk	αχνη
Physalis	bladder	φυσα		
Anthemis	flourish	ανθεω		
Aster	star	αστηρ		
Astrantia	constellation	αστρον	opposite	αντιος
Antholyssa	of a flower	ανθος	rage	λυσσα
Anthyllis	of a flower	ανθος	down	
Amorpha	shape	μορφη	privative	α
Hesperis	evening	εσπερος		
Anacardium	without	ανα	heart	καρδια
Hydrangea	for water	υδωρ	vessel	αγγος
Hypecoum	resound	υπηχεω		
Ceratonia	siliqua	κερατιον		
Lepidium	scale	λεπις		
Thlaspi	press	θλαω?		
Tribulus	caltrops	τριβολοι		
Tirglochin	point	γλωχιν	three	τρεις
Erigeron	of spring	ηρ	old man	γερων
Eriophorum	wool	εριον	bear	φερω
Phalaris	shining	φαλος		
Tordylium	lathe	τορνος	revolve	ιλλω
Elymus	roll in	ελυω		
Raphanus	easily	ραδιως	appear	φαινω
Selinum	moon	σεληνη		
Chamaerops	copse	ρωψ	low down	χαμαι

5. MEDICINAL effects applied as names.

Alcea	relief	αλκη		
Althaea	heal	αλθεω		
Erysimum	protect	ερυω		
Panax	all	παν	medicament	ακος
Pancratium	all	πας	strong	κρατυς
Heracleum	Hercules	Ηρακλης		
Jatropha	medicament	ιατρον	eat	φαγω[sic]
Bromus	food	βρωμα		
Olax	furrow	ωλαξ		
Galium	milk	γαλα		
Polygala	milk	γαλα	much	πολυ
Poterium	cup	ποτηριον		
Draba	sharp	δραβη		
Capsicum	bite	καπτω		
Glycine	sweet	γλυκυς		
Oxalis	acid	οξυς		
Picris	bitter	πικρος		
Xanthium	yellow	ξανθος		
Lapsana	drain	λαπτω		
Carthamus	clean	καθαιρειν?		
Rheum	flow	ρεω		
Corchorus	clean	κορεω	place	χωρος
Ischaemum	restrain	ισχω	blood	αιμα
Peganum	freeze	πηγνοω		
Aristolochia	midwife	λοχεια	best	αριστος
Dictamnus	derive	τιπτειν [sic]		
Horminum	be impelled	ορμαινω		
Thymus	spirit	θυμος		
Symphytum	grow together	συμφυειν		
Holcus	dragging	ολκος		
Parthenium	maidenly	παρθενιον		
Cnyza	itch	κνυζα		
Argemone	eye-ulcer	αργεμα		
Alyssum	rage	λυσσον [sic: λυσσα]	privative	α
Rhexia	rupture	ρηξις		
Antidesma	bond	δεσμος	against	αντι
Catananche	violence	αναγκη	against	κατα
Anisum	unequal	ανισος		
Trachelium	kneck	τραχηλος		
Anchusa	dye	ανχουσειν [sic: αγχουσα, αγχειν]		
Phytolacca	pigment		plant	φυτον
Anagallis	laugh	αναγελαω		
Briza	sleep	βριζω		

continued

Dipsacus	thirst	διψαω			
Alisma	anxiety	αλυσμα			
Butomus	carving	τομος	beef	βους
Helleborus	tie up	ειλεω	food	βορα
Orobanche	strangle	αγχω	vetch	οροβος
Phellandrium	cork	φελλος	man's	ανδριον
Trichomanes	madness	μανια	hair	θριξ, τριχος
Hypericum	image	εικων	above	υπερ

6. Native LOCATION taken as a name.

Ephedra	water	υδωρ	above	επι
Origanum	of a mountain	ορος	gladness	γανος
Aconitum	sharp stone	ακονη		
Crambe	dry	κραμβος		
Azalea	arid	αζαλεος		
Bunias	hill	βουνος		
Bunium	hill	βουνος		
Empetrum	stone	πετρος	in	εν
Gypsophila	chalk	γυψος	girl-friend	φιλη
Ammi	sand	αμμος		
Agrostis	field	αγρος		
Cichorium	field	χωριον	go	κιω
Diodia	on the way	διοδιος		
Monotropa	alone	μονος	turn	τρεπω
Lathraea	clandestine	λαθραιος		
Mandragora	fold	μανδρα		
Anthericum	flower	ανθος	of the hedges	ρηχος
Alsine	grove	αλσος		
Hydrocharis	water	υδωρ	grace	χαρις
Hydrocotyle	water	υδωρ	vessel	κοτυλη
Typha	marsh	τιφος		
Potamogeton	river	ποταμος	neighbour	γειτων
Pistia	trough	πιστηρ		

7. SEVERAL miscellaneous things from which names have been borrowed.

Dodecatheon	gods	θεος	twelve	δωδεκα
Theobroma	gods'	θεων	food	βρωμα
Ambrosia	mortal	βροτος	privative	α
Baccharis	Bacchus	Βακχος		
Cypripedium	Venus	Κυπρις[20]	slipper	ποδιον
Jasione	god's	σιος for θεος	violet[s][21]	ια
Neottia	nestling	νεοττος		

continued

Cucubalus	shot	βολος	bad	κακος
Euonymus	name	ονομα	good	ευς
Hemerocallis	day	ημερα	beautiful	καλος
Heliotropium	sun	ηλιος	turn	τρεπειν
Andryala	man's	ανδρος	wandering	αλη
Androsace	man's	ανδρος	shield	σακος
Allophylus	alien	αλλοφυλος		
Æthusa	beggar-woman	αιθουσα[22]		
Isoëtes	year	ετος	like	ισος
Jasminum	violet[21]	ιον	smell	ιοασμη
				[*sic*: οσμη]
Leucojum	violet	ιον	white	λευκον
Ipomœa	caterpillar	ιψ, ιπος	likeness	ομοιος
Microcos	kernel	[κοκκος]	little	μικρος
Calla	cock's wattle	καλλαιον		
Arachis	harm	αρα		
Arum	harm	αρα	privative	α
Ballota	throw	βαλλω		
Phyteuma	beget	φυτευω		
Brassica	boil	βρασσω		
Hemionitis	mule	ημιονος		
Cachrys	parched barley	καχρυς		
Cardamine	cress	καρδαμον		
Glechoma	pennyroyal	γληχων		
Hedysarum	of ointment [?]	αρον	sweetness	ηδυσμα
Fraxinus	fence	φραξις		
Hypochæris	pig	χοιρος	diminutive	υπο
Chondrilla	lump	χονδρος		
Ornithogalum	milk	γαλα	birds	
Peucedanum	fir	πευκη	lowly	δαυος [*sic*]
Caryota	nut	καρυον		
Conium	dust	κονια		
Erica	break	ερεικω		
Hamamelis	apple-tree	μηλις	at once	αμα

241. Plant names that were commonly used by the ancients are read either
as GREEK in the works of HIPPOCRATES (H.), THEOPHRASTUS
(T.), and DIOSCORIDES (D.); or as LATIN in the works of PLINY
(P.) and those of the agriculturists and the poets.

The authority for the words is to be sought in the works of the fathers, Section 9.
GREEK names commonly used by the Greeks.

Acalypha	ακαληφη T.D.
Acanthus P.	ακανθος T.
Achillea P.	αχιλλειος T.D.
Achras	αχρας D.
Aconitum P.	ακονιτον T.D.
Acorus P.	ακορον D.
Adiantum P.	αδιαντον H.
Ægilops P.	αιγιλοψ T.D.
Æschynomene P.	αισχυνομενη
Agaricum P.	αγαρικον D.
Ageratum P.	αγηρατον D.
Agrostis	αγρωστιςT.D.
Aira P.	αιρα H.T.
Aizoon P.	αειζωον T.D.
Alcea P.	αλκεα D.
Alisma P.	αλισμα D.
Aloë P.	αλοη D.
Alopecurus	αλωπεκουρος T.
Alsine P.	αλσινη D.
Althaea P.	αλθαια D.
Alyssum P.	αλυσσον D.
Amarantus	αμαραντος D.
Ambrosia P.	αμβροσια D.
Ammi	αμμι D.
Amomum	αμωμον H.T.D.
Amygdalus P.	αμυγδαλη H.
Anagallis P.	αναγαλλις D.
Anagyris P.	αναγυρις D.
Anchusa P.	αγχουσα H.T.D.
Anethum P.	ανηθον H.T.D.
Anemone P.	ανεμωνη H.T.D.
Andrachne P.	ανδραχνη H.T.D.
Androsace P.	ανδροσακες D.
Anisum P.	ανισον D.
Anthemis P.	ανθεμις D.
Anthericum P.	ανθερικος T.D.
Anthyllis P.	ανθυλλις D.
Antirrhinum P.	αντιρρινον T.D.
Aparine P.	απαρινη T.D.
Apocynum P.	αποκυνον D.
Arabis	αραβις D.
Arachos P.	αραχος T.

continued

Arctium P.	αρκτιον D.
Argemone P.	αργεμωνη D.
Aristolochia P.	αριστολοχια D.
Artemisia P.	αρτεμισια H.D.
Arum P.	αρον H.D.T.
Aruncus	ηρυγγος A.
Asarum P.	ασαρον D.
Asclepias P.	ασκληπιας D.
Ascyrum	ασκυρον D.
Asparagus	ασπαραγος D.
Asphodelus P.	ασφοδελος D.
Asplenum P.	ασπληνον D.
Aster P.	αστηρ D.
Astragalus P.	αστραγαλος D.
Athamanta	αθαμαντικον D.
Atractylis P.	ατρακτυλις T.D.
Atraphaxis	ατραφαξις H.T.
Baccharis P.	βακχαρις D.
Ballote P.	βαλλωτη D.
Borassus	βορασσος D.
Briza	βριζα
Bromus P.	βρωμος T.D.
Bryonia P.	βρυωνια T.D.
Bryum P.	βρυον T.D.
Bubon	βουβωνιον H.
Bulbocodium	βολβοκωδιον T.
Bunias	βουνιας D.
Bunium P.	βουνιον D.
Buphtalmum P.	βουφθαλμον D.
Bupleurum P.	βουπλευρον H.
Butomus	βουτομος T.
Byssus P.	βυσσος Poll.
Cachrys P.	καχρυς D.
Cactus P.	κακτος T.
Canna O.	καννα Arist.
Cannabis P.	κανναβις D.
Capparis P.	καππαρις T.D.
Cardamine	καρδαμον D.
Carpesium	καρπησιον Gal.
Carum	καρος[καρον] D.
Caryophyllus P.	καρυοφυλλον Æ
Caryota P.	καρυωτας D.
Cassia	κασσια D.

continued

Catananche P.	καταναγκη D.
Caucalis P.	καυκαλις H.T.
Ceanothus	κεανωθος T.
Cenchrus	κεγχροςT.D.
Centaurea P.	κενταυριον H.D.
Cerasus	κερασια T.D.
Ceratonia	κερατωνια T.
Cercis	κερκις T.
Cerinthe	κηρινθος T.
Cestrus	κεστρον D.
Chaerophyllum	χαιρεφυλλον O.
Chelidonium P.	χελιδονιον D.
Chondrilla P.	χονδριλλη D.
Chrysanthemum	χρυσανθεμον D.
Chrysocoma P.	χρυσοκομη D.
Chrysogonum	χρυσογονον D.
Cichorium P.	κιχωριον T.
Circaea P.	κιρκαια D.
Cissampelos	κισσαμπελος D.
Cissus P.	κισσος D.
Cistus (Cisthus) P.	κιστος D.
Citrus	κιτρος D.
Clematis P.	κληματις D.
Clethra	κληθρα T.
Clinopodium P.	κλινοποδιον D.
Cneorum P.	κνεωρον D.
Cnicus P.	κνικος [κνηκος] H.T.D.
Coix	κοιξ T.
Colchicum	κολχικον D.
Colutea	κολουτεα T.
Comarum	κομαρος T.
Conium	κωνειον T.D.
Conyza P.	κονυζα H.D.
Corchorus P.	κορχορος T.
Coriandrum P.	κοριαννον T.D.
Coris P.	κορις D.
Corymbia P.	κορυμβιον
Costus P.	κοστος T.D.
Cotinus P.	κοτινος T.
Cotyledon P.	κοτυληδων D.
Crambe P.	κραμβη D.
Crataegus	κραταιγος D.
Crinum P.	κρινον D.
Crithmum	κριθμον D.
Crocus P.	κροκος H.T.D.

continued

Croton	κροτων Nic. D.
Cupressus P.	κυπαρισσος D.
Cyclamen P.	κυκλαμινος T.D.
Cyminum P.	κυμινον T.D.
Cynanchum	κυναγχη
Cynoglossum P.	κυνογλωσσον D.
Cyperus P.	κυπειρος H.T.D.
Cytisus P.	κυτισος H.T.D.
Daphne	δαφνη T.D.
Daucus P.	δαυκος D.
Delphinium	δελφινιον D.
Dictamnus P.	δικταμνος T.D.
Dipsacus P.	διψακος D.
Dolichus P.	δολιχος T.
Dorycnium P.	δορυκνιον D.
Draba	δραβη D.
Dracontium P.	δρακοντιον T.D.
Drosera	δροσιον [sic]
Drypis	δρυπις T.
Echium P.	εχιον D.
Elæagnus	ελαιαγνος T.
Elatine P.	ελατινη D.
Elymus	ελυμος D.
Empetrum P.	εμπετρον D.
Epimedium P.	επιμηδιον D.
Eranthemum	ηρανθεμον D.
Erica P.	ερεικη T.
Erigeron P.	ηριγερων T.D.
Erinus	ερινος D.
Eriophorum P.	ηριφορον T.
Eryngium P.	ηρυγγιον D.
Erysimum P.	ερυσιμον T.D.
Erythronium	ερυθρονιον D.
Eupatorium P.	ευπατοριον D.
Euphorbia P.	ευφορβιον D.
Euonymus	ευωνυμος T.
Exacum	εξακον D.
Galeopsis P.	γαλεοψις D.
Galium	γαλιον D.
Gentiana P.	γεντιανη D.
Geranium P.	γερανιον D.
Glaux P.	γλαυξ D.

continued

Glechoma	γληχων D.
Glycyrrhiza P.	γλυκυρριζα D.
Gnaphalium P.	γναφαλιον D.
Hamamelis	αμαμηλις Ath.
Hedysarum	ηδυσαρον T.D.
Helenium P.	ελενιον D.
Heliotropium P.	ηλιοτροπιον D.
Helleborus P.	ελλεβορος D.
Helxine P.	ελξινη D.
Hemerocallis P.	ημεροκαλλις D.
Hemionitis	ημιονιτις D.
Hibiscus	ιβισκος D.
Hieracium	ιερακιον D.
Hippomane P.	ιππομανες
Hippophaë P.	ιπποφαες
Hippuris P.	ιππουρις D.
Holcus P.	ολκος
Holosteum P.	ολοστιον [sic: εον] D.
Horminum	ορμινον H.D.
Hyacinthus P.	ιακινθος T.D.
Hydnum	υδνα [υδνον] D.
Hyoscyamus	υοσκυαμος
Hypecoum P.	υπηκοον D.
Hypericum P.	υπερικον D.
Hypnum	υπνον D.
Hypochaeris P.	υποχοιρις T.
Hyssopus P.	υσσωπος D.
Jasione P.	ιασιωνη T.
Jasminum	ιασμινον D.
Iberis P.	ιβηρις D.
Iris P.	ιρις T.D.
Isatis P.	ισατις D.
Ischaemum	ισχαιμον
Isopyrum	ισοπυρον D.
Itea	ιτεα T.D.
Ixia P.	ιξιας D.
Lampsana P.	λαψανη D.
Lathyrus	λαθυρος T.
Lemna	λεμνα T.
Lepidium P.	λεπιδιον T.
Leucojum	λευκοιον T.D.

continued

Lichen P.	λειχην D.
Ligusticum P.	λιγυστικον D.
Linum P.	λινον D.
Lithospermum P.	λιθοσπερμον D.
Lonchitis P.	λογχιτις D.
Lotus P.	λωτος T.D.
Lychnis P.	λυχνις T.D.
Lycium	λυκιον D.
Lycopsis P.	λυκοψις D.
Lysimachia P.	λυσιμαχιον D.
Lythrum	λυθρον D.
Malope P.	μαλοπη
Mandragora P.	μανδραγορας T.D.
Melampyrum P.	μελαμπυρον T.
Melanthium	μελανθιον T.D.
Melia	μελια D.
Melothrion P.	μηλωθρον T.
Memecylon P.	μεμηκιλον D.
Menta P.	μινθη [μιντη]T.
Mespilus P.	μεσπιλον D.
Mnium	μνιον
Morus P.	μορεα D.
Myagrum P.	μυαγρος D.
Myosotis P.	μυος ωτιον D.
Myrica P.	μυρικη T.D.
Myriophyllum P.	μυριοφυλλον D.
Myrsinum P.	μυρσινη D.
Myrtus P.	μυρτος Arist.
Nama P.	ναμα T.
Narcissus P.	ναρκισσος T.D.
Nardus P.	ναρδος T.D.
Nepenthes P.	νηπενθες H.
Nerium P.	νηριον D.
Nymphaea P.	νυμφαια
Ochna P.	οχνας Ath.
Ocimum P.	ωκιμον H.T.D.
Oenanthe P.	οινανθη T.D.
Oenothera P.	οινοθηρη [sic: -ρας] T.
Ononis P.	ονωνις D.
Ophioglossum P.	οφιογλωσσον R.
Orchis P.	ορχις T.D.
Origanum P.	οριγανον T.D.

continued

Ornithogalum P.	ορνιθογαλον D.
Orobanche P.	οροβαγχη T.D.
Orobus	οροβος T.D.
Oryza P.	ορυζα T.D.
Osyris P.	οσυρις D.
Othonna P.	οθοννα D.
Oxalis P.	οξαλις D.
Paeonia P.	παιονια [sic: ? -ω-] H.D.
Panaces P.	παναξ T.D.
Pancratium P.	παγκρατιον D.
Parthenium P.	παρθενιον D.
Peganum	πηγανον T.D.
Peplis	πεπλις D.
Peucedanum P.	πευκεδανον D.
Phaca	φακος D.
Phalaris P.	φαλαρις D.
Phaseolus	φασιολος T.D.
Philadelphus	φιλαδελφος A.
Phillyrea	φιλλυρεα D.
Philyca	φιλυκα [–κη] T.
Phleum	φλεον [? εως] T.
Phlomus	φλομος D.
Phlox P.	φλοξ T.
Phœnix	φοινιξ T.D.
Phyteuma P.	φυτευμα D.
Picris	πικ ρις D.
Pimpinella	πιμπινελε M.
Piper	πεπερι T.D.
Pistacia	πιστακια T.D.
Pisum	πισον T.
Platanus P.	πλατανος T.D.
Poa	ποα T.
Polemonium	πολεμονιον D.
Polycnemum P.	πολυκνημον D.
Polygala P.	πολυγαλον D.
Polygonum P.	πολυγονον D.
Polypodium	πολυποδιον T.D.
Polytrichum P.	πολυτριχον
Potamogeton P.	ποταμογειτων D.
Poterium P.	ποτηριον D.
Pothos P.	ποθος T.
Prasium	πρασιον D.
Psidium	σιδιας H. [sic: σιδιον]
Ptelea	πτελεα T.D.

continued

Pteris P.	πτερις T.D.
Rhamnus P.	ραμνος T.D.
Rhaphanus P.	ραφανις T.D.
	ραφανος T.
Rheum	ρηον D.
Rhododendron	ροδοδενδρον D.
Rhus P.	ρους D.
Samyda	σημυδα T.
Satyrium P.	σατυριον D.
Scandix P.	σκανδιξ T.D.
Schinus	σχινος Athen.
Schœnus	σχοινος D.
Scilla P.	σκιλλα T.D.
Scolymus P.	σκολυμος D.
Selinum	σελινον T.D.
Sesamus P.	σησαμος D.
Seseli P.	σεσελι T.D.
Sicyos	σικυος T.
Sida	σιδη T.
Sideritis P.	σιδηριτις D.
Silphium P.	σιλφον D.
Sinapi P.	σινηπι T.D.
Sison	σισων D.
Sisymbrium P.	σισυμβριον T.
Sisyrinchium P.	σισυριγχιον T.
Sium P.	σιον D.
Smilax P.	σμιλαξ D.
Smyrnium	σμυρνιον D.
Sonchus P.	σογχος T.D.
Spartium	σπαρτιον T.D.
Splachnum	σπλαχνον D.
Sparganium	σπαργανιον D.
Sphagnum P.	σφαγνον [sic: ? σφαγνος] H.D.
Spongia P.	σπογγος D.
Stachys P.	σταχυς D.
Statice P.	στατικη
Stoebe P.	στοιβη D.
Stratiotes	στρατιωτης D.
Struthion P.	στρουθιον D
Strychnon P.	στρυχνος D.
Styrax	στυραξ T.D.
Symphytum P.	συμφυτον D.

continued

Taxus P.	ταξος Aet.
Telephium P.	τηλεφον D.
Tetragonia	τετραγωνια T.
Teucrium P.	τευκριον D.
Thalictrum P.	θαλικτρον D.
Thapsia P.	θαψια D.
Thlaspi P.	θλασπι D.
Tridax	θριδαξ D.
Thya P.	θυα T.
Thymus	θυμος T.D.
Tordylium P.	τορδυλιον D.
Tragacantha P.	τραγακανθα T.D.
Tragopogon P.	τραγοπωγων D.
Tribulus	τριβολος D.
Trichomanes	τριχομανες D.
Trifolium	τριφυλλον H.D.
Typhe P.	τυφη T.D.
Vella Gal.[23]	
Xanthium	ξανθιον D.
Xylon P.	ξυλον Poll.
Xyris P.	ξυρις D.
Zea P.	ζεια D.
Zizania	ζιζανιον S.S.

LATIN NAMES that were received by the Romans.

Abies P.	*Avena* P.
Abiga P.	*Bellis* P.
Acer P.	*Berberis*
Actaea P.	*Beta* P.
Aeschynomene P.	*Betonica* P.
Alnus P.	*Betula* P.
Allium P.	*Boletus* P.
Alga P.	*Brassica* P.
Amarantus P.	*Buxus* P.
Anabasis P.	*Caepa* P.
Anacampseros P.	*Calla* P.
Apium P.	*Callitricha* P.
Arbutus P.	*Caltha* P.
Arundo P.	*Carduus* P.
Asperugo P.	*Carex* V.
Atriplex P.	*Carica* P.
Atropha P.	*Carpinus*

continued

Celtis P.

Centunculus P.

Cerasus P.

Chamaerops P.

Chenopus P.

Chironia P.

Cicer P.

Cicuta P.

Cinara P.

Cleome Hor.

Coccus P.

Conferva P.

Convolvulus P.

Cornus P.

Corylus P.

Crepis P.

Cucubalum P.

Cucumis P.

Cucurbita P.

Cunila P.

Cupressus P.

Cycas P.

Dactylos P.

Diospyros P.

Dodecatheos P.

Ephedra P.

Equisetum P.

Erigeron P.

Ervum P.

Aesculus P.

Ervum P.[24]

Exacum P.

Faba P.

Fagus P.

Ferula P.

Ficus P.

Filix P.

Foeniculum P.

Fragaria P.

Fraxinus P.

Fucus P.

Fungus P.

Genista P.

Geum P.

Gladiolus P.

Gnidium P.

Gossypium P.

Gramen P.

Hedera P.

Helianthe P.

Heraclion P.

Hesperis P.

Hordeum P.

Hyosiris P.

Ilex P.V.

Illecebra P.

Inula P.

Isoëtes P.

Juncus P.

Juglans P.

Juniperus P.

Lactuca P.

Lamium P.

Laserpitium P.

Laurus P.

Leontice P.

Ligustrum P.

Lilium P.

Lolium P.V.

Lupinus P.

Malva P.

Marrubium P.

Melitis P.

Mercurialis P.

Milium P.

Mimmulus P.

Minyanthes P.

Mollugo P.

Mucor Col.

Nepeta P.

Olea P.

Onopordum P.

Ophrys P.

Palma P.

Panicum P.

Papaver P.

Pentapetes P.

Pezica P.

Pharnaceum P.

Phellandrium.[25]

continued

Phyllanthes P.
Pinus P.
Pirus P.
Populus P.
Porrum P.
Portulaca P.
Proserpinaca P.
Prunus P.
Punica P.
Quercus P.
Ranunculus P.
Reseda P.
Rhexia P.
Ricinus P.
Rosa P.
Rosmarinus.
Rubia P.
Rubus P.
Rumex P.
Ruscus V.
Ruta P.
Saccharum P.
Salix P.
Salvia P.
Sambucus P.
Samolus P.
Sanguinaria P.
Satureja P.
Saxifragum P.
Scirpus P.
Scorpiurus P.
Secale P.
Sedum P.

Selago P.
Sempervivum P.
Senecio P.
Serratula P.
Spireon P.
Solanum P.
Sorbus P.
Syringia [sic: *Syringias*] P.
Tamarix P.
Tamus P.
Thelygonum P.
Thesium P.
Tilia P.
Tinus P.
Triticum P.
Vaccinium P.V.
Valeriana.
Veratrum P.
Verbascum T.
Verbena T.
Viburnum V.
Vicia P.
Vinca (pervinca) P.
Viola P.
Viscus P.
Vitex.
Vitis.
Ulex P.
Ulmus V.
Ulva V.O.
Urtica P.
Zoster P.

242. An ancient generic name (241) is appropriate for an ancient genus.

Translation of the Latin *language* into Greek.

Dens leonis	*Leontodon*	[lion's tooth]
Ferrum equinum	*Hippocrepis*	[horse's iron (shoe)]
Nidus leporis	*Lagoecia*	[hare's form]

Slight changes

Acacia robini	*Robinia*	
Gramen Parnassi	*Parnassia*	[Parnassus grass]
Lilium convallium	*Convallaria*	[lily of the valleys]
Jan-Raja	*Rajania*	

Abbreviations

1.	Achyracantha	*Achyranthes*
2.	Calophyllodendron	*Calophyllum*
	Staphylodendrum	*Staphylea*
	Tetragonocarpus	*Tetragonia*
	Leontopetalon	*Leontice*
	Heleniastrum	*Helenia*
	Partheniastrum	*Parthenium*
	Arachidna	*Arachis*
	Sicyoides	*Sicyos*
3.	Oreoselinum	*Selinum*
	Melocactus	*Cactus*
	Anapodophyllum	*Podophyllum*
	Hydroceratophyllum	*Ceratophyllum*[26]
4.	Ananthocyclus	*Anacyclus*

The genus *Aster* formerly comprised Enula campana and those connected with it; but since its essential character has been discovered, this very common plant cannot assume another name; accordingly the very ancient name of *Inula* has been accepted.

243. A generic name that is satisfactory (213–242) may not be changed for another, even if the latter is more appropriate.

Asclepias catches flies with its flowers; so the name *Myagrum* would suit this genus very aptly.

On account of its woolly flower, *Menyanthes* would be more appropriately called by the essential name *Erianthus* or *Lasianthus*.

But we must abstain from such innovations, which would never end, since more appropriate names would be discovered every day for ever and ever.

244. New generic names should not be contrived, so long as adequate synonyms are readily available.

But when new genera are discovered, new names are very appropriately composed and applied to them.

If an old genus is divided into several, it is advisable not to contrive new generic names, so long as there are adequate names in the ranks of the synonyms for the several species.

245. Unless it is superfluous (215–17), the generic name of one genus ought not to be transferred to another, even if it suits the latter more aptly.

Who nowadays would change names that have been accepted for a long time, for
those used by the fathers.

The Hyacinthus of the ancients is *Delphinium*.

The Tribulus of the ancients is *Fagonia*.

The Opulus of the ancients is *Humulus*.

The botanists of the sixteenth century nearly destroyed botany by seeking out the
names used by the ancients.

We must take careful note of the genera A, B, C, and D, according to our *Genera
plantarum*.

    A. *Lithospermum*

    B. *Myosotis*

    C. *Alsine*

    D. *Cerastium*

Tournefort included genus B under A, and accordingly transferred the name [of]
B to D.

Ruppius separated genera A and B, and recalled the name Myosotis T. from D to
B; and he subordinated genus D to C.

Dillenius separated genus C from D, with Tournefort; and he did not unite D
with C, as Ruppius did; moreover, he kept genera A and B separate, with
Ruppius, and did not combine A and B, as Tournefort did; and so he allowed the
name Myosotis for plant B, and applied a new one (Cerastium) to D.

The beginner, who at first attached the name Myosotis to genus D, according to
Tournefort's opinion, and is now obliged to unite it to B, according to Ruppius',
is always in doubt about whether he should attach it to B or to D.

246. If a genus that has been accepted ought to be divided into several,
according to the law of nature (162) and art (164), then the name that
was previously shared should be left for the plant that is most
widespread and common.

Let us suppose that the genus CORNUS is to be divided into three:

    A. A tree with flowers that have involucres and umbels.

    B. A herb with flowers that have involucres and umbels.

    C. A tree with flowers that do not have involucres, but do have cymes.

So A should be called *Cornus*, B *Mesomora*, and C *Ossea*; and it is not permissible for
A to be called Mesomora or Ossea.

247. Generic names (229) should be written in Latin letters.

αι	becomes	æ	ανδροσαιμον	*Androsaemum*
ει	"	e	ποταμογειτων	*Potamogeton*
	"	i	αειζωον	*Aizoon*

continued

η	"	a	οθοννη [sic][27]	*Othonna*
	"	e	νηπενθης	*Nepenthes*
o	"	o	οροβος	*Orobus*
	"	u	υπηκοον	*Hypecoum* [in the] final [syllable]
ου	"	u	αγχουσα	*Anchusa*
οι	"	oe	φοινιξ	*Phœnix*
ω	"	o	σισων	*Sison*
θ	"	th	βουφθαλμον	*Buphthalmum*
φ	"	ph	φιλαδελφος	*Philadelphus*
χ	"	ch	χαιρεφυλλον	*Chaerophyllum*
κ	"	c	ωκιμον	*Ocimum*
γχ	"	nch	αγχουσα	*Anchusa*
γγ	"	ng	ηρυγγιον	*Eryngium*
	"	h	αντιρρινον	*Antirrhinum*

248. The endings and the pronunciation of generic names should be made easy, as far as possible.

Unusual endings			Ambiguous [names]	
in	e	*Ballote*	*Alpina* Pl.	Alpinia
"	i	*Sinapi*	*Phyllum*	Phyllis
		Seseli	*Meum* T.	Athamanta
		Thlaspi		
"	ois	*Hedypnois*	Inverted [names]	
"	t	*Tetrahit*	*Anthoceros* M.	Ceranthus
"	n	*Triglochin*	*Caraxeron* V.	

249. Generic names $1\frac{1}{2}$ feet long, those that are difficult to pronounce, or are disgusting, should be avoided.

.... *Words 1 ½ feet long*
Are actually painful to pronounce, and liable to damage the throat of the speaker.[28]
I regard as *1 ½ feet long* words that contain more than 12 letters.

17.	*Kalophyllodendron* V.	Calophyllum
18.	*Titanoceratophyton* B.	Isis
19.	*Leuconarcissolirion*	Galanthus
21.	*Coriotragematodendros* Plk.	Myrica
22.	*Hypophyllocarpodendron* B.	Protea

I regard as *difficult* those that are composed with several consecutive consonants.

Acrochordodendros Pl.	Cephalanthus
Stachyarpogophora V.	Achyranthes
Orbitochortus Kn.	Fagonia

And I regard as *disgusting* those that suggest something or other unusual.

Caraxeron V.	Gomphrena
Galeobdolon D.	Leonurus
Myrobatindum V.	Morinda

250. It is unwise to misuse technical TERMS (199) instead of generic names.

Latin		Greek
Tuberosa H.	Polianthes	*Polyanthes* Pt.
Graminifolia R.	Subularia	*Phyllon*
Spica Hk.	Lavandula	*Hexapetala* Pk.
Siliqua T.	Ceratonia	
Nux T.	Juglans	
Odorata R.	Scandix	

251. The same argument that concerns generic names applies equally to those of CLASSES (160) and ORDERS (161); (204).

> The names of classes are subject to the laws given for genera.
>> Particular [names]: Sections 213, 214, and 217.
>> [They must consist of]
>> a *single* word: Sections 215 and 221.
>> the *same* word: Section 216.
>> not a *primitive* word: Section 220.
>> not a *hybrid* word: Sections 223–227.
>> not a *barbarous* word: Section 229.
>> not an *ambiguous* word: Sections 230–1.
>> not, an *inconsistent* word: Section 232.
>> not *from [names of] men:* Section 236.
>> not $1^1/_2$ *feet long*: Section 249.
>> not *difficult*: Section 248.
> *Cesalpino* made use of definitions as the names of classes.
> *Tournefort, Rivinus*, and their predecessors quite often allowed several words.

252. Names of classes and orders derived from the potencies, root, herbage, or habit of the plant are wrong.

> Systematic names ought to be essential ones, taken from the fruit-body.
> *Fragments of the natural method* have taken names from accidental circumstances as surrogates; whereas, in an absolute system, it is necessary that they should be changed, as required by the laws of classification in the system that is to be adopted.

I have always applied erroneous names to plants of doubtful genera—names in –*oides* or else diminutives—when the plant needed to be numbered in the catalogue, and so far nothing had been established about the fruit-body, so that it was impossible to give it its true name yet; thus readers, warned by the single erroneous word, would ask me about the uncertain genus, and give more attention to the fruit-body.

From the potencies	The root	The leaves	The habit
Cordialis [affecting the heart]	*Bulbous*	*Rough-leaved*	*Whorled*
Capillaris [affecting the hair]	*Tuberous*	*Succulent*	*Star-shaped*
	Fibrous	*Ridge-backed*	
	Trees		
	Shrubs		

253. The names of classes and orders should contain a feature that is essential (187) and *characteristic* (189).

Reasons: because methods are numerous, variable, and modern.
because names are the unavoidable burdens of technique;
because names must be ready and prompt.

Right:	Good:	Wrong:
Calyciflorae [with flowers formed by the calyx]	*Siliquosae [with siliquae]*	*Discoidae [disk-shaped]*
Papilionaceae [butterfly-shaped]	*Leguminosae [with pods]*	*Corymbiferae [bearing clusters]*
Cruciformes [cross-shaped]		
Difformes [irregular]	*Multisitiquae [with many siliquae]*	
Syngenistae [with united anthers]		

254. Names of classes and orders that are taken from a particular plant, and which are inadmissible for a genus, may be included only for natural classes.

Palm	[natural order]		2.
Fern	"	"	64.
Moss	"	"	65.
Alga	"	"	66.
Fungus	"	"	67.
Grass	"	"	14.
Lily	"	"	10. Lirium *Roy.*
Poppy	"	"	30.
Corydalis	"	"	28.
Reed	"	"	13.
Orchid	"	"	4.
Gourd-like	"	"	45.
Like the gillyflower	"	"	42.
Like the white violet	"	"	45 Leucojiformes Kr.
Pepper-like	"	"	1.

They are to be included as names for classes and orders, inasmuch as they are the names of plants of the same kind, and never otherwise; or else, as new methods never cease, so also botanists would never stop excluding names of plants as generic names, and including them as names for artificial classes; however, this should cause less anxiety in the case of the natural method, which is unique and will always be so.

255. The names of classes and orders must consist of a *single* (215) word.

> Radiati T. [rayed]: *Compounds consisting of irregular [florets] on the periphery, and of regular ones in the middle.* Riv.
> *With a perfect simple flower, and seeds uncovered and solitary, or single seeds in single flowers.* Raj.
> Those consisting of a single word are fine.

Campaniformes	[bell-shaped].
Infundibuliformes	[funnel-shaped].
Personati	[mask-like].
Labiati	[lip-like].
Cruciformes	[cross-shaped].
Rosacei etc.	[rose-like etc.].
Monopetali	[with a single petal].
Dipetali	[with two petals].
Tripetali	[with three petals].
Tetrapetali	[with four petals].
Pentapetali &c.	[with five petals etc.].
Monandria	[with a single stamen].
Diandria	[with two stamens].
Triandria	[with three stamens].
Tetrandria	[with four stamens].
Pentandria etc.	[with five stamens etc.].

> Care should be taken to prevent these words from turning out to be too long or compounded[29] of too many parts.

VIII. DIFFERENTIÆ.

256. Perfecte nominata est planta nomine *generico* &
specifico (212) instructa.

Botanices Tyro novit Classes, *Candidatus* omnia Genera, *Magi-
ster* plurimas species.

Quo plures Botanicus noverit species, eo etiam præstantior est.

Cognitione specierum innititur omnis solida eruditio Physica,
Oeconomica, Medica; immo omnis vera cognitio humana.

Speciei notitia consistit in nota essentiali, qua sola ab omnibus
congeneribus distinguitur.

Sine notitia Generis nulla certitudo speciei.

   Cæsalpinus: Ignorato genere nulla descriptio, quamvis ad-
      curate tradita, certam demonstrat, sed plerumque fallit.

Differentia specifica continet notas, quibus species a congeneribus differt.

Nomen specificum autem continet Differentiæ notas essentiales.

Notæ in nomine specifico sint

     non *lubricæ incertæ* aut *falsæ* §. 259 - 274. 281. 283.
     sed *firmæ, certæ, mechanicæ* §. 275 - 280. 257. 287.
     quæ *caute, caste, judiciose* §. 284 - 305.

257. Nomen specificum *legitimum* plantam ab *omnibus* congeneribus (159) distinguat; *Triviale* autem nomen legibus etiamnum caret.

Fundamentum est hic canon nominum specificorum, quo negle-
cto, lubrica erunt omnia.

Nomina specifica omnia, quæ plantam a congeneribus non distinguunt, falsa sunt.

Nomina specifica omnia, quæ plantam ab aliis, quam congeneribus distinguunt, falsa sunt.

Nomen specificum est itaque Differentia essentialis.

NOMINA TRIVIALIA forte admitti possunt modo, quo in
Pane suecico usus sum; constarent. hæc

     Vocabulo unico;

     Vocabulo libere undequaque desumto.

Ratione hac præcipue evicti, quod differentia sæpe longa
evadit, ut non ubique commode usurpetur, & dein mu-
tationi obnoxia, novis detectis speciebus, est. e. gr.

                        PYRO-

❧ VIII. DEFINITIONS[1]

256. A plant is completely named, if it is provided with a generic name and a specific (212) one.

> A *beginner* in *botany* knows the classes, a *candidate* all the genera, and a *master* most of the species.
>
> The more species a botanist knows, the more outstanding he is.
>
> All substantial education in science, trade, and medicine, indeed all true human knowledge, relies on a knowledge of the several species.
>
> The concept of a species consists of an essential feature, by which alone it is distinguished from all others in the same genus.
>
> Without the concept of a genus, there is no certainty of the species.
>
> *Cesalpino*: 'Without knowledge of the genus, no description, however accurately reported, can indicate a definite species, but is usually deceptive.'
>
> A specific definition contains features in which the species differs from those in the same genus.
>
> But the specific name contains the essential features of the definition.
>
> The features [contained] in a specific name should not be
>
> *elusive, uncertain*, or *deceptive*, Sections 259–74, 281, and 283;
>
> but *firm, certain* and *mechanical*, Sections 275–80, 257, and 287;
>
> and they [should be used] with care, integrity, and judgement; Sections 284–305.

257. The *legitimate* name for a species should distinguish the plant from *all* those of the same genus (159); but a *trivial* name is still free from any laws.

> This rule is the foundation of all specific names, and if it is disregarded, they are all elusive.
>
> All specific names that do not distinguish a plant from those of the same genus are deceptive.
>
> All specific names that distinguish a plant from plants other than those of the same genus are deceptive.

Therefore the specific name is the essential definition.

TRIVIAL NAMES can perhaps be allowed in the manner in which I have used them in *Pan Suecicus*; these would consist of

a single word;

a word freely taken from any source.

We are convinced by this argument, that the definition often turns out to be long-winded, so that it cannot conveniently be used in all cases, and is liable to be changed later on, as new species are discovered; for example:

PYROLA *irregularis*	PYROLA with stamens going up, and pistil bent down. [My] *Flora Suecica 330*.
PYROLA *Halleriana*	PYROLA with scattered flowers in racemes, stamens and pistils upright. *Flora Suecica 331*.
PYROLA *secunda*	PYROLA with one-sided racemes, *Flora Suecica 332*.
PYROLA *umbellata*	PYROLA with flowers in umbels. *Flora Suecica 333*.
PYROLA *uniflora*	PYROLA with a scape bearing a single flower. *Flora Suecica 334*.

But in this work we lay trivial names aside, since we are concerned only with definitions.

258. The specific name must declare its own [particular] plant at first sight, since it contains the definition (257) that is *inscribed on the plant itself.*

The names used by our predecessors were trivial, and those used by the most ancient botanists were the most trivial.

A description is the *natural character* of a species, and a definition is the *essential character* of a species.

I was the first to begin to establish specific names; before me, no satisfactory definitions existed.

This method has been acknowledged by the most intelligent of the more recent botanists, *Royen, Gronovius, Guettard* and *Dalibard* in all cases; and by *Haller, Gmelin,* and *Burman* in most.

My specific names have extricated the definitions from the descriptions; from the definitions, they have traced the most obvious essential character by which they are established.

We should exclude from the specific name all accidental features that do not exist in the actual plant or are not palpable: for example, *place, time, duration,* and *use.*

Erroneous specific names are all derived from the ranks of ideas or from supposition.

The ranks of ideas	*Supposition*
Tinus *prior* [the first].	Hyoscyamus *peculiaris* [special].
Tinus *alter* [the second].	Meum *spurium* [false].
Tinus *tertius* [the third].	Acorus *verus* [true].
	Campanula *pulchra* [beautiful].

259. The specific name ought to be derived from parts of the plant *that do not vary.*

> In the animal kingdom, no sensible person would readily say that *varieties* are distinct species.
> *White, black, red, grey, and variegated cows; small and large, thin and fat, smooth and hairy cows;* no one has said that there are so many distinct species.
> Excrescences, crowns of the head, and sutures of the skull have demonstrated that dogs, whether Melitean,[2] spaniels, mastiffs, Greek, poodles, etc., are all of the same species.[3]

> *Species* derived from varieties were multiplied by our ancestors because of:
> Fear of confusing different species;
> Lack of essential definitions;
> Ignorance of the continued generation of species, Sections 79 and 132.
> Contagious madness among lovers of flowers.
> Zeal for subtleties.

> *Colour, smell, taste, hairiness, curliness, fullness, and deformity* are mostly variable and rarely constant.

> The advocates of species, who have added varieties to the number of the species, have been principally the most recent botanists before our time: *Barrelier, Tournefort, Boerhaave, Pontedera,* and *Micheli.*

> The introduction of varieties has done more to contaminate botany than any other thing. It confused those who used synonyms to such an extent that the science of botany would have been done for, but for a quick remedy.

> *Definitions* that pass off varieties as species are *erroneous.*
> TRIFOLIUM with the heads almost round, the florets pedunculate, the pods containing four seeds each, and the stem bending down: [my] Hortus Cliffortianus, 375.

> Trifolium pratense album [white]: *Bauhin, pinax, 327.*
> 1. Trifoliastrum pratense bearing corymbs, greater creeping; with the leaves quite rounded and marked with an arrow-shaped white spot; and siliquae with four seeds each[4]: Micheli gen. 26.
> 2. Trifoliastrum *pratense bearing corymbs, greater creeping; with the leaves quite rounded, and marked with an arrow-shaped white spot; the corymbs of flowers placed on very long pedicels; and siliquae with four seeds [each]:* Michieli gen. 26. pl. 25 f. 1.

3. Trifoliastrum *pratense bearing corymbs, greater creeping; with the leaves more rounded, and marked with an arrow-shaped white spot extended further into an acute angle; and siliquae with four seeds each*: Micheli gen. 26. pl. 25 f.4.

4. Trifoliastrum *pratense, bearing corymbs, greater creeping; with quite blunt oblong leaves, with a whitish spot that is pyramid-shaped on the upper side, and on the lower side beautifully hollowed out in the shape of a heart; and siliquae with four seeds each*: Micheli gen. 26.

5. Trifoliastrum *pratense, bearing corymbs, greater creeping; with an almost round leaf marked with a half-moon-shaped white spot, which is a little hollowed out at the back; and a siliqua with four seeds*: Micheli gen. 26.

6. Trifoliastrum *pratense, bearing corymbs, greater creeping; with a heart-shaped leaf ensigned with a white spot of the same shape; and a siliqua with four seeds*: Micheli gen. 26.

7. Trifoliastrum *pratense bearing corymbs, greater creeping; with a blunt leaf ensigned with two white spots, whereof the upper, which is the smaller, is triangular, and the lower, which is the larger, is heart-shaped; and a siliqua with four seeds*: Micheli gen. 27.

8. Trifoliastrum *pratense bearing corymbs, greater creeping; with blunt leaves which are as it were heart-shaped, and not spotted; and siliquae with four seeds [each]*: Micheli gen. 27.

9. Trifoliastrum *pratense bearing corymbs, medium creeping, with oblong leaves which are quite pointed, with a broad arrow-shaped spot; and siliquae with four seeds each*: Micheli gen. 27.

10. Trifoliastrum *pratense bearing corymbs, medium creeping, with a round leaf, a very narrow arrow-shaped spot; and siliquae with four seeds [each]*: Micheli gen. 27.

11. Trifoliastrum *pratense bearing corymbs, lesser creeping: with a leaf that is almost round, with a very small arrow-shaped spot*: Micheli gen. 27.

12. Trifoliastrum *pratense bearing corymbs, least creeping; with blunt leaves which are not spotted; and siliquae with four seeds [each], level on the upper side and as it were knotty on the lower side; and yellowish seeds*: Micheli gen. 27. pl. 25 f.6.

13. Trifoliastrum *annuum, bearing corymbs, white and bending down; with heart-shaped leaf shining vivid black on the under side; and a siliqua with four seeds, with a sickle-shaped division in the lower part*: Micheli gen. 27 pl. 25 f. 6.

14. Trifoliastrum *pratense bearing corymbs, not creeping, but prostrate on the ground, with deep roots; with leaves that are almost round, lightly marked with a sickle-shaped white spot; flowers quite small and delicately ruddy; and siliquae with four seeds each, with only the upper side bordered, and brown seeds*: Micheli gen. 27.

15. Trifoliastrum *pratense bearing corymbs, upright, annual and very tall; with a stem quite thick and full of pipes; a leaf that is quite long, and heart-shaped; a white flower; and a siliqua that is curved, broad and compressed, with two seeds*: Micheli gen. 28 pl. 25 f.2.

16. Trifoliastrum *supinum bearing corymbs, greater white annual, with a leaf that is quite long and blunt; and a siliqua that is curved; broad, and compressed, with two seeds*: Micheli gen. 28. pl. 25 f. 5.

Botanists should learn from this horrendous example, that a very small circumstance should not constitute a material variation, against the law of Nature; for the Creator has entrusted the generation of species to Nature, and not to men. Indeed, the very observant Micheli offended against Section 262 with his leaves *quite rounded, long, pointed, or blunt*; with his *quite thick* stem; and with his *very long* peduncles: against Section 260 with his *very tall* stem: against Section 266 with his *spot on the leaves*, that is different in colour and shape; his *delicately ruddy* flowers; and his *brown and yellowish seeds*.

260. *Size* does not separate species.

The size varies with the *location, soil* and *climate*; it varies with the quantity of the nourishment, in plants no less than in animals.

The size, if it is variable but does not alter the species, cannot alter the essential definition so as to supply a specific name.

All *specific names* that are derived from the size of *the plant, the root, the herbage or the fruit-body* are *erroneous*.

Plukenet's *largest* Polytrichum is smaller than his *smallest* Thalictrum.

Plantago major [greater] is called by some *media* [medium]. J.B.

From the size of the *plant*.
Alsine *altissima* [tallest]
Alsine *major* [greater]
Alsine *media* [medium]
Alsine *minor* [lesser]
Alsine *minima* [least]
Alsine *exigua* [minute]
Sedum *majus* [greater]
Sedum *minus* [lesser]
Sedum *parvum* [little]
Sedum *minimum* [least]
Boletus *magnus* [great]
Galeopsis *procerior* [larger]
Gramen *elatius* [taller]
Fraxinus *excelsior* [taller]
Trachelium *giganteum* [gigantic]
Jasminum *humile* [lowly]
Virga aurea *humilior* [lowlier]

Salix *pumila* [pygmy]
Betula *nana* [dwarf]
Melampyrum *perpusillum* [tiny]

From the size of the *leaves*.
Nicotiana *angustifolia* [narrow-leaved]
Nicotiana *latifolia* [broad-leaved]

From the size of the *fruit-body*.
Magnolia flore *ingenti* [with a *huge* flower]
Aster flore *ingenti* [with a *huge* flower.]

261. Features that draw *comparisons* with other species of a *different genus* are deceptive.

Our predecessors presupposed an empirical knowledge—in the beginner—of most European plants, [as if that knowledge was gained] from innate ideas, and accordingly they wrote for those who were learned in the art; but our business is only to teach the unlearned.

A plant should be recognized from its name, and the name from the plant, according to technical rules: and both should be recognized by the peculiar character, which is written in the former case and drawn in the latter; a third case should not be allowed.

Names that presupposed other plants have eventually made people *giddy*.

We denounce, as *erroneous, specific names* that emphasise a similarity to another plant, in *herbage, fruit-body*, or *habit*, in Greek or Latin.

From the leaves—
Jacobaea
with a *Betonica* leaf.
with a *Glastum* leaf.
with a *Chrysanthemum* leaf.
with a *Rosmarinus* leaf.
with an *Absinthium* leaf.
with a *Hieracium* leaf.
with a *Horminum* leaf.
with a *Sonchus* leaf.
with a *Dens leonis* leaf.
with a *Helenium* leaf.
with a *Limonium* leaf.
with a *Senecio* leaf.
Through a giddy cycle—
Jacobaea with a *Hieracium* leaf.
Hieracium with a *Blattaria* leaf.
Blattaria with a *Verbascum* leaf.
Verbascum with a *Conyza* leaf.

Conyza with a *Salvia* leaf.
Salvia with a *Horminum* leaf.
Horminum with a *Betonica* leaf.
Betonica with a *Scrophularia* leaf.
Scrophularia with a *Melissa* leaf.
Melissa with a *Plantago* leaf.
Plantago with a *Coronopus* leaf.
Coronopus with a *Senecio* leaf.
Senecio with a *Jacobaea* leaf.
From imagination—
Clinopodium resembling *Origanum*.
Clinopodium resembling *Ocymum*.
Adonis with a *Buphthalmum* flower.
Cirsium with the root of a *Helleborus niger*.
Through the use of Greek—
Acer *platanoides*.
Brassica *asparagoides*.
Carduus *centauroides*.
Adonis *helleboroides*.
Vicia *lathyroides*.
Pseudo-Helleborus *ranunculoides*.

262. Features that draw *comparisons* with other species *of the same genus* are wrong.

Except in the presence of all the other plants in the same genus, a specific name is not soundly constructed when it contains a feature that exists in none of the other plants of the genus.

It is for the master to establish the name, and it is for the beginner to recognize the plant by it.

The beginner cannot collect various species; but he ought to recognize them successively when they do not grow at the same time or when specimens of each do not exist close together.

Specific names that presuppose another known species are erroneous.
Orchis *with the whitest* flower.
Campanula *with a narrow leaf*, with a *large* flower, *lesser*. T.
Campanula with a *smaller* flower, *more branchy*. Morison.

263. The *name* of the *discoverer* or anyone else should not be applied in the definition.

Names are the plant's hands; the generic name is the right hand and the specific the left; the hands of plants have eyes, and they believe only what they see; the plant should proffer them to the botanist to see if he will believe in the facts.[5]

So plant science, once recognized, brings out all things that are hidden—historical, synonymous and the rest.

We maintain that all *specific names* called after men, whether they are contrived from the *discoverer*, the *describer*, or as a *memorial*, are *erroneous*.

From the discoverer:

Trifolium *gastonium* Moris.

From the describer:

Gramen cyperoides *Boelii* Lob.

Conyza tertia *Dioscoridis* C.B.

Conyza media *Matthioli* I.B.

Narcissus *Tradescantii* Rudb.

From history:

Sideritis *Valerandi Douroz.* I.B.

Campanula prepared [as a drink] by *Toussaint Charles* I.B.

Mimosa [sent] by Herr *Hermans*, [from the garden] of his excellency the Lord of *Syen*. Breyn.

As memorials:

Chamaepithys flore plusquam eleganti, sive *Plusqueneti* [with a flower that is more than elegant, or *Plukenet's*]. Pluk.

Eriocephalus *Bruniades*. Pluk.

Amanita *Divi Georgii* [*St George's*]. Dill.

264. The native *location* does not indicate separate species.

There are many reasons to persuade us that the location ought not to enter into the specific name.

1. No one could easily go to Japan, the Cape of Good Hope, or Peru to recognize a plant.
2. The location quite often changes; all alpine plants, especially the mountain ones, turn out to be marsh plants outside the Alps.
3. The location of plants of the same species is not unique: Lapland, Siberia, Canada, Asia, and America quite often produce the same ones.
4. A Paradise garden often includes plants from all over the world.
5. Who would not have to work hard to identify a plant that was proffered to him without its location?
6. Botanists are glad to recognize species in the herbarium; so are physicians, in the dispensary.

It is impossible to imagine a plant without a location; but it is adventitious, even though next door; and it is very variable; therefore it cannot come into the specific name.

We maintain that all *specific names* taken from the location, whether from the soil, the region or the frequency, are *erroneous*.

From the soil:

Valeriana	*sylvestris*	[*wood*]
	palustris	[*marsh*]
	campestris	[*field*]
	montana	[*mountain*]
	alpina	[*alpine*]
Mentha	*arvenis*	[*arable land*]
	aquatica	[*water*]
Alsine	*nemorum*	[*grove*]
	pratensis	[*meadow*]
	littoralis	[*coast*]
Vicia	*segetum*	[*corn*]
	sepium	[*hedge*]
	dumetorum	[*thicket*]
Glaux	*maritima*	[*sea*]
Sedum	*rupestre*	[*rock*]
Muscus amans	*uvida*	[liking *wet*]
Erisimum juxta	*muros*	[by *walls*]

From the region:

Sagittaria	*Europaea*	[*European*]
Acrostichum	*Septentrionalium*	[*Northern*]
Calceolaria maritima	*Lapponum* Rd.	[*Lapland*]
Pentaphylloides	*Suecicum* Pluk.	[*Swedish*]
Bugula	*Suecica* Mor.	[*Swedish*]
Acetosa	*Moscovitica* Mor.	[*Russian*]
Cochlearia	*Danica* C.B.	[*Danish*]
Cochlearia	*Batavica* Hr.	[*Dutch*]
Cochlearia	*Britannica* Dod.	[*British*]
Cochlearia	*Anglica*	[*English*]
Cytisus	*Germanicus*	[*German*]
Pulmonaria	*Gallica*	[*French*]
Tamariscus	*Narbonensis*[6]	[*Southern French*]
Iris	*Florentina*	[*Florentine*]
Salicornia	*Cretica*	[*Cretan*]
Aster	*Atticus*	[*Attic*]
Ranunculus	*Turcicus*	[*Turkish*]
Iris	*Chalcedonica*	[*Kadiköy*]
Iris	*Damascena*	[*Damascus*]
Lilium	*Persicum*	[*Persian*]
Stoechas	*Arabica*	[*Arabic*]
Fritillaria	*Capitis Bonae Spei.*	[*Cape of Good Hope*]

continued

Virga aurea	*Novae Angliae*	[*New England*]
Virga aurea	*Marilandica.*	[*Maryland*]
Filix	*Brasiliensis*	[*Brazilian*]

From the frequency:

Oenanthe *rara* R.	[rare]
Clematis *peregrina* C.B.	[exotic]
Valeriana *hortensis*	[garden]
Scabiosa *communior* J.B.	[commoner]
Hydrocotyle *vulgaris*	[common]
Muscus *vulgatissimus*	[commonest]

265. The *season* of flowering and thriving is a very deceptive definition.

The season is adventitious to the plant and does not exist in the plant; but rather, the plant exists in the season; the seasons of plants are liable to change, and are not constituent parts of them.

Plukenet and his contemporaries introduced an extremely large harvest of plants from India, not defined by genus or species; I do not know if this resulted in a greater benefit or burden to botany. A house that is built on an unreliable foundation should be demolished and rebuilt soundly; anything from the old house that is serviceable should be used again, and the rest rejected, even if the work would then be finished late: likewise too with specific names, so that Science may stand on a firm footing.

I conclude that *specific names* taken from the seasons, whether from the year, the month, or the time of day, are *erroneous*.

From the year:

Tulipa *praecox* [early].

Tulipa *serotina* [late].

Crocus *vernus* [spring].

Geranium *aestivale* [summer].

Crocus *autumnalis* [autumn].

Aconitum *hyemale* [winter].

From the month:

Rosa *omnium calendarum* [all months].

Vila *Martia* [March].

Rosa *Majalis* [May].

Boletus *Julii mensis* D. [July].

Boletus *Augusti mensis* D. [August].

From the time of day:

Lychnis *Noctiflora* [night-flowering].

Althaea *Horaria* [of a certain time of day].

266. *Colour* within the same species is remarkably sportive, and so is of no value in definitions.

This is very clearly shown in the variations of colour in domestic animals.

Nothing is more variable than colour in flowers; *red* and *blue* flowers, above all, change into *white* very easily and very frequently.

The flowers of *Mirabilis* and *Dianthus barbatus* bear corollas of different colours in the same plant.

Colour especially attracts and delights the most refined of the senses.

And so the attention of the ancients was invited back to the colours by the heads as they opened: but do not put too much trust in colour.

The zeal that *lovers of flowers* derived from that results in dishonour to the art of botany, so that it seems that no [other] mortal men wander further beyond the olives:[7] this is obvious in *Tulipa, Pulsatilla, Ranunculus, Hyacinthus,* and *Primula.*

Tournefort entered the camp of the lovers of flowers and saw, as in a kaleidoscope, 63 species in one Hyacinthus and 93 in a single Tulipa—more than existed in reality.

We maintain that all *specific names* taken from the colour, whether from the *flower,* the *fruit,* the *seeds,* the *root,* the *plant,* or the *leaves,* or through *imagination,* are *erroneous.*

Leaves are described as *coloured* when they take on a colour other than green. These are very variable, and quite often display a colour that is foreign to them.

WHITE-SPOTTED: *Cyclamen, Acetosa* italica, *Ranunculus* repens, *Trifolium* album, and *Amarantus* emarginatus.

BLACK-SPOTTED: *Ranunculus* ficaria *and* hederaceus, *Arum, Galeobdolon* D., *Hypochoeris, Persicaria* ferrum equinum referens [resembling a horseshoe] T., and the *Orchises.*

RED-SPOTTED: *Ranunculus* acris, *Nymphoides* folio maculis purpureis notato [with a leaf marked with purple spots] T., *Amarantlus* tricolor.

CHEQUERED: *Satyrium* [my] Flora Suecica 732, *Cypripedium* Flora Suecica 736.

WITH DOTS ON THE UNDERSIDE: *Anagallis, Plantago* maritima Flora Suecica 127.

WITH A WHITE LINE: *Arundo* indica cornuta, *Phalaris* gramen pictum [painted grass], and, on the underside, *Empetrum.*

WITH A SILVER FRINGE: *Ilex* T., *Buxus* T., *Caprifolium* T., and *Glechoma* T.

From the flower:
Primula veris with a *yellow* flower.
Primula veris with a *red* flower.
Primula veris with a *rusty* flower.
Auricula ursi with a *scarlet* flower.
Auricula ursi with a *purple* flower.
Auricula ursi with a *violet* flower.
Auricula ursi with a *variegated* flower.

From the plant:
Brassica, *green.*
Brassica, *red.*
Brassica, *white.*
Marrubium, *white.*
Marrubium, *black.*
Hyoscyamus, *black.*
Martagon, *gory.*

From the fruit:
Melo with *yellow* fruit.
Cucumis with *white* fruit.
Pepo with *variegated* fruit.
Prunus with *blue-black* fruit.
Prunus with *golden-yellow* fruit.
Prunus with *corn-coloured* fruit.

From the seeds:
Papaver with *white* seed.
Papaver with *black* seed.
Sinapi with *red* seed.
Sinapi with *yellow* seed.

From the root:
Daucus with *ruddy-black* root.
Daucus with *orange-coloured* root.
Daucus with *yellow* root.

From the leaves:
Agrifolium with leaves *variegated* from a *yellow* background.
Agrifolium with leaves *variegated* from a *white* background.
Agrifolium with leaves with *silver* fringes and prickles.
Agrifolium with leaves with *golden* fringes and prickles.
Ocymum maculatum [spotted].
Esula marked with saffron-coloured dots.
Malva foliis margine superius *micis sulphureis ad solem splendentibus* donata
    [furnished, on the upper side, with a margin *with sulphur-coloured grains,*
    which shine in the sunlight]. Moris.

From imagination:
Alypum or Frutex [shrub] *terribilis* [*frightening*].
Campanula *pulchra* [*beautiful*] I.B.
Filix scandens [climbing fern] *perpulchra* [*extremely beautiful*] Br.
Poinciana flore *pulcherrimo* [with a *very beautiful* flower] T.
Filix saxatilis [rock fern] *elegantissima* [*very elegant*].

267. *Scent* never clearly distinguishes a species.

> The sense of smell examines very fine emanations; it is the most obscure of the senses, and it gives names to very few genera.
>
> The scent of all things very easily varies, and is different in different individuals.
>
> It is clearly shown, by the recent[8] testing of dogs in a crowd of humans, that there are as many distinct scents as there are scented bodies.
>
> Scents do not allow for fixed boundaries and cannot be defined; therefore they should have no place among the miscellaneous characters that we use for characteristic features.
>
> We rightly denounce as *erroneous* all those *specific names* that allow for smell as a definition.
>
> *From kinds of scent*:
>> Hypericum *hircinum* [goat].
>> Melo *moschatus* [musk].
>> Agrimonia *medio modo* odorata [*moderately* scented]. M.
>> Arbor [tree] *merdam* olens [smelling of *dung*].
>
> *From the time*:
>> Hesperis *noctu* olens [smelling *at night*].
>> Caryophyllus *inodorus* [*without scent*].
>
> *From other plants*:
>> Ocimum *caryophyllatum* [*like Caryophyllus*]. C.B.
>> Ocimum with the scent *of Citrus* [*citron*].
>> Ocimum with the scent *of Foeniculum* [*fennel*].
>> Ocimum with the scent *of Melissa* [*balm*].
>> Ocimum with the scent *of Cinnamomum* [*cinnamon*].
>> Ocimum with the scent *of Ruta* [*rue*].
>> Ocimum with the scent *of Styrax liquidus* [*liquid storax*].

268. *Taste* is often variable, according to the person doing the tasting; therefore it should be excluded from definitions.

> Different ages make different judgement about the taste of things.
>
> Differences of soil and climate change tastes.
>
> Cultivation elicits acid and sour [qualities].
>> CICHOREUM bitter wild CHICORY.
>> LACTUCA poisonous wild LETTUCE.
>> ALLIUM unscented GARLIC in Greece.
>> APIUM disagreable marsh CELERY.
>> MALA extremely sharp APPLES in the woods.

The technique of the cultivation of fruit has increased the seasonal produce of the
PEAR (PYRUS), the APPLE (MALUS) etc. by such a numerous progeny that it
has produced 172 different kinds of pears and 200 kinds of apples (according to
Boerhaave); and each one is customarily honoured by its own particular name
because of its special taste.

Ludicrous *specific names* derived from taste are all *erroneous*; therefore we conclude
that they should be altogether excluded from definitions.

From the plant:
Apium [celery] *ingratius* [*quite disagreable*].
Apium [celery] *dulce* [*sweet*].
Lactuca [lettuce] *with the poisonous juice of opium*.
Lactuca [lettuce] *mitis* [*mild*].
From the fruit:
Pisum [pea] *with edible outer tissue*.[9]
Pyrus [pear] with *sugary* fruit, which *melts* in the mouth.

269. *Potency* and *use* provide definitions that are worthless to a botanist.

To be recognized, a species ought to become [the subject of] an experiment on a life
of little value, according to the method of the ancients.

If you taste *Hippomane*, you will observe some very dangerous experiments. *Arum
renders dumb those who taste it with the tips of their lips*. Sloan.

The pharmacists' plants and their names should be placed among the synonyms:
they should no more prescribe names to botanists than they themselves would
withdraw from their customary usage because of botanical principles.

Are we to make of one genus, which comprises *Convolvulus, Turbith, Mechoacanna,
Scammonium, etc.*, as many genera as names, when they would be contrived
contrary to the laws of the Creator, merely to please the pharmacists.?

Are several genera of plants to be established from the one and only *Punica*, namely
Balaustium from the flowers, *Granatum* from the fruit and *Malacorium* from the
outer tissue of the fruit? One should keep a sound mind in a sound body.[10]

We declare *erroneous* all *specific names* that comprise *potencies* or use in themselves,
whether they take their distinguishing mark from shops, health, diet, trade or
historians.

From the pharmacy:
Agrimonia *officinarum* [*of the shops*] C.B.
Calamintha *officinarum Germaniae* [*of the shops in Germany*] C.B.
Martagon *alchimistarum* [*of the alchemists*] Lob.
Hieracium *usuale* [*useful*] Rd.
From life:
Solanum *lethale* [*deadly* nightshade].
Aconitum *salutiferum* [*health-giving*].
From housekeeping:
Genista *scoparia* [broom *for sweeping*].

Rubia *tinctoria* [madder *for dyeing*].

Dipsacus *fullonum* [*fullers'* teasel].

Ricinoides, ex qua paratur *Tournesol* Gallorum [from which the French *Tournesol* is prepared]. T.

From the potencies:

Menyanthes *antiscorbutica* [against scurvy].

Rhamnus *catharticus* [purgative].

Solanum *somniferum* [soporific].

Solanum *furiosum* [maddening].

From diet:

Pisum cortice eduli [pea with *edible* outer tissue].[9]

Pisa, quae simul cum *folliculis* comeduntur [peas, that are eaten together with the *shells*][9] Volk.

From history:

Punica, which bears the *pomegranate*.

Canellifera [cinnamon] tree, which has *a quite insignificant bark* and a leaf that is the *malabathrum* [*base cinnamon*] of the shops.

270. SEX nowhere ever establishes different species.

Here is to be understood the dioecious sex of male and female, but not the sex that is monoecious, hermaphrodite etc., for example in RUMEX, the *Acetosa* species are male and female, or dioecious; *Beta spinosa* is androgynous or monoecious; *Lapathums* are hermaphrodite or monoclinous; *Acetosa alpina* has two pistils.[11] No one has denied that these features serve as specific definitions.

Several authors, including J. Bauhin, Ray, and Tournefort, have established different species from the sex of the male and female; but, as they differ only in sex, we say that they are not distinguishable species.

For example:

Urtica, *male* and *female*.

Cannabis, *male* and *female*.

Humulus, *male* and *female*.

The ancients made an *erroneous* distinction of species on account of the distinction of sex, where the sex was no different, but the plant was very different, for example:

Male and female:

Anagallis	Cistus	Orchis
Aristolochia	Cornus	Pæonia
Abrotanum	Christa galli	Pulegium
Abies	Ferula	Quercus
Amaranthus	Filix	Symphytum
Balsamina	Mandragora	Tilia
Caltha	Nicotiana	Veronica

271. ABNORMAL flowers (150) and plants always betray their origin from natural ones.

> In Sections 119–22, we have stated that multiple, full and prolific flowers are abnormal; and in Section 150, that they are derived from simple ones.
>
> We distinguish species, which are by the work of the Creator, from varieties, which are the result of Nature's tricks.
>
> These abnormalities most usually occur as a result of display in cultivation and excessive nourishment.
>
> No one ever takes abnormalities among animals for distinct species; therefore plants should not be so taken either.
>
> Enlarged, multiple, full and prolific flowers should be removed from botany, and the numerous crowd, which has long been a burden to botany, must be banished.

272. PUBESCENCE (163:VIII) is a ludicrous definition, since it is often shed as a result of cultivation.

> We understand *thorns* and *hairiness* as included in pubescence, which plants quite often doff as a result of location or cultivation.
>
> It is an achievement of cultivation that the fiercest animals become wonderfully tame; and we observe every day that the same thing happens in plants too.
> The *fierce aurochs* becomes a *mild bull*.
> The *fierce wild dog* becomes a *tame dog*.
>
> Thorny trees often shed their thorns in gardens as a result of cultivation, and produce mild instead of bitter fruit, for instance *Pyrus [pear]*, *Citrus [citron]*, *Limon [lemon]*, *Aurantium [orange]*, *Mespilus [medlar]*, *Oxyacantha [sharp-thorn]*, *Grossularia [gooseberry]*, and *Cynara [artichoke]*.
>
> Wild *Cichorium [endive* or *chicory]* has leaves that have recesses, and are toothedged and rough with a disagreable bitter taste.
>
> HAIRINESS is very easily shed owing to location or age.
>> *Fagus [beech]* is at first very hairy, whether grown from seed or buds; soon it turns out to be smooth.
>> A tender young *Heliocarpus* has leaves with matted hair, but a full-grown one has them smooth.
>> A little *Triumfetta* has matted hair, a full-grown one is shaggy.
>> *Asperula odorata* in the woods is shaggy, but in sunny places it becomes smooth.
>> *Persicaria amphibia* in water is very smooth, in dry places it is rough.
>> *Thymus serpyllum [thyme]* in the open fields is smooth, and it is hairy in sandy places by the sea.
>> *Scabiosa succisa* [devil's-bit scabious] in sunny places is smooth, in the woods it is slightly hairy.

Plantago [plantain] coronopus in a damp place has smooth entire leaves, in a dry one they are hairy and tooth-edged.

Lilium martagon in the woods is hairy, in gardens it is very smooth.

Alchemilla palmata is smooth and yellowish in arid sunny places, and in marshy shady places it is green and hairy.

A milder CLIMATE makes the plants milder, just as a harsher climate renders them harsher, and as it were clothes them with hide. Likewise they do not have to resort to hairiness and thorns unless they are compelled by dire necessity: for example:

Pentaphylloides palustre rubrum, *crassis et villosis foliis,* suecicum et hibernicum [red marsh, *with thick and shaggy leaves,* Swedish and Irish]. Pluk.

Bugula non crenata, *tomentosa,* suecica [without scalloped edges, and *with matted hair,* Swedish.] Pluk.

273. The DURATION is often related more to the place than to the plant; so it is not acceptable to use it in definitions.

The warm regions, which enjoy perennial summer, nourish plants that will scarcely die throughout the year. So very many plants in these regions turn out to be perennial and tree-like in growth, whereas with us they are annual: such as *Tropaeolum, Beta, Majorana, Malva arborea,* etc.

The cool regions make annual plants of perennials, such as *Mirabilis, Ricinum,* etc. Therefore no definition is to be looked for in duration, unless it is very obvious.

274. The QUANTITY of herbage often varies according to the location.

A creeping stem usually multiplies itself very extensively by putting out radicles.

Plants are multiplied either in the soil, or in the root, stem, leaves, or fruit-body.

A plant is described as *frequent* and *common,* if it is found growing naturally and copiously in a suitable soil.

A plant becomes caespitose, when numerous stems grow from the same root; this does not happen entirely consistently, because, in a poor soil, a plant that would otherwise be caespitose has difficulty in propagating a single stem: on the other hand, if, in a plant that usually produces a single stem, that stem is cut off towards the root, then it puts forth numerous stems, like a hydra.

A plant is usually called *bundled* when several stems grow together, so that one compressed stem, in the shape of a bundle, is formed out of many. The same thing can be produced artificially, if several growing stems are forced to enter a confined space, and as it were to be born from a narrow womb: so frequently with *Ranunculus, Beta, Asparagus, Pinus, Celosia, Tragopogon, Scorzonera* and *Cotula foetida.*

Beta *lato caule [with a wide stem]* C.B.

Amarantus *cristatus [crested].*

A plant is said to be *pleached*, when a tree or branch grows out with very small interlaced twigs, like a Polish plait of hair, or in the manner of a magpie's nest, about which the common people believe that it is made by an evil spirit; with us it is frequent in the birch, especially in Norrland, and in the hornbeam in Skåne; and not rare in the pine.

Curly (83:63) leaves are formed when the periphery of the leaves is enlarged, so that it waves about like a billowing selvage.

Studded leaves are formed from wrinkled ones, when the disk (but not the outer edge) grows, so that matter rises up between the wrinkles, in the shape of cones which are concave on the underside; such as most *Salvias* and *Ocimum*.

It is clear, from what has been said before (119–22, 150, and 271), that multiple, full, and prolific flowers develop out of simple ones.

Varieties instead of species are variable plants, which have increased in number.

Ophioglossum lingua bifida [*with cleft tongue*]. Barth.

Plantago spica bifida [*with cleft spike*]. Barth.

275. The ROOT (81) often provides a real definition, but recourse should not be had to it unless all other ways are blocked.

If any other constant feature remains, the root should not be used, for we cannot easily obtain permission to uproot plants in gardens; in herbaria, the root is not easily transplanted; in living plants, the root withdraws itself from our sight.

The more easily plants can be identified, the better; but no necessary law is laid down.

Scillas are distinguished with difficulty by the herbage, but very easily by the bulbs, which may be *tunicate, solid*, or *scaly*.

Orchises are not safely distinguished unless the roots, *fibrous, almost round*, or *testiculate* are called in to help.

Bulbous *Fumarias with hollow roots*, greater and lesser, and those *with a root that is not hollow*, greater and lesser, are only varieties, as is shown by the habit of the whole plant: moreover the calyx is hardly visible to our eyes, and the bracts differ from the rest [of the leaves.]

276. The features of the TRUNK often produce excellent definitions.

In many plants, the stem offers such essential definitions that without it there would be no certainty about the species.

An angular stem distinguishes several species that could hardly be distinguished otherwise.

Hypericum androsaemum, [my] Flora Suecica 626, has a rounded stem.

Hypericum perforatum, Flora Suecica 625, has a two-edged stem.

Hypericum ascyron, Flora Suecica 624, has a four-cornered stem.

Convallaria polygonatum vulgare, Flora Suecica 274, a, has a two-edged stem.

Convallaria polygonatum maximum, Flora Suecica 274, b, has a rounded stem.

The latter of these should be described as *Convallaria with alternate leaves, rounded stem, and peduncles with many flowers*; the former as *Convallaria with alternate leaves, two-edged stem, and peduncles with single flowers.*

Hedysarum, [my] Flora Zeylanica 286, is remarkable for its three-cornered stem.

The several species of *Lupinus* could hardly be distinguished, except by the stem, simple in some, compound in others.

The five-cornered and six-cornered *culm* is the best definition of *Eriocaulon*, Flora Zeylanica 48–50.

A three-cornered *scape* distinguishes [some] *Pyrolas* from the rest, [my] Flora Suecica 337. 332.

Petioles that are winged or enlarged on either side by a membrane distinguish *Aurantium* from others of the same genus, and *Hedysarum*, Flora Zeyanica 286.

Peduncles that are *bifoliate* or furnished with two opposite leaflets, and those placed beneath a head, neatly define *Gomphrena*, Flora Zeylanica 115.

277. The LEAVES (83) show the neatest and most natural definitions.

Nowhere has nature been more multiform than in the leaves; and therefore their extremely numerous shapes must be assiduously learnt by beginners.

The leaves commend themselves for definitions because they are very shapely, produce shapes that are very different in appearance, and readily provide definitions; for this reason, it is from the leaves that I have borrowed very many of my definitions in specific names; this is obvious to anyone who consults *Hortus Cliffortianus, Hortus Upsaliensis, Flora Suecica, Flora Lapponica,* or *Flora Zeylanica*; to these add *Gronovius' Flora Virginica, Royen's Flora Leydensis, Guettard's Flora Stampensis* and *Dalibard's Flora Parisina.*

We have described the characters of the leaves above, Section 83.

Very many leaves occur that are comparatively rare, and so are not included in a general survey; for example:

A *hood-shaped* leaf, the sides of which are very close together towards the base, and are spread out at the top: *Geranium africanum.*

A *glandular* leaf, which displays glands that are placed on it;

On the *saw-toothed edges*: *Salix, Persica.*

On *the back*: *Urena.*

Sharply pointed leaves display the shape of a needle and are like an awl; they are usually inserted into a branch by articulation at the base, as in *conifers.*

Rooted leaves are those that send radicles down from the substance of the leaf.

Conjoined leaves are those that are attached to each other, so that they grow together at the base.

Crossed leaves issue from opposite ones (83:112) so that, if the plant is viewed from directly above, the leaves extend in four directions.

We pass over several synonyms for describing leaves.

Chopped or *cut up*, see *slashed*, p. 57, No 28.

Umbilicate [as used] by some, see *shield-shaped* p. 61, No. 119.

Cusped, see *tapering*, p. 57, No 38, but with the point somewhat more rigid.

Rough, see *scabrous* p. 58, No. 55.

Bristling, see *prickly* p. 58, No. 54.

Pubescent, see *shaggy*, p. 58, No 53, but not very shaggy.

Hoary see *tomentose*, p. 58, No. 51. Leaves that have a grey, almost silvery colour because it arises from a particular kind of surface.

Straight up see *upright*, p. 62, No. 131.

Direct see *straight*, but it enhances the meaning; to wit, absolutely straight.

Rising in an upward curve, but at first sloping down, and eventually upright towards the tips.

Sloped down, bent downwards like the keel of a boat.

Leaves of several kinds are known intrinsically from the leaf characters that have already been described.

Pairs, just like sets of three or four, p. 61, No. 111.

Nerveless, the opposite of *nervous*, p. 58, No. 67.

Veinless, the opposite of *veiny*, p. 58, No. 66.

Turned over or *inverted*, with the base relatively narrow, so that one thinks of the base as being where the tip actually is: for example:

Egg-shape turned over, or *heart-shape turned over*.

Inversely egg-shaped or *inversely heart-shaped*.

Latin [words], which are understood intrinsically in their proper and most usual linguistic sense:

Trapeziform and *rhomboid* from mathematics.

Perennial, *biennial*, and *annual* from the duration of the plant.

278. The SUPPORTS (84) and the WINTER-BUDS (85) usually yield excellent definitions.

The botanist, if he is deprived of these features, will scarcely—or not even scarcely—define the [several] species easily and safely; and this is clearly shown by several examples:

SHARP POINTS in *Rubus*. THORNS are especially noticeable in *Prunus*.

BRACTS in *Fumaria*, *Hedysarum indicum*, and *Dracocephalum*.

A TOP-KNOT consists of bracts that are conspicuous for their size at the end of the stem, as in *Corona imperialis*, *Lavandula* and *Salvia*.

GLANDS provide the essential features in *Padus*, *Urena*, *Mimosa*, and *Cassia*.

Glandular saw-toothed edges at the bases of the leaves in *Heliocarpus*, *Salix*, and *Amygdalus*.

A glandular back to the leaves in *Padus*, *Urena* and *Passiflora*.

Glandular points secreting liquid at the tip, in *Bauhinia aculeata*.

Anyone who was deprived of knowledge of the glands could never certainly and safely distinguish the several species in very many genera, especially in *Cassia, Mimosa*, and others.

Amygdalus is distinguished from *Persica* only by the glands of the saw-toothed edges.

No one can define the several species of *Urena* before he has examined the glands of the leaves.

The several species of *Convolvulus, with the calyx shaped like a tubercle*, would be divided into even more if they were not kept together by the glands.

Monarda with a corolla sprinkled with glands turns out to be very obviously distinct from others of the same genus.

STIPULES are valuable in very large genera, where doubt may exist about a species.

One *Melianthus* is distinguished by single stipules, another by pairs of them.

Cassia with kidney-shaped barbed stipules, Flora Zeylanica 151, turns out to be absolutely distinct from all others of the same genus.

BUDS often differ greatly within the same genus, as is shown by the genus *Rhamnus* in which *cervispina, alaternus, paliurus*, and *frangula* differ in their buds.

The family of the *Salixes* is divided with great ease, and certainty into absolutely constant species, by the very full and intricate structure of the buds and of the foliation.

The genus *Scilla* is divided excellently and almost solely by the BULBS.

The bulbs that are placed in the axils of the leaves define *Dentaria, Lilium, Ornithogalum, Saxifraga*, and *Bistorta*, by this peculiar feature.

279. The INFLORESCENCE (163:XI) is the most certain definiton.

The inflorescence is the means by which the peduncle produces the fruit-body (p. 125), with respect to its structure or [rather] its location; some have described it as *the site of the fruit-body*, and, going on from there, have formed new genera.

The inflorescence has almost seemed to me the most perfect among the definitions, in most genera.

In *Spiraea*, there are some species with flowers in double racemes, others with flowers in clusters, and others with umbellate flowers, so that if this feature is removed, there is no certainty of the species.

The PEDUNCLE, which produces the flowers, puts them forth in various ways.

It is *flabby*, when it is so weak that it hangs downwards with the weight of the flower itself.

It is *drooping*, when it is curved at the tip, so that the flower nods towards one side or towards the gound; and it cannot be raised owing to the fixed curvatureof the peduncle; as in *Carpesium, Bidens radiata, Carduus nutans, Scabiosa alpina, Helianthus annuus, and Cnicus sibiricus*.

The flowers are *erect and bundled*, when the petioles raise the fruit-bodies into a
bundle, so that they turn out to be equal in height at the top, as if they had been
shorn off horizontally: this is shown by the examples of *Dianthus*, and *Silene*.

A *spreading* petiole disperses small branches in all directions, so that it bears flowers
at a distance; the opposite of a *constricted* one.

Flowers become *conglomerate*, when a branching petiole bears closely congested, and
compacted flowers without any order; so they are the opposite of a *diffused panicle*.

A *jointed* petiole is one that is furnished with a single joint: *Oxalis*, *Sida*, and *Hibiscus*.

Pairs of peduncles issue forth together in *Capraria*, and *Oldenlandia zeylanica*.

Sets of three peduncles in the same axil in *Impatiens zeylanica*.

Bending or wavy peduncles in *Aira*, *Flora Suecica* 64, and *Poa zeylanica* 46.

Peduncles *that remain* in the plant after the fall of the fruit-body; in *Jambolifera*,
Ochna, and *Justicia*.

Peduncles *that are thickened* towards the flower; in *Cotula, and Tragopogon*, and
generally in drooping flowers.

280. The parts of the fruit-body quite often provide the most reliable
definitions.

Once I thought otherwise, and I did not go to the fruit-body, unless other ways were
blocked, because the flower lasted only for a short time, and its parts were quite
often very small.

There are more parts in the fruit-body than in all the rest of the plant; and therefore
more features are extracted from it.

All certainty, throughout the whole of nature, depends on very small parts; and
anyone that shies away from them, shies away from nature.

The features of the fruit-body should be divided into the essential, the natural, and
the specific ones.

The beginner, when he first dissects, observes special features, and forms new,
and false, genera. He believes that he was the first to see the flowers; but after
he has become a mature botanist, he will think that he has quite often been
misled.

Gentianas cannot be distinguished if the flower is removed, as established by those
observed by the illustrious Haller; but in different samples, the corollas are *bell-
shaped, wheel-like*, or *funnel-shaped*; being *split into five, four, or eight*, they very
readily make a distinction.

The *Hypericum* with flowers *with three pistils* should be properly distinguished from
that *with five*.

The *Geraniums* described as *African*, with irregular corollas, and conjoined stamens,
ought to be separated from the European ones of the same genus.

In *lichens*, the *tubercle* is a constant fruit-body, with scabrous dots as it were heaped
up out of dust.

The *scutellum* [little shield] is a circular concave fruit-body, with the edge raised all round.

The *pelta* [light shield] is a flat fruit-body, usually stuck to the edge of a leaf.

In *mosses*, the *head* is the anther.

In *funguses*, the *cap* is a circle spread out horizontally, which displays the fruit-bodies on the underside.

In *grasses*, the *spikelet* is a partial spike, which some have called the locusta.

The *tortilis* is an awn, so called because it is noted for its node, which is twisted in the middle: for example, *Avena*.

The *joint* is a part of the culm that is placed between two nodes.

A *compound rayed* flower consists of a disc, and a ray.

The *ray* consists of irregular small corollas on the periphery.

The *disc* consists of still smaller, mostly regular, corollas.

A *decompound* flower, or one compounded of compounds, contains, within a common calyx, smaller common calyces with many flowers: for example *Sphaeranthus*.

A *corolla* is *even*, when the parts of the corolla are equal in shape, size, and proportion.

> It is *uneven*, when the parts do not correspond in size, but do so in proportion, so that the flower turns out to be regular: for example, *Butomus*.

> A *regular* corolla is even in the shape, size, and proportion of the parts.

> An *irregular* corolla is one that is different in the parts of the edge, in shape, and in size or proportion.

> A *gap* is an opening between two lips.

> An *orifice* is an opening in the tube of the corolla.

> A *palate* is a prominent swelling in the orifice of the corolla.

> A *spur* is a nectary extended, from the corolla backwards, into a cone.

> An *urn-shaped* corolla is one that is blown up in the shape of an urn or jug, and swollen all round.

> It is *cup-shaped* when it begins as a cylinder, and is slightly widened at the top.

> A *closing* corolla is one in which the lobes at the edge turn towards each other.

> A *torn* corolla is one with the edge finely divided.

A *swinging*, and *reclining anther* is one that is fixed by the side.

> An *upright anther* is one that is fixed by the base.

A *pericarp* is *inflated*, when it becomes hollow like a bladder, and is not filled with seeds: for example, *Fumaria cirrhosa*.

> It is *prismatic* when it becomes a linear polyhedron, with flat sides.

> It is *top-shaped* when the fruit is narrowed at the base: *Pyrus*.

> It is *twisted*, when it becomes twisted spirally: *Ulmaria*, *Helicteres*, and *Thalictrum*.

> It is *scimitar-shaped*, when the fruit is compressed into the likeness of a bill-hook, with one lengthwise corner blunt, and the other sharp: *Mesembryanthemum*, Dill.

> It is one with *nestling seeds* T. when the seeds are dispersed throughout the pulp in a berry-like pericarp.

It is a *hedgehog-like one* when it is dotted all over with prickles or thorns, like the animal hedgehog.

It is a *bulging* one, when it turns out to be lumpy with little bulges or protuberances on this side and that: *Lycopersicon* T., and *Phytolacca*.

281. Generic features (192) are absurd if used in definitions.

Here we understand the generic features of the natural character; and they never distinguish the several species. For these features are in agreement throughout all the species in the genus; and those that agree do not disagree.

So we have established as erroneous all definitions that take their specific features from the features of the character.

Polygala *with small siliquae, four petals, and two capsules*. Moris.

Aponogeton *with solitary stamens*. P.

Gujacum *with fruit of Acer or Bursa pastoris*. Br.

282. Every definition must necessarily be taken from the number, shape, relative size, and site of the various parts of the plants (80–6).

We have described above the sources of features that are deceptive, and of those that are reliable.

We have established that very many features are deceptive; such as

Colour,	Section 266.
Scent,	Section 267.
Taste,	Section 268.
Use,	Section 269.
Sex,	Section 270.
Pubescence,	Section 272.
Time,	Section 265.
Place,	Section 264.
Number,	Section 274.
Size,	Section 260.
Abnormality,	Section 271.
A variant [part,]	Section 259.
An accidental [feature],	Section 258.
Authority,	Section 263.
Insufficiency,	Section 257.
Comparison,	Sections 261 and 262.

And [we have established] that trustworthy features are to be taken only from parts of the plant, such as the

The means, by which a definiton is selected, are four:

    number, shape, site, and *relative size;*

    so they are the same as in the case of the genus; Section 167.

These are constant everywhere, in a plant, a herbarium, or a picture.

283. One must be careful everywhere, that a variety (158) is not taken for a species (157).

> That is the job, the trouble;[12] therefore it must be investigated with the greatest care.
>
> The following causes, primarily, give rise to mistakes, so that we quite often become blind.
>
> 1. *Nature,* which has many forms, and never ceases its operations.
> 2. The differing, and peculiar nature of *regions,* and *climates.*
> 3. Native *places* that are very remote.
> 4. The *brevity* of human *life,* which perishes by a fate that comes too soon.
>
> The following discover certainty in species, which are to be distinguished from varieties.
>
>   *Cultivation* in very different, and varying soils.
>
>   Very attentive examination of all the *parts* of the plant.
>
>   Examination of the *fruit-body,* extending to all its parts, even the smallest.
>
>   Inspection of species *in the same genus.*
>
>   The constant laws of *nature,* which never make leaps.
>
>   The far-fetched modes of *varieties.*
>
>   The *placing* of the species in the nearest distinct genus.

284. The generic name ought to be applied to each of the several species.

> When the species have been referred to genera, the species should take the name of the genus, so that the genus of the plant named is established by the name.
>
> The generic name is valid as current coin in the Republic of Botany.
>
> Ray, and Morison quite often referred species to genera, but they did not adopt the name of the genus.
>
> STOECHAS, Ray, hist. 280.
>
> 1. *Stoechas citrina germanica, with leaf quite broad.*

2. *Chrysocome aethiopica, with plantain's leaf.*
3. *Helichrysum, with leaves of a female Abrotanum.*
4. *Elichrysum creticum.*
5. [*A plant*] *related to Lobel's second scentless Stoechas citrina.*
6. *Gnaphalium montanum album.*

Thus, anyone who hears Ray's name [for the plant] has no idea of the genus, unless he looks in the book.

285. The specific name must always follow the generic name.

Since there is no certainty when the genus is unknown, it is necessary that the name of the genus should begin the expression, if it is to distinguish the things that need to be distinguished.

In this, LOBEL offended very greatly, and quite often:
minus Heliotropum *repens* [*lesser creeping*] *Lob.*
Matthioli secundum [*Matthioli's second*] Limonium. *Lob.*
aquatica Plantago *foliis Betae* [*with leaves of beet*]. *Lob.*

286. A specific name without a generic one is like a bell without a clapper.[13]

A definition is merely a distinction within a genus; therefore no definiton can be conceived without the genus.

Names have been technically composed to identify plants.

Definitions without a specific name are creatures with their heads cut off.
_____ *a herb related to myagrum, with almost round capsules.* J.B.
_____ *somewhat similar to Linaria, hairy, and not lappeted.* C.B.
_____ *somewhat similar to Linaria, with a daisy leaf.* J.B.
_____ *resembling Periclymenum, being a plant with a single flower.* Mor.
_____ *a native of the Alps of Savoy.* Bocc.

287. The specific name must not be part of the generic name.

There have been botanists that have tried to divide genera by means of tails, by inflecting the name at the end, but this has produced very great confusion.

The genus, and the definition should be distinct from each other.
Gentian*ella* for Gentiana parva [small].
Aceto*sella* for Acetosa parva.

288. A specific name is either *synoptic* or *essential*.

Specific names distinguish the several species quickly, safely, and agreeably.

A selection should be made from all the possible definitions of a species, and from this the best ones should be chosen, so that eventually one may be in a position to recognize the species safely.

The mode of specific names is either synoptic or essential, or a mixture of both.

289. A specific name that is synoptic (288) imposes features, that are partially divisive, on plants of the same genus (159).

When the essential features of a species cannot be traced, it is necessary that the definition should be made by synopsis; therefore synopsis is a substitute for essential definition.

In the very largest genera, we are quite often compelled to adopt synopsis.

Salix with leaves that are saw-toothed, smooth, egg-shaped, pointed, and almost sessile. Roy: *synoptically.*

Salix with five-stamened flower. Flora Lapponica: *essential.*

Salix with leaves that are almost entire, lanceolate-linear, very long, pointed, and silky on the under-side, and with rod-like branches. Flora Suecica: *synoptically.*

Salix with leaves that are linear, and rolled back: essential.

290. An essential specific name (288) shows a feature of definition that is particular, or peculiar only to its own species.

The essential specific name is settled by one or another single word, or single expression.

When we have identified stable genera, and species by means of essential definitions, we have attained the highest point in botany. If botanists were to arrive eventually at the position in which they could identify all species by essential names, there would be no possibility of further progress.

The excellence of a name commends itself by brevity, ease, and certainty.

Essential names are valid without citation of the authority, but other names are never so.

Once the essential name has been discovered, synopsis cannot be allowed in the specific definition; therefore we must work on, so that we may eventually arrive at the goal that is set before us.

Eriophoron *with pendulous spikes;* [my] Flora Lapponica 22.

Plantago *with single-flowered scape;* Flora Lapponica 64.

Alchemilla *with simple leaves;* Flora Lapponica 66.

Alchemilla *with digitate leaves;* Flora Lapponica 67.

Menyanthes *with leaves in sets of three;* Flora Lapponica 80.

Convallaria *with bare scape;* Flora Lapponica 112.

Convallaria *with whorled leaves;* Flora Lapponica 114.

Pyrola with *single flowered scape*; Flora Lapponica 167, and [my] Flora Suecica 334.
Betula *with circular, scalloped leaves*; Flora Lapponica 342, and Flora Suecica 777.

291. The shorter the specific name the better, provided that it is such (257).

> Artistic beauty demands brevity; for the simpler the better too; and it is foolish to do in more [words] what can be done by fewer; even nature itself is very compendious in all its actions.
>
> The total of the words used in a definition should never admit more than twelve; just as generic names must consist of twelve letters at the most, Section 249, so too a definition must consist of twelve words at the most, so that limits are set eventually.
>
> From the following calculation, it is clear that twelve words are enough for a specific definition. Suppose that a genus consists of a hundred species—and we know of no genus hitherto recognized that has reached this number—and suppose that the following species are distinguished: a. 50, b. 25, c. 13, d. 7, e. 3, f. 2, and g. 1; in that case, six substantives with as many adjectives would be needed for the total; yet one and the same substantive often receives several adjectives, with the result that twelve words will hardly ever be necessary for the definition[s] of a genus of a hundred species.
>
> Therefore the names used by the ancients, that are $1\frac{1}{2}$ feet long, and which constitute descriptions instead of definitions, are to be abhorred.
>
> > Cenchramidea, *a tree which grows on rocks, with a roundish, thick leaf, a pome-shaped fruit divided into very many capsules containing grains like those of a fig, (which grains adhere to a very hard hexagonal columnar stylus); and which tree produces balsam.*
> >
> > Gramen, [a grass] *myloicophoron carolinianum [from Carolina] or altissimum [very tall]; with a very large and beautiful tuft composed of quite large spikes, which are compressed, and pinnate on both sides, and in some way resemble a mill-blade; with leaves that are rolled up, pointed, and pungent.* Pluk. alm. 173.
> >
> > *A spiny American tree, in some way resembling* Acacia, *similar to Vesling's* Myrobalanus chebulae, *with* Ceratonia *leaves in pairs on the pedicle, a siliqua with two valves, which is compressed, and horn-shaped or curved like the horns of snail-shells or rams' horns, or like a cat's claws.* Breyn. prodromus 2, p. 29.

292. A specific name should include no words except those needed to distinguish it from others of the same genus.

> There must not be any superfluous word in a specific definition.
>
> If the same feature can be expressed in comparatively few words, the best definition will be the one that is the shortest.
>
> Contradiction, tautology, and rhetorical flourishes must be banished.

Betula *nana pumila* [*diminutive dwarf*]. Franken. *Tautology.*
Lamium *caule folioso* [with *leaf-like stalk*]. Lind. *Contradiction.*

293. No specific name may be given to a species that is the only one in its genus (203).

> There are some who rule that a specific name should be added to the species of new genera, even if they are the only ones, so that an idea of the plant may be formed from this.
> We do not deny that the habit serves well for the formation of notions about the plant; but in a specific name it is erroneous, inasmuch as it should contain nothing else but only the feature by which it may be distinguished from others of the same genus.
> When there is no specific definition of the plant named, then it is assumed that no species in the genus, other than the solitary one, has been discovered.
> Therefore specific names that impose definitions on plants that are the only ones in their genera, are erroneous, see p. 148: for example:
> Morina *orientalis, with Carlina leaf.* Tournef. cor. 48.
> Dalechampia *scandens, with Lupulus leaves, and prickly fruit with three kernels.* Plum. amer. 17.
> Matthiola *with rough, almost round leaf, and blackish fruit.* Plum. amer. 16.
> Maranta *arundinacea, with a Cannacorus leaf.* Plum. amer. 16.
> Valdia *with thistle leaf, and with somewhat blue fruit.*

294. Anyone who comes upon a new species, should give it a specific name, provided that such a name is needed (293).

> Anyone who discovers a new species, should not only provide a definition of it, but also add to the definitions given for the other or others in the same genus, so that in the future the several species may be distinguished by adequate definitions.
> CLAYTONIA Gron. virg. 25 first became known in Virginia, then another species called Limnia became known in Siberia; and so the Siberian one should be called CLAYTONIA *with egg-shaped leaves, and* the Virginian, CLAYTONIA *with linear leaves.*

295. The words in a specific name must not be compound, or similar to generic manes; and they must not be Greek but Latin, for the simpler they are, the better too.

> The character determines the generic name, whereas the specific definition determines itself; and so the former can be a foreign name, but the latter must be very clear in itself; therefore the latter must be pure Latin, and not Greek.

It follows that all Greek specific definitions are erroneous.

Lathyrus *distoplatyphyllos.*	Mimosa *platykeratos.*
Myrrhis *conejophyllon.*	Mimosa *brachyplatolobos.*
Potamogeton *lejophyllon.*	Pisum *leptolobon.*
Potamogeton *iteophyllon.*	Lotus *tetragonolobus.*
Potamogeton *malacophyllon.*	Trifolium *katoblebs.*
Potamogeton *ulophyllon.*	Clematis *bucananthos.*
Pilosella *monoclonos.*	Ficus *aizoides.*
Pilosella *polyclonos.*	Asclepias *aizoides.*
Lotus *oligokeratos.*	Hieracium *piloselloides.*
Lotus *polykeratos.*	Oreoselinum *anisoides.*

[Pierre Richar] de BELLEVAL, Professor at Montpellier about the end of the sixteenth century, had some very rare pictures engraved; but they were not published. These were communicated to me by the illustrious Sauvages, and I understood that it had been the intention of the author to print all the definitions in Greek, using compound words.[14]

Alsine	μυοσωτις ανθομηλινος
Androsace	ορεκατογ καυλον [sic]
Auricula muris	ορεα[ν]θολευκος
Auricula muris	ορεομικρανθολευκος
Betonica	ορεοπιζοδοντωδης and λευκανθος
Campanula	αλπικυβυυγλωσσυφυλλυς
Campanula	κυανανθοκαλος
Campanula	οπωρομειζορειος
Campanula	ορεομικροπωρος
Carduus	ανακανθος 11
Carduus	λευκεριοκεφαλος C.M.
Condrilla	μικρομηλινοπολυκαυλος
Corruda	λευκοκαυλοσφαιρορρεπης
Cynoglossum	αλπικοχαμαιανθερυθρον
Doronicum	ορεοπολυκλωνανθομηλινον
Gentianella	εαρακαυλορειος
Glycyrrhiza	μακρορριζοπολυσχιδης
Hieracium	μακροστενοφυλλον B.
Hieracium	πλατυανθον
Jacea	ορεανθοκαλος
Jacea	ορειπετρομονοκαυλος
Jacea	ορεακαυλοπορφυρος
Χαμαιλειμωνιον	*Montis Ceti.*
Nardus	ορειψιλοκαυλος
Plantago	πλατυφυλλορρεπης

continued

Polygonum	ανθαλσινοειδες
Pulsatilla	ανθυπομηλινος
Quercus	εναλιοπλατυφυλλος
Racemus marinus	σφυροειδης
Tanacetum	ανοσμον C.M.
Thlaspi	ορεοκαυλοφυλλοστεφης *Horti dei.* [of God's garden]
Trachelium	αλπικοπυραμιδοειδες
Trachelium	ορεογκωπυκνοφυλλον [sic]
Tulipa	ανθεντομωτατος
Tulipa	οξυμεσανθος

296. The specific name must not be decorated with rhetorical figures; still less may it be erroneous, but it should faithfully express what nature dictates.

> *Synecdoche* of the whole for a part is very frequent in botanical works, where something valid only for a part is predicated of the whole; and in our opinion, this should never be allowed.
>
> *Synecdoche* of the singular number for the plural is very common, and just as erroneous.
>
> A *metaphor* is always obscure; for this reason bare simplicity is better.
>
> *Irony*, being a sort of lie, must be excluded for that reason.
>
> > *Synecdoche of the whole for a part.*
>
> Salicaria *purpurea (purple)* for *with purple corollas.*
>
> Quinquefolium *with silver leaf* for *white on the underside.*
>
> Molucca *spinosa [spiny]* for *with spiny calyxes.*
>
> Pimpinella *with pure white umbel* for *with pure white corollas.*
>
> > *Synecdoche of the singular number for the plural.*
>
> Lupinus *with a yellow flower,* for *with yellow flowers.*
>
> Ranunculus *with a round, hairy leaf,* for *with round hairy leaves.*
>
> > *Metaphor.*
>
> Limon *incomparabilis,* for *maximus [greatest].*
>
> Caryophyllus *superbus [proud],* for *with very many flowers.*
>
> Majorana *gentilis [exotic],* for *odoratissima [strongly scented].*
>
> Tragon *improbus [outrageous],* for *aculeatus [stinging].*
>
> Ranunculus *sceleratissimus [very wicked],* for *vesicatorius [blistering].*
>
> Urtica *fatua* for *inermis [unarmed,* i.e. *dead*-nettle].
>
> Cannabis *erratica,* for *mas [male].*
>
> Hesperis *melancholica,* for *scented by night, unscented by day.*
>
> Urtica *mortua [dead]* for *inermis [unarmed].*
>
> Bulbonac *with reviving root,* for *perennis.*
>
> Meum *adulterinum [bastard],* for *not a genuine species.*
>
> Orchis *abortiva,* for *with a peculiar form of flower.*

Pinus *incubacea* [*one lying on another*], for *fasciata* [*with parts grown together*].

Fucus *haemorrhoidalis*, for *remedy for haemorrhoids*.

Pepo *strumosus* [*with tumours*], for *with fruit covered with pimples*.

Caryophyllus *barbatus* [*bearded*], for *with bristling scales on the calyx*.

Mentha *cataria* [*catmint*], for *with scent that is agreeable to cats*.

Gramen *leporinum* [*hare*-like], for *trembling like a hare*.

Lactuca *agnina* [*lamb's* lettuce], for *a species that is agreeable to sheep*.

Aparine *with sugared seed*, for *with fruit covered with warts*.

A tree *for fixing boundaries*,[15] for *with dye for marking out maps in colour by means of it.*
 Irony

Lysimachia *bifolia flore globoso* [*with two leaves, and a spherical flower*], for *with opposite leaves, and egg-shaped bunches* [*of flowers*].

Ornithogalum *with smaller flower growing inside*, for *with flat filaments*.

Narcissus *with yellow cup*,[16] for *with yellow nectary*.

Dentaria *baccifera* [*bearing berries*], for *bearing bulbs in its axils*.

Dracunculus *with a very long pistil*, for *with a very long receptacle*.

Fragaria *sterilis* [*barren* strawberry], for *with a dry receptacle*.

Sabina *sterilis* [*barren*] for *male*.

297. A specific name should not be a comparative or a superlative.

The comparatives *greater* or *lesser*, and the comparative, and superlative degrees of adjectives are not to be allowed, since they presuppose knowledge of another plant.
 Alsine *altissima* [*the tallest*] is smaller than Betula nana [dwarf].

But the opposite is to be understood, where the superlative degree is used of a part of the plant that is the subject, and indicates that the part is the largest of all the parts of the plant; then it is an outstanding feature.
 Lobelia *with very short peduncles, and a very long tube of the corolla.* Roy.

All specific names that draw a comparison with something outside the plant are erroneous.
 Equisetum *laevius* [smoother]. Raj.
 Pilosella *major, minus hirsuta* [*greater, less hairy*]. C.B.
 Pilosella *minor* [*lesser*], *with narrower, less shaggy, leaf*. I.B.

298. A specific name should make use of positive, not negative, terms.

Negative terms tell us nothing, or rather tell us what is not, not what is.

So long as positive terms are available, it is never permissible to make use of privatives; thus there will always be words ready to express the opposite meanings.

Almost round	and	oblong.
Rounded <rotundatum>	and	angled.
Split	and	undivided.
Blunt	and	sharp.
Stinging	and	unarmed [dead, of nettles].
Saw-toothed	and	absolutely entire.
Nervous	and	nerveless.
Tufted	and	smooth.
Covered	and	bare.
Veiny	and	veinless.
Rounded <teres>	and	angular.
Hollow	and	filled in.
Simple	and	compound.
Petioled	and	sessile.
Upright	and	rolling.
Rounded <teres>	and	angled.
Absolutely simple	and	branchy.
Awned	and	cut off.
Loose (flabby)	and	tight.
Separated	and	crowded.
Spread out	and	confined.
Stemmed	and	stemless.
Herbaceous	and	woody or shrubby.
Thinned down	and	thickened up.

If the description of the plant were to be greatly enlarged with negative terms, no one could form the slightest idea of the plant from it: see the *Examen epicriseos* of Dr. *Bishop Browallius*.

Therefore all specific names that include negative words or particles are erroneous.

Lysimachia *non papposa* [*without pappus*] Mor. for *with bare seeds*.

Hippuris *non aspera* [*not rough*] I.B., for *smooth*.

Bidens *folio non dissecto* [*with a leaf that is not divided*], T., for *entire*.

Phalangium *non ramosum* [*not branchy*] Wehm., for *with single stem*.

Lychnis *petalis non bifidis* [*with petals that are not divided into two*] Mor. for *entire*.

299. Any simile that is misused in a specific name, may be more familiar than your right hand; but even so, it is unacceptable.

A simile exhausts in a single word what would otherwise have to be explained by a whole speech; but every simile is lame, and therefore every obscure simile, which is not absolutely obvious to everyone, is used to the disgrace of the art of botany.

Therefore no simile is to be used, other than those derived from the external parts of the body, such as the ear, finger, navel, eye, scrotum, penis, vulva or breast; and not from the internal parts of the body, which are well known only to anatomists.

A simile is valid without limitation, in a case where there is no other technical term.

Botanists have introduced obscure similes, and indeed very many of them, that are not understood by all: for example:

Agaricus *with a tube like a Fallopian [tube]*. T.

Orchis *bearing [the likeness of] a man <anthropophora>*.

Orchis *resembling an ape*. C.B.

Orchis *representing a long-tailed monkey*. Col.

Orchis *resembling a fly*.

Mesembryanthemum *resembling a heron's beak*. D.

Mesembryanthemum *resembling a dog's open mouth*. D.

Mesembryanthemum *resembling a cat's open mouth*. D.

Lotus *with siliquae resembling crows' feet*. C.B.

Fungus *with hairs precisely similar to a roebuck's*. Loes.

Atriplex *with the seed of a bucephalon*. Col.

Tree with prickly crooked branches, *forming topiary*. Pluk.

Fungus *with Daedalian[17] [labyrinthine] recesses*. T.

Fungus *resembling an ear-pick*.

Hemionitis *with a leaf in the shape of a Roman axe*. Pluk.

Medica *caseiformis [cheese shaped]*. Rudb.

300. A specific name should not make use of any adjective without a corresponding substantive.

In a specific name, there should be no adjective, and therefore no attribute, without a previous substantive with which it agrees; but where no part is previously named, there it is assumed that it is predicated of the whole plant.

Substantives regularly used in this context must always be parts of the plant.

All specific names that contain adjectives without any corresponding substantives are erroneous.

Adjectives without a veritable substantive.

Millefolium *cornutum [horny]* C.B.: with .... leaves.

Nigella *cornuta* C.B.: with a .... capsule.

Thlaspidium *cornutum* T.: with a .... calyx.

Lysimachia *corniculata [small-horned]* C.B.: with a .... capsule.

Allium *bicorne [two-horned]* C.B.: with a .... spathe.

Viola *tricolor [three-coloured]* C.B.: with a .... corolla.

Myrtus *cristata [crested]*: with .... leaves.

Amaranthus *cristatus*: with a .... spike.

Gramen *cristatum [crested* grass]: with .... bracts.

Solanum *vesicarium [bladder-like]*: with a .... calyx.

Colutea *vesicaria*: with a .... pericarp.

Ranunculus *vesicarius [to cure the bladder]*: with power to ....

Millefolium *vesicatorium [blistered]*: with a root furnished with blisters.

Mesembryanthemum *vesicatorium*: with leaves sprinkled [with blisters].
　　Erroneous substantives.
Sederitis with spiny *bags. Herm.* for whorls.
　　Adjectives and substantives should agree in gender.
Juniperus *alpinus* Clus.: alpina.
Hippuris *muscosus* Mor.: muscosa.

301.　Every adjective (300) in a specific name ought to follow its substantive.

Just as the part or predicate should always come first in a character, so in a definition
　　should the substantive with which the adjective agrees; so that the meaning may
　　be absolutely clear, and that a different sense may not creep in as a result of a
　　printer's error over a comma or a fullstop.
Corona solis *parvo* flore, *tuberosa* radice (with a *small* flower and a *tuberous* root]. T.
　　BETTER: *flore* parvo, *radice* tuberosa.
Sinapistrum aegyptium heptaphyllum, *flore carneo*, majus spinosum [with seven
　　leaves, *flesh-coloured flower*, greater spiny]. *Herm.*
　　BETTER: aegypt. heptaphyll. majus spinosum, flore carneo.
Orchis aethiopica, greatest, spotted; a little bird, snow-white, [which is] with a spot,
　　blood-red, on its back marked, resembling; with a crest, blue, very large, with
　　powder, silvery, sprinkled, [which is] conspicuously splendid.[18]
　　BETTER: aethiopica, greatest, spotted; resembling a little bird, snow-white: on
　　　　its back a blood-red mark; with a crest which is blue, very large and sprinkled
　　　　with silvery powder.

302.　Adjectives (300) used in a specific name should be sought among the
　　　　special technical terms (80–6), provided that they are appropriate.

If botanists agree about technical terms, and are consistent in their writings,
　　knowledge will turn out to be very easy to acquire.
A botanist should never allow *circumlocution,* so long as clearly defined technical
　　terms are available.
Synonyms of terms should be excluded, and a single, consistent selected term
　　should be used.

Synonyms		
Lychnis *viscosa*	}	glutinosa.
Lychnis *glutinosa* Rd		
Caryophyllus *supinus* C.B.	}	procumbens
Malva *procumbens* [lying down].		
Ligustrum *foliis pictis* [*with decorated leaves*] Weh.	}	variegatis.
Laurocerasus *foliis variegatis* Weh.		

continued

Pilosella repens *hirsuta [hairy]*
Pilosella repens *folio piloso [with a shaggy leaf]* } pilosa (hairy)
Pilosella repens *folio villoso [with a villous leaf].*
Quinquefolium *pubescens [downy]*

Hieracium *radice succisa [with root cut back]* I.B.
Hieracium *radice praemorsa [with root bitten off]* Mor. } praemorsa
      By circumlocution

Quinquefolium *molli lanugine pubescens (pubescent*
   with soft down) I.B. villosum [villous].

Conyza *humidis locis proveniens (occurring in damp*
places) I.B. palustris [marsh].

Muscus squamosus *in aquis nascens (growing in*
water) Moris. aquaticus [water].

303. **A specific name should not include particles that connect adjectives or substantives.**

> Particles of this kind are either
>> Conjunctive:, *and, and also, together with;*
>> Disjunctive: *or, or else.*
>
> All features in a definition should be set forth in the *ablative* case, without any preposition.
>
> And where two different things are to be indicated in the same plant, we use the suffix *ve* [or] or *que* [and] at the end of the second word,[19] so as not to increase the number of words: for example:
>> Carduus with *leaves that are lanceolate, ciliate, entire, and lappeted <laciniatisque>.* Hortus Cliffortianus 391.
>> Juncus *with leaves that are flat, and a spike that is sessile, and [both] are pedunculate <pedunculatisque>.* Flora Suecica 288.
>
> All definitions that include disjunctive particles are banished: for example:
>> Medica *silv[estris] frut[escens],* or Trifolium *falcatum,* or else Medica *with a twisted silique.* Moris.
>> Absinthium *ponticum,* or else *romanum officinarum [of the shops in Rome],* or else *Dioscoridis,* Moris. C.B.
>> Aster *montanus* or else *like, if not identical with, Oculus Christi* or else Conyza 3.[20] I.B.

304. **Punctuation marks should separate parts of the plant (80), and not adjectives, in a specific name.**

> The separating pieces, in this context, are the *comma (,), comma with point* [semicolon] *(;), colon (:),* and *point* [full stop]*(.).*

The specific definition is brought out very clearly by means of punctuation properly inserted.

I use a *comma* to separate the parts; I make use of a *colon*, where there is a subdivision of a part; and I finish the definition with a *full stop*. A.

OTHERS use a *semi-colon* to separate the parts; they separate all adjectives with a *comma*. B.

OUR PREDECESSORS mostly separated both the parts and the adjectives with commas. C.

    A. Bauhinia inermis, *with leaves* that are heart-shaped [and] half-divided into two: *with lappets* that are acuminate-ovate [and] vertically dehiscent. H[ortus] C[liffortianus].

    B. Bauhinia inermis, *with leaves* that are heart-shaped, half divided into two; *with lappets* that are acuminate-ovate, vertically dehiscent. H.C.

    C. Bauhinia inermis, *with leaves* that are heart-shaped, half-divided into two, *with lappets* that are acuminate-ovate, vertically dehiscent. H.C.[21]

305. A specific name should never include a parenthesis.

A parenthesis, whether just understood or enclosed in brackets, is rejected in the same way in either case.

Either indicates either an exception or a failure of arrangement, and so is not admissible.

Boerhaave, and his contemporaries, to avoid a radical alteration of the names, put the title of the new genus before the old generic name, with the insertion of a syllable *qui, quae* or *quod* [*which (is)*]: but we do not adopt this practice, on account of the view expressed above.

    a. Parenthesis that is just understood.

Sinapistrum pentaphyllum, with flesh-coloured flower, lesser. H.

    b. Parenthesis that is marked in writing.

Androsaemum maximum (*as it were fruiting*) bearing berries. Mor.

A climbing relative of the Viola, with a navel-shaped (*or shield-shaped*) leaf. Br.

    c. Qui, quae, quod.[22]

Dens leonis *which is* <*qui*> a Pilosella with a less villous leaf. T.

Doria *which is* <*quae*> a Jacobaea orientalis with a lemon leaf. T. cor. *B.*

Titanokeratophyton *which is* <*quod*> a Lithophyton marinum (a marine mineral-vegetable), that is whitish. Gesn. *B.*

IX. VARIETATES.

306. Nomini Generico (vii.) & Specifico (viii.) etiam varians, si quod (158), addi potest.

Varietates sunt plantæ ejusdem speciei, mutatæ a caussa quacunque occasionali.

Usus varietatum in Oeconomia, Culina, Medicina necessariam reddidit earum cognitionem in vita communi; ad Botanicos cæteroquin non spectant varietates, nisi quatenus Botanici curam gerant, ne Species multiplicentur aut confundantur.

Evidentiores varietates, ob usum publicum, ad finem differentiæ inserat Botanicus, ubi necesse est.

307. Nomina Generica, Specifica & Variantia literis diversæ magnitudinis scribenda sunt.

Nomen *Genericum* pingatur literis majoribus Romanis.
    Specificum literis mediocribus & vulgatissimis.
    Varians autem literis minoribus currentibus vulgo dictis.

Fiat hoc ut Varietas distinctissima tradatur a Differentia.
    CONVALLARIA scapo nudo; *corolla plena.*
    CONVALLARIA scapo nudo; *corolla rubra.*
    SAXIFRAGA alpina ericoides; *flore purpurascente.*
    SAXIFRAGA alpina ericoides; *flore cæruleo.*
    PENTAPHYLLOIDES palustre rubrum; *crassis & villosis foliis.*

308. Sexus (149) Varietates naturales constituit; reliquæ omnes monstrosæ sunt.

Dioicæ plantæ constituunt unicum modum varietatum vere naturalem, in *Mares* & *Feminas* distinctum, quas nosse & differentiis addere Botanicis perquam necessarium est.

Veteres, ignari fundamenti fœcundationis plantarum, *Mares pro Feminis*, & *Feminas pro Maribus* assumsere, quod sedulo vitandum. e. gr.

    Mercurialis *mas.* T.
    Cannabis *mas.* I.B.   sunt *feminæ.*
    Lupulus *mas.* T.

309. Varie-

✳ IX. VARIETIES

306. Some further varietal name, if any (158), may be added to the generic (VII) and specific (VIII) names.

> Varieties are plants of the same species, changed by some adventitious cause or other.
>
> The use of varieties in gardening, cookery, and medicine makes it necessary to recognize them in ordinary life; otherwise, varieties do not concern botanists, except in so far as the botanists bother about them, so that the several species shall not be multiplied or confused.
>
> Where it is necessary, the botanist should insert the more obvious varieties at the end of the definition, for the benefit of the public.

307. Names of genera, species, and varieties are to be written in letters of different sizes.

> The name of a genus should be represented in Roman capital letters; that of a species in ordinary common letters; that of a variety in small letters, commonly called italics.
>
> This should be done, in order that a variety may be shown as absolutely distinct from a definition of a species.
>
> CONVALLARIA scapo nudo; *corolla plena.*
> CONVALLARIA scapo nudo; *corolla rubra.*
> SAXIFRAGA alpina ericoides; *flore purpurascente.*
> SAXIFRAGA alpina ericoides; *flore caeruleo.*
> PENTAPHYLLOIDES palustre rubrum; *crassis & villosis foliis.*

308. SEX (149) constitutes natural varieties, all the rest are abnormal.

> Dioecious plants constitute the only sort of variety that is truly natural, separated into males and females; and it is very necessary for botanists to recognize them and add them to the definitions.

The ancients, who did not understand the basis of the fertility of plants, took *males for females* and *females for males*; but this should be carefully avoided: for example:-

Mercurialis *mas* [*male*] T.
Cannabis *mas* I.B.   } are *female*
Lupulus *mas* T.

309. Abnormal varieties (308) are constituted by *flowers* that are mutilated (120), filled in (121), or prolific (222), and herbage that is luxuriant, bundled, folded, or mutilated: in number, shape, relative size, and position of all the parts, and quite often in colour, scent, taste, or season.

The principal sorts of varieties are the following:

Mutilation of the corolla	Sections 119 and 184.
Multiplying of the corolla	Sections 120, 126 and 127.
Filling in of the corolla	Sections 121 and 124.
Proliferation of the corolla	Sections 123 and 124.
Bundled development of the stem	Section 274
Pleached development of the stem	Section 274.
Curled development of the leaves	Section 311.
Puckered development of the leaves	Section 311.
Colour of the herbage	Section 266.
Scent of the herbage	Section 267.
Taste of the herbage	Section 268.
Quantity of the herbage	Section 260.
Season of the herbage	Sections 265 and 273.

Colouring of the leaves.
Buxus *with leaves that are golden at the edges*. T.
Aquifolium *with silver thorns and edges*. T.
Alaternus *with leaves that are variegated with yellow*. H.R.P.
Salvia *with leaves that are variegated with green and yellow*. H.R.P.
Aloe sobolifera *with elegantly variegated leaves*. Herm. prodromus.
Caprifolium perforatum *with leaves that are uneven at the edge and variegated*. T.
Urtica urens minor *with leaves that are elegantly variegated in green and red*. Rudb.
Gramen paniculatum aquaticum, with Phalaris seed and *with variegated leaves*. T.

310. A botanist does not care about very small variations (7).

LOVERS OF FLOWERS, with their excessive industry and careful inspection, have observed, in the corollas of flowers, such marvels as the inexpert eye cannot descry; their object is the very beautiful flowers of *Tulipa, Hyacinthus,*

Pulsatilla, Ranunculus, Dianthus, and *Primula*: they have called obscure varieties of these by words which give rise to astonishment; they pursue their own peculiar kind of knowledge of flowers, which is clear only to adepts; therefore no sane botanist will enter their camp.

Phoebus.
Apollo.
Astraea.
Daedalus.
Cupido [Cupid].
Aurora.
Gratiosa [Gracious].
Pretiosa [Precious].
Triumphus Florae [Triumph of Flora].
Pompa Florae [Procession of Flora].
Splendor Asiae [Splendour of Asia].
Corona Europae [Crown of Europe].
Gemma Hollandiae [Gem of Holland].
Sponsa Amstelodami [Bride of Amsterdam].
Alexander Magnus [Alexander the Great].
Carolus Duodecimus [Charles XII (of Sweden)].
Julius Caesar.
Imperator Augustus [The Emperor Augustus].
Tartar Cham [The Tartar Khan].

The multitude of GARDENERS have applied to trees, that bear pomes or stone-fruits, varietal names that are not, and cannot be, defined, and are almost infinite in number.

Poma (Apples)		Pyra (Pears)	
	Paradisiaca		*Falerna*
	Prasomila		*Favonia*
	Rubelliana		*Boni Christiani*[1]
	Borstorphiana		*Crustamina*
	Appiana		*Picena*
	Melimela		*Libraria.*

To the disgrace of the art of botany, the order of the FUNGUSES is even now in confusion, since the botanists do not know what is a species and what is a variety.

311. Luxuriation of the leaves can very easily occur when they are opposite or compound. All CURLED or PUCKERED leaves are abnormal.

OPPOSITE leaves, Section 83:112, often turn out to be in sets of three 83:111 or four 83:111; and then a polygonal stem is made out of a quadrangular one. *Lysimachia lutea major, foliis ternis* [*with leaves in sets of three*]. T.

Lysimachia lutea major, *foliis quaternis* [*sets of four*]. T.

Lysimachia lutea major, *foliis quinis* [*five*]. T.

Anagallis caerulea, foliis binis *ternisve ex adverso nascentibus* (with leaves in sets of two, *or three, growing opposite* [*each other*]). Raj.

Anagallis phoenicea, foliis amplior[ibus] *ex adverso quaternis* [*with quite large leaves, opposite each other in sets of four*]. T.

Salicaria *trifolia, caule hexagono* [*with three leaves, and a hexagonal stem*]. T.

DIGITATE leaves often have an additional leaflet.

Trifolium [three-leaved] with four leaves.

All plants with CURLED leaves are abnormal varieties, like filled-in corollas in the flowers; and so plants furnished with these leaves are no natural plants, but acknowledge something other [than Nature] as their mother.

Apium or Petroselinum *crispum* [curled]. C.B.

Heracleum: [?or] Sphondylium *crispum*. [curled]. C.B.

Rumex: [?or] Lapathum *folio* acuto *crispa*. [curled, with pointed leaf] C.B.

Rumex acetosa *foliis crispis* [with *curled leaves*]. C.B.

Reseda *crispa* gallica [French *curled*]. Bocc.

Reseda luteola lusitanica pumila *crispa* [diminutive Portuguese, *curled*] T.

Brassica *laciniata* rubra. [red, *lappeted*]. J.B.

Nasturtium hortense *crispum* [garden, *curled*]. C.B.

Malva *crispa* J.B.

Lactuca *crispa*. C.B.

Cichorium *crispum*. T.

Lapsana folio amplissimo *crispo* [with a very large *curled* leaf]. B.

Tanacetum foliis *crispis*. C.B.

Matricaria *crispa*.

Asplenium: [?or] Lingua cervina maxima, *undulato folio* [greatest harts tongue *with a wavy leaf*]. H.R.P.

Leonurus; [?or] Cardiaca *crispa*. Raj.

Mentha *crispa* danica [*curled* Danish]. Park.

Ocymum latifolium maculatum vel *crispum* [spotted or *curled*]. C.B.

In *Tanacetum, Mentha, Ocimum* and *Matricaria*, the SCENT increases with the curliness; which is peculiar.

PUCKERED leaves, Section 374, are mostly formed from wrinkled ones, 83:64, when the substance of the leaf is enlarged and multiplied, and so pushed up between its vessels.

Ocimum foliis *bullatis* [with *puckered* leaves]. C.B.

Brassica *undulata* [*wavy*]. Renealm.

Lactuca capitata, *foliis magis rugosis* [*with leaves more wrinkled*]. B.

Lactuca capitata major, *foliis rugosis & contortis* [greater, *with leaves wrinkled and twisted*]. B.

Lactuca capitata omnium maxima, *verrucosa* [greatest of all, *with warts*]. B.

SAPONARIA *concava anglica* [*concave English*], which J.B. elegantly described as Gentiana folio convoluto [with leaf rolled together] J.B., grows a puckered leaf in a peculiar manner, in the absence of wrinkles: the edge is contracted, and the leaves turn out to be concave, like a spoon.

Plants *with thin leaves* sometimes develop out of broad-leaved ones, but this variation is relatively infrequent.

Heracleum hirsutum, *foliis angustioribus* [*with quite narrow leaves*]. C.B.

Lycopus *foliis in profundas lacinias incisis* [*with leaves carved into deep lappets*]. T.

Brassica *angusto apii folio* [*with a narrow celery leaf*]. C.B.

Veronica austriaca, *foliis tenuissine laciniatis* [*with leaves very thinly lappeted*].

Sambucus *laciniato folio* [*with a lappeted leaf*]. C.B.

Sonchus *asper laciniatus* [*rough, lappeted*]. C.B.

Valeriana sylvestris, *foliis tenuissime divisis* [*with leaves very thinly divided*]. C.B.

312. **It is usually superfluous to include diseased plants, or even the ages [of plants], in the names of varieties.**

Diseased plants that are recognized by botanists are diverse, like their diseases.

Erysiphe[2] Th. is a white mould with brown sessile heads, with which leaves are sprinkled, frequent in *Humulus, Lamium* [my] Flora Suecica 494, *Galeopsis* Flora Suecica 411, Lithospermum Flora Suecica 152, and *Acer* Flora Suecica 303.

Rust is iron dust, sprinkled on leaves on the underside, frequent in *Alchemilla* Flora Suecica 135, *Rubus saxatilis* Flora Suecica 411, *Esula degener* R., and especially in *Senecio* or *Jacobaea, with a grey perennial Senecio leaf*, Hall. jen. 177, principally in woodland soil that has been scorched.

Clavus [*nail*], when the seeds grow out into quite large horn-like projections, which are black on the outside, as in *Secale* and the *Carexes*.

Ustilago [*scorching*], when the fruits produce black powder instead of seeds.

Ustilago of Hordeum [*barley*]. C.B. *Ustilago of Avena* [*oats*]. C.B.

Scorzonera pulveriflora (with dusty flowers). H.R.P.

Tragopogon abortium [*miscarriage*]. Loes.

Insects' nests, caused by insects that have deposited their eggs in plants, from which there are various excrescences.

Galls of *Quercus, Glechoma, Cistus, Populus tremula*, the *Salixes*, and *Hieracium myophorum*.

Bedeguar of *Rosa*.

Follicles of *Pistacia* and *Populus nigra*

Contortions of *Cerastium, Veronica*, and *Lotus*.

Scaly surfaces of *Abies* and *Salix rosea*.

Insects often cause filling in and proliferation of flowers.

Matricaria Chamaemelum vulgare, Flora Suecica 702, is made prolific by very small insects.

Carduus caule crispo [*with curled stem*], Flora Suecica 658, displays, as a result of the attention of insects, flowers that are larger, grey, and filled in, or rather prolific leaves, with pistils growing out on the leaves.

313. Colour changes very readily, especially from blue or red to white.

The following colours count as the chief ones with botanists:

Crystal, *watery* and *glassy*.

White, *milk-* and *snow-*.

Grey, *hoary, livid,* and *leaden*.

Black, *dusky*

    Swarthy, *lurid*.

    Sable, *pitch-black*.

Yellow.

    Flaxen, *sulphurous*.

    Tawny, *saffron* and *fiery*.

    Pale, *brick-colour* and *rusty*.

Red, *blood-*.

    flesh-colour.

    scarlet, *Phoenician*.[3]

Purple, *Phoenician*.[3]

    violet, *blue-purple*.

Blue.[4]

Green, *leek-*.[4]

In PLANTS, different colours are more appropriate to different parts.

    Black occurs frequently in the *root*, often in the *seeds*, more rarely in the *pericarp*, very rarely in the corolla.

    Green is appropriate to the *herbage* and the *calyx*, and very rare in the corolla.

    Crystal is frequent in the *filaments* and the *pistil*.

    Yellow is frequent in the *anthers*, and not infrequent in *corollas*, especially *autumnal* ones, and those with *demi-florets*. T.

    White is frequent in *spring-time corollas* and *sweet berries*.

    Red is frequent in *summer flowers,* and in *berries* that grow in the shade and are acid.

    Blue too is not infrequent in flowers.

The colours of FLOWERS quite often change.

Red *into* white:

    Erica, Serpyllum, Betonica, Galeopsis, Pedicularis.

    Dianthus, Silene, Cucubalus, Agrostemma, Coronaria.

    Trifolium, Orchis, Digitalis.

    Carduus, Serratula, Gnaphalium,

    Rosa, Papaver.

    Fumaria, Geranium.

Blue *into* white:

    Campanula, Polemonium, Convolvulus.

    Hepatica, Aquilegia, Viola.

    Vicia, Galega, Polygala.

Echium, Anchusa, Symphytum, Borrago.
Hyssopus, Dracocephalum, Scutellaria.
Scabiosa, Jasione, Cyanus, Cichorium.

Yellow *into* white:

Melilotus, Agrimonia, Verbascum.
Tulipa, Blattaria, Alcea.
Cyanus turcicus, Chrysanthemum.

White *into* purple

Oxalis, Datura.
Pisum, Bellis.

Blue *into* yellow:

Commelina, Crocus.

Red *into* blue:

Anagallis.

Multiple changes.

Aquilegia blue into red and white.
Polygala blue into red and white.
Hepatica blue into red and white.
Cyanus blue into red and white.
Mirabilis red into yellow and white.
Impatiens yellow into red and white.
Tulipa yellow into red and white.
Anthyllis yellow into red and white.
Primula red into yellow and white.
Cheiranthus yellow into blue and white.

PERICARPS that consist of berries are at first green, then whitish; when ripe, they show various colours, white, red, and blue, especially in pears, plums and cherries.

Solanum guineense, fructu nigerrimo [with very black fruit]. B.
Solanum annuum, baccis luteis [with yellow berries]. Dill.
Solanum judaicum, baccis aurantiis [with orange berries]. Dill.
Rubus vulgaris major, fructu albo [with white fruit]. Raj.
Ribes vulgare acidum, albas baccas ferens [bearing white berries]. J.B.

SEEDS change their colours too, though more rarely.

Papaver	*hortense, nigro semine* [with a black seed]. C.B.	
	hortense, semine albo [with a white seed]. C.B.	
Avena	*vulgaris & alba* [common and white]. C.B.	
	nigra [black]. C.B.	
Phaseolus	*vulgaris, fructu violaceo* [with violet fruit]. T.	

           vulgaris fructu ex rubro & nigro variegato [with fruit variegated red and black]. T.

           fructu albo, venis nigris & lituris distincto [with white fruit marked with black veins and smudges]. T.

Pisum   *maximum, fructu nigra linea maculato* [with fruit mottled by a black line]. H.R.P.

           hortense, flore fructuque variegato [with variegated flower and fruit]. C.B.

Faba   *ex rubicundo colore purpurascente* [with the colour changing from reddish to purple].

The ROOT quite rarely varies its colour.

    Daucus sativus, radice alba [with white root]. T.

    Daucus sativus, radice lutea [with yellow root]. T.

    Daucus sativus, radice aurantii coloris [with orange-coloured root]. T.

    Daucus fativus, radice atro-rubente [with reddish-black root]. T.

    Raphanus niger [black]. C.B.

The LEAVES quite rarely change their green colour, as in the *Amaranti*; yet sometimes they put on spots.

    Persicaria cum maculis ferrum equinum referentibus [with spots resembling horseshoes]. T.

    Ranunculus hederaceus, atra macula notatus [marked with a black spot].

    Orchis palmata palustris maculata [marsh, spotted]. C.B.

    Hieracium alpinum maculatum [spotted]. T.

    Lactuca maculosa [with spots]. C.B.

The whole HERB sometimes take on an extraneous colour.

    Eryngium latifolium planum, caule ex viridi pallescente, flore albo [with stem changing from green to sallow, and white flower]. T.

    Abrotanum cauliculis albicantibus [with the smaller stems turning white]. T.

    Artemisia vulgaris major, caule ex viridi albicante [with stem turning from green to white]. T.

    Atriplex hortensis rubra [red]. C.B.

    Amarantus sylvestris maximus novae angliae, spicis purpureis [from New England, with purple spikes]. T.

    Portulaca sativa, foliis flavis [with flaxen leaves]. Moris.

    Lactuca capitata rubra [red]. B.

314. A *wet* location quite often causes the division of the lower leaves, a *mountainous* one that of the upper leaves.

    Irregular leaves, or leaves of various shapes on the same plant, are comparatively rare.

    Tithymalus *heterophyllus* [with different leaves]. Plum. Pluk. alm. 112. f.6.

Rudbeckia *foliis inferioribus trilobis, superioribus indivisis* [with three-lobed lower leaves and upper leaves undivided]. [My] Hortus Upsaliensis.

Hibiscus *foliis inferioribus integris, superioribus trilobis* [with entire lower leaves and three-lobed upper leaves]. Hortus Cliffortianus.

Lepidium foliis caulinis pinnato-multifidis, rameis cordatis amplexicaulibus integris [with leaves on the stem that are pinnate with many divisions, and leaves on the branches that are heart-shaped, clasping the stem and entire]. Hortus Cliffortianus.

AQUATIC plants quite often make their lower, submerged leaves capillary.

Ranunculus *aquaticus, folio rotundo & capillaceo* [with round, capillary leaves]. C.B.

Sisymbrium *foliis simplicibus dentatis serratis* [with simple, toothed leaves]. Hortus Cliffortianus.

Cicuta, Sium, Phellandrium, Oenanthe. &c.

MOUNTAIN plants form their lower leaves comparatively entire, and their upper leaves more divided: for example, *Pimpinella, Petroselinum, Anisum* and *Coriandrum*.

Thin-leaved plants, Section 311, belong here. This is confirmed by the fact that they are all formed by a dry soil.

315. A natural plant (157) ought not to be recorded by a name that distinguishes it from its varieties (158).

Since varieties are superfluous in the forum of botany, this rule must be observed to prevent definitions from being multiplied indefinitely; and it is not necessary to distinguish a natural plant from abnormalities.

316. Cultivation is the mother of very many varieties and is the best means of testing varieties.

Horticultural display has produced flowers that are *filled in*, fruit *in* [*different*] *seasons, turions* on stems, *nutritive* herbs, herbs *that form heads*, and tender *vegetables*; these plants, if left to themselves in poor soil, assume a wild and natural nature.

I have observed that [*vegetables*] *that have been selected for a long time and looked after with much labour, will degenerate even so, unless they are selected by hand every year, with very great human effort; thus all things are doomed to go to ruin and be carried back in a retrograde slide.* Virgil, Georgics Bk I [ll. 197–200.]

Thus the sweetest *vines* would become acid, the pleasantest *apples* harsh, the most agreable *pears* tart, the blandest *almonds* bitter, the juiciest *peaches* dry, the smoothest *lettuces* spiny, the fleshiest *asparagus* woody, and the tastiest *cherries* very sharp; and all cereal *vegetables* and fruits would become worthless.

Soil changes plants, and as a result varieties are produced; and when the soil is changed they revert.

Buxus *arborescens* [*growing like a tree*]. C.B. Buxus humilis [lowly]. Dod.

Xanthium. Dod. Xanthium *canadense majus* [*greater Canadian*]. T.

Acanthus *mollis* [*soft*]. C.B. Acanthus *aculeatus* [*prickled*]. C.B.

Cinara *aculeata*. C.B. and *non aculeata*. C.B.

Brunella. Dod. Brunella *caeruleo magno flore* [*with a large blue flower*]. C.B.

Myosotis *foliis hirsutis* [*with hairy leaves*]. H.C. and *foliis glabris* [*with smooth leaves*]. H.C.

Crista galli *femina* [*female*]. I.B. and *mas* [*male*]. I.B.

Cerinthe *flore ex rubro purpurascente* [*with flower turning from red to purple*]. C.B. and *flavo flore* [*with flaxen flower*]. C.B.

317. It is no less important to assemble the different varieties under their proper species, than to place species together under their proper genus.

The consistency of the ancients, in handing species down to posterity separately, has been defeated by the zeal of more recent botanists, at the end of the last century, for increasing the number of the plants; and it has infected the science of botany by the introduction of varieties as species, when a new species was created on account of an unimportant feature, to the detriment of botany. This notion progressed so far that varieties came out as species, and species as genera. Vaillant was the first to oppose this heresy, then myself, then Jussieu, Haller, Royen, Gronovius, and not a few others, to save the science from ruin.

Most varieties are very easily explained and dealt with by comparing the variant features with the natural plant; but there are several varieties that demand both ability and experience.

HELLEBORUS *aconiti folio, flore globoso croceo* [*with aconite leaf and spherical saffron flower*]. Amm. ruth. 101. Trollius *humilis flore patulo* [*with wide-open flower*]. Buxb. cent. 1. p. 15 pl. 22. The Helleborus variety *Trollii*, [My] Flora Suecica 474, with nectaries as long as the corolla.

GENTIANA *corolla hypocrateriformi: tubo villis clauso, calycis foliis alternis majoribus*, [*with salver-shaped corolla: with tube closed by villi, and alternate leaves of the calyx larger*]. Flora Lapponica 94. The Gentiana variety *fauce barbata* [*with bearded throat*], Flora Suecica 203, with the flower divided into four, and with alternate lappets of the calyx doubled in width.

FUMARIA *bulbosa with the root hollow, and with the root not hollow, greater and lesser* are shown to be of the same species by the very small perianth, by the several species of the same genus, the scales on the bud, the structure of the leaves, the position of the branch, the location of the bract, by the corolla, the silique, the seeds, and the stigma; but it varies in the divided bracts, in the root, and in being more or less hollow.

VALERIANA *arvensis praecox humilis, semine compresso* [*early, low-growing, with compressed seed*]. T.

Valeriana arvensis praecox humilis, foliis serratis [*… with saw-toothed leaves*]. T.

Valeriana arvensis serotina altior, semine turgidiore [*late, quite tall, with seed quite swollen*]. Mor.

Valeriana semine umbilicato nudo rotundo [*with seed that is umbilicate, uncovered, and round*]. Moris.

Valeriana semine umbilicato nudo oblongo [*… and oblong*]. Moris.

Valerianella semine umbilicato hirsuto majori [*… hairy and relatively large*]. Moris.

Valerianella semine umbilicato hirsuto minori [*… and relatively small*]. Moris.

Valerianella cretica, fructu vesicario [*with bladder-shaped fruit*]. Tournef. cor.

Valerianella semine stellato [*with star-shaped seed*]. C.B.

That these plants, differing greatly in the fruit, with leaves quite often relatively more divided, are of the same species, is proved by the stem divided into two, the annual root, and the structure of the leaves, corolla, and seed.

SCORPIURUS H.C., with a pod that differs greatly in separate specimens, is nevertheless the same species.

Scorpioides siliqua campoide hispida [*with a curved prickly silique*]. I.B.

Scorpioides siliqua cochleata et striata ulyssiponensis [*with the siliqua shaped like a snail-shell and streaked, of Lisbon*]. T.

Scorpioides bupleuri folio, siliquis levibus [*with a Bupleuron leaf, and smooth siliques*]. Park.

Scorpioides siliqua crassa [*with a thick silique*] Boelius' Ger.

MEDICAGO *leguminibus cochleatis, stipulis dentatis, caule diffuso* [*with shell-shaped pods, saw-toothed stipules, and spreading stem*]. H.C.

Scutellata [*shaped like a small shield*], *coronata* [*crown-shaped*], *hirsuta* [*hairy*], *polycarpos* [*with many fruits*], *orbiculata* [*circular*], *doliata* [*barrel-shaped*], *lupulina* [*hop-like*], *dicarpos* [*with two fruits*], *echinata* [*prickly like a sea-urchin*], *ciliaris* [*with eyelashes*], *spinosa* [*spiny*], *arabica* [*Arabic*], *turbinate* [*top-shaped*], *tornata* [*rounded*], *rugosa* [*wrinkled*], *cretica* [*Cretan*].

In its fruit, the separate varieties are of as many shapes as the actual shells which nature imitates in these plants by the sea-shore.

The botanist who chooses to exercise himself over varieties can hardly come to the end of the playfulness of nature in its numerous shapes.

X. SYNONYMA.

318. Synonyma funt diverfa Phytologorum (6) nomina, eidem plantæ impofita, eaque *Generica*
(VII), *Specifica* (VIII) & *Variantia* (IX).

Patres §. 9. conveniebant plerumque in plantarum nominibus,
genericis folum contenti.

Commentatores §. 10 ob defectum Defcriptionum & Figurarum
in Patrum fcriptis, varie eorum nomina plantis applica-
bant.

Defcriptores §. 11. 12. detectis numerofioribus plantis, pro ar-
bitrio nomina imponebant.

C. Bauhinus quadraginta annorum labore, in Pinace 1622, An-
teceflorum nomina conjungebat & reduxit ad 6000 Species.

Curiofi §. 4. conquirentes undique novas plantas, reddidere nu-
merum earum duplo auctiorem.

Syftematici §. 53. in Generum conftructione olim maxime diffen-
tientes, fecundum falfa genera, falfiffima nomina plantis
imponebant.

Legibus differentiarum fpecificarum non datis, Botanici Differentias partim triviales, partim variabiles, omnes lubri-
cas, fpeciebus imponebant.

G. Sherardus, Anglus, Botanicus infignis, Bauhinii Pinacem
continuare allaboravit, fed fato præventus 1728 reliquit
opus Dillenio.

Dillenius, Profeffor Sherardinus Oxoniis, continuavit opus Sherar-
di in annum, quo obiit, 1747.

Sibthorpius, Succeffor Dillenii, Pinacem Sherardianum & Dille-
nianum, etiamnum ineditum, fervat & fupplet.

Hallerus variis in Operibus Synonomiam abfolutam plantarum
Helveticarum elaboravit.

Synonymiæ abfolutæ opus maxime neceffarium eft Botanicis:
Unico auctoris nomine detecto, innotefcunt omnium.
Evolvi ideoque poffunt Defcriptiones & Figuræ omnes.
Innotefcunt inde omnia quæ beneficio feculi innotuere de
planta.
Plura nomina non reddant amplius ideam plurium plantarum.

Specierum fynonyma ad Botanicos præcipue fpectant; *Varieta-*
tum autem, quæ fæpius fuperflua, addat qui lubet, ut mi-
nuatur fpecierum falfarum numerus.

319. In

❧ X. SYNONYMS

318. Synonyms are variant names given to the same plant by phytologists ([6]), and they refer to *genera* (VII), *species* (VIII), and *varieties* (IX).

> *The fathers*, Section 9, mostly agreed on the names of plants, content with those of the genera only.
>
> *The commentators*, Section 10, owing to he lack of descriptions and figures in the writings of the fathers, have applied their names to plants in various ways.
>
> *The describers*, Sections 11 and 12, discovered a greater number of plants, and gave them names arbitrarily.
>
> *Caspar Bauhin*, in a labour of forty years, put together the names given by his predecessors and reduced them to 6,000 species in his Pinax of 1622.
>
> *The meticulous*, Section 14, have doubled the number of plants, by finding new ones on all sides.
>
> *The systematists*, Section 53, who formerly disagreed a very great deal in the formation of genera, used to give plants utterly false names in accordance with their false genera.
>
> As no fixed rules exist concerning the definitions of species, botanists have devised definitions that are partly trivial, partly variable, and always uncertain.
>
> *William Sherard*, an Englishman and a famous botanist, worked on a continuation of Bauhin's Pinax, but was overtaken by death in 1728, and left the work to Dillenius.
>
> *Dillenius*, Sherardian Professor at Oxford, continued Sherard's work until the year when he died, 1747.
>
> *Sibthorpe*, Dillenius' successor, keeps up and adds to the *Pinax* of Sherard and Dillenius, which is still unpublished.
>
> *Haller* worked out a complete system of synonyms for Swiss plants, in his various works.
>
> A complete *system of synonyms* is a thing very necessary to botanists:
>
>> When the name used by a single author is found, the names used by all of them become known.
>>
>> And so all the descriptions and pictures can be produced.[1]
>>
>> As a result, everything that has been made known about the plant by earlier good work, becomes known to us.
>>
>> More names should no longer express the idea of more plants.

Synonyms of *species* especially concern botanists; those of *varieties*, which are quite often superfluous, can be added by anyone, to reduce the number of false species.

319. In the case of synonyms, the best name should be at the head of the column; such a name should be either one used by another botanist and *chosen*, or one *peculiar* to the author.

Among synonyms, the first will be that used by the author, whether peculiar to him or taken from another.

The first name will be the one that is chosen for the species, and foremost among the synonyms.

It is a bad practice of some authors to place their peculiar name at the end.

Alsine minor *Fuchs. Lon. Tab.* minor *Rudb.* minor multicaulis (many-stemmed). C.B. Morsus gallinae minor *Brunsf.* Spec. 4. *Trag.* ALSINE vulgatissima *Ourself, Brom. chlor.*

Others put genuine definitions, which are accepted in a name, after specific names that are uncertain and illegitimate:

Veronica mas supina vulgatissima (male, reclining, very common) *C.B. Seguier. veron.* 1, *p.* 333. *Berg. viadr* 76. *Ludolf. berol* 212. Veronica supina vulgaris *Moris.* VERONICA mas serpens (male, creeping) *Dod.* VERONICA caule repente, scapis spicatis, foliis ovatis strigosis (with creeping stem, spiked scapes and leaves that are egg-shaped and covered with strigae). *Lin. [Hortus] Cliffortianus* 8. VERONICA, foliis siccis ovatis serratis, caule procumbente, ex aliis ramosa (with dry, egg-shaped, toothed leaves, and prostrate stem branching from others) *Haller, helv.* 530.

320. The said synonyms should be collected together.

Synonyms are presented in two ways, either descending from the most ancient to the modern, or ascending from the modern to the primitive.

In descent, the discoverer's name leads the line, and anyone who is keen to work it out will have to proceed backwards: see Haller and Dillenius.

Those who start from the more recent generic name, and end with the most ancient present the names *in ascent*.

Ascending	Descending
LIMOSELLA. [*My*] *Fl. lapp.* 249. [my] *Fl. su.* 521. *Hall. helv.* 609. *Wachend. ultraj.* 144. *Guett. stamp. I.p.*195. *Dalib. paris.* 193. Limosella annua. *Lind. alsat.* 156. *t.I.*	LIMOSELLA. *Fl. lapp.* 249. *Fl. su.* 521. *Hall. helv.* 609. *Wachend. ultraj.* 144. *Guett. stamp. I. p.*195. *Dalib. paris.* 193.

continued

Menyanthoides vulgaris. *Vaill. par.*
126. *Fabreg. V.p.*91.

Plantaginella *Rupp. jen.*18. *Dill. gen.*
113. *Buxb. act.* 3. *p.*271.

Plantaginella palustris. *Bauh. pin.*
190. *Moris. hist.* 3. *p.*605.
s.15.t.2.f.I. Raj. angl. 237. *hist.*
1077. *syn.* 3. *p.*278. *Mapp. alsat.*
242.

Plantago aquatica minima *Clus. hist.*
2. *p.*110. *Park. theatr.* 1244.
Merret. pin. 95. *Boerh. lugdb.* 1.
*p.*45.

Spergula perpusilla, lanceolatis foliis
(with lanceolate leaves). Loes.
pruss. 261 pl. 81.

Alsine palustris repens, foliis
lanceolatis, floribus albis
perexiguis (with lanceolate leaves
and very small white flowers).
Pluk. alm. 20. *pl.*74 *f.*4. *Volk*
norib. 22.

Alsine palustris exigua, with
lanceolate leaves like a
Plantaginella aquatica, and white
florets which are not very
conspicuous. *Mentz. pug.* 2 *pl.* 7.
*f.*6. *Comm holl.* 8. *Ray supplem.*
498. *Tournef. paris.* 381.

Ranunculus palustris minimus, with
a Plantago leaf. *Herm. lugdb.* 517.

Plantago aquatica minima. *Clus.*
hist. 2. *p.* 110. * *Park. theatr.*
1244. *Merrett. pin.* 95. *Boerh.*
*lugdb. I. p.*45.

Plantaginella palustris. *Bauh. pin.*
190. *Moris. hist.* 3. *p.* 605.*or* 15.
*pl.*2. *or* 1. *Raj. angl.* 237 *hist.*
1077. *syn.* 3. *p.*278. *Mapp.*
alsat. 242.

Spergula perpusilla, lanceolatis
foliis. *Loes. pruss.* 26 1. *pl.*81.

Alsine palustris enigua, with
lanceolate leaves like a
Plantaginella aquatica, and
white florets which are not very
conspicuous. *Mentz. pug.* 2.
*pl.*7. *f.*6. *Comm. holl* 8. *Ray*
supplem. 498. *Tournefort paris.*
381.

Alsine palustris repens, with
lanceolate leaves and very small
white flowers. *Pluk alm.* 20.
*pl.*74 *f.*4. *Volk. norib.* 22.

Ranuncalus palustris minimus
with a Plantago leaf. *Herm.*
lugdb. 517.

Plantaginella. *Rupp. jen* 18. *Dill.*
gen. 113. *Buxb. act.* 3. *p.*271.

Menyanthoides vulgaris. *Vaill.*
paris. 126. *Fabregou V. p.*91.

Limosella annua. *Lind. alsat.* 156
pl.1.

321. The synonyms are set out each on a fresh line.

Synonyms are usually rehearsed by authors in five ways.

a. Synonyms according to genera.

PARTHENIUM with leaves that are egg-shaped and scalloped.

[My] *Hortus Cliffortianus* 442. *Gronovius virg.* 115. *Royen lugdb.* 86.

Partheniastrum with a Helenium leaf. *Dillenius elth.* 302. *pl.* 225. *f.* 292.

Ptarmica virginiana, with Helenium leaves. *Morison bles.* 297.

Ptarmica virginiana, with the divided leaves of Scabiosa austriaca. *Pluk. alm.* 308. *pl.*
53 *f.* 5. *and pl.* 219. *f.* 1.

PseudoCostus virginiana or Anonymos corymbifera virginiana, with a white
flower. *Ray hist.* 363.
Dracunculus latifolius or Ptarmica virginiana with a Helenium leaf. Morison hist. 3.
p. 41.

b.   Synonyms each beginning on fresh lines.

PARTHENIUM with leaves that are egg-shaped and scalloped. *Hortus
Cliffortianus* 442. *Gronovius virg.* 115. *Royen lugdb.* 86.
Partheniastrum with a Helenium leaf. *Dillenius elth.* 302. *pl.* 225. *f.* 292.
Ptarmica virginiana, with Helenium leaves. *Morison bles.* 297.
Ptarmica virginiana, with the divided leaves of Scabiosa austriaca. Pluk. alm. 308.
pl. 53. f. 5. and pl. 219. f. 1.
PseudoCostus virginiana or Anonymos corymbifera virginiana, with a white
flower. *Ray hist.* 363.
Dracunculus latifolius or Ptarmica virginiana with a Helenium leaf. *Morison hist.* 3.
p. 41.

c.   Synonyms in a continuous series.

PARTHENIUM with leaves that are egg-shaped and scalloped. *Hortus Cliffortianus*
442. *Gronovius virg.* 115. *Royen lugdb.* 86. Partheniastrum with a Helenium leaf.
Dillenius elth. 302. *pl.* 225. *f.* 292. Ptarmica virginiana, with Helenium leaves.
Morison bles. 297. Ptarmica virginiana, with the divided leaves of a Scabiosa
austriaca. *Pluk. alm.* 308. *pl.* 53. *f.* 5. *and pl.* 219. *f.* 1. PseudoCostus virginiana or
Anonymos corymbifera virginiana, with a white flower. *Ray hist,* 363.
Dracunculus latifolius or Ptarmica virginiana with a Helenium leaf. *Morison hist.*
3. *p.* 41.

d.   Synonyms, excluding the generic name in later instances.

PARTHENIUM with leaves that are egg-shaped and scalloped. *Hortus Cliffortianus*
442. *Gronovius virg.* 115. *Royen lugdb.* 86. Partheniastrum with a Helenium leaf.
Dillenius elth. 302. 225. *f.* 292. Ptarmica virginiara, with Helenium leaves. *Morison
bles.* 297. virginiana, with the divided leaves of a Scabiosa austriaca. *Pluk alm.* 308
t. 53. *f.* 5 *and pl.* 219. *f.* 1. PseudoCostus virginiana or Anonymos corymbifera
virginiana, with a white flower. *Ray hist.* 363. Dracunculus latifolius or Ptarmica
virginiana with a Helenium leaf. *Morison hist.* 3. *p.* 41.

e.   Synonyms, abridged, with parenthesis.

PARTHENIUM with leaves that are egg-shaped and scalloped. *Hortus Cliffortianus*
442. *Gronovius virg.* 115. *Royen lugdb.* 86. Partheniastrum with a Helenium leaf.
Dillenius elth. 302. *pl.* 225. *f.* 292. Dracunculus latifolius (or Ptarmica virginiana
(with the divided leaves of Scabiosa austriaca. *Pluk. alm* 308 *pl.* 53. *f. and pl.* 219.
f. 1) with a Helenium leaf. *Morison bles.* 297) hist. 3. p. 41. PseudoCostus
virginiana or Anonymos corymbifera virginiana, with a white flower, *Ray hist.*
363.

322. In the case of synonyms, the *author* and the *page* should always be indicated at the end.

> The citation of the work, in the case of the names of plants, must consist of the genus and the definition, or else the *author* and the *book*.
>
> The *author*'s name is not enough, since the same author has quite often published several writings; also, there have quite often been two or more of the same name, or in the future there might be several who become famous by the same name; already there have been cases of two of the same name becoming well known.

The Gesners,	the Ammans,	the Rudbecks,
the Bauhins,	the Knauts,	the Commelins,
the Camerarii,	the Volkamers,	the Magnols,
the Hermanns,	the Horsts,	the Trionfettis,
the Jussieus,	the Millers	the Montis,
the Scheuchzers,	the Cordii,	the Beslers,
		the Hoffmanns.

> The *work* alone is not enough for citation, since very many have either appeared under the same name already, or will so appear in future; and they will provide a handle for confusion to future ages, unless the author is added.

Hortus Lugduno-Batavus	Hortus Patavinus	Hortus Parisinus
Vorst's	Cortusi's	Cornut's
Paaw's	Guilandini's	Tournefort's
Schyylius'	Schenck's	Vaillant's
Hermann's	Vesling's	Fabregou's
Boerhaave's	Marcellus'	Dalibard's
Royen's	Turreus'	

> Particular authors have the habit of excluding their own name and mentioning only

(Dill.)	Catalogus Gissensis,	or C.G.
	Hortus Elthamensis	or H.E.
	Historia muscorum	or H.M.

> the work, or, what is still less acceptable, only the initials.
>
> The work may be named *very briefly* in a single word, the initial letter of which must be small, whereas that of the author must be a capital, so that the citation shall turn out to be none too extensive, but clear.
>
> Lastly, the *page* should be added, so that the plant can quite easily be looked up.

323. In a complete company of synonyms, it is acceptable to note the discoverer with an asterisk.

> This may be difficult before a recent and complete list of the synonyms is available. A note of this kind would bring about a great deal of enlightenment in the dating of plants, so that a plant discovered by quite modern botanists would not be sought in the works of the ancients.

324. Vernacular names of regions should either be excluded, or else placed at the end of the list of synonyms.

> The vernacular names of each and every place generate a great deal of light in specialized Floras, not only so that the names of the plants may be more easily learnt from the inhabitants, but also that the nature of the plant may become known from the name used by the common people, which is often apt.
> Barbarous[2] names should be placed at the end of the list of synonyms, such as RHEEDE'S Malayalam, HERMANN'S Sinhalese, KAEMPFER'S Japanese, HERNANDEZ'S Mexican, MARCGRAAF'S Brazilian, and RUMPF'S Amboinese.

XI. ADUMBRATIONES.

325. ADUMBRATIONES Hiſtoriam plantæ continent, uti *Nomina* (VII), *Etymologias* (234-242), *Claſſes* (II), *Characteres* (VI), *Differentias* (VIII), *Varietates* (IX), *Synonyma* (X), *Deſcriptiones* (326), *Icones* (332), *Loca* (334), *Tempora* (335).

Methodus demonſtrandi propoſita in *Syſt. Nat. 6. p. 222.* ſiſtit ordinem, ſecundum quem plantæ hiſtoria concinnari debet.

Continebit Adumbratio omnia, quæ ad *Hiſtoriam plantæ* pertinent, uti ejusdem Nomina, Signa, Faciem, Naturam & Uſum.

326. DESCRIPTIO (325) eſt totius plantæ character naturalis, qui deſcribat omnes ejusdem partes externas (80. 81. 82. 83. 84. 85. 86.).

Perfecta Deſcriptio non adquieſcat more. recepto in *Radice*, *Caule*, *Foliis*, & *Fructificatione*, ſed etiam probe obſervabit *Petiolos*, *Pedunculos*, *Stipulas*, *Bracteas*, *Glandulas*, *Pilos*, *Gemmas*, *Foliationem* & *Habitum* omnem.

RICINUS foliis peltatis ſerratis, petiolis glanduliferis *Hort. cliff. 450.*

Radix *ramoſa*, *fibroſa*.

Caulis *erectus, teres, viridis, articulatus, inanis, lævis:* ſtriis *ſparſis longitudinalibus*, *ſuperne flexuoſus*, *altus orgyam unam alteramve.*

Rami *ſolitarii, ex axillis ſuperioribus foliorum, cauli ſimiles, altiores;* ex inferioribus axillis *breviores*, *vel marceſcentes*, *vel ſeriores* Rami.

Folia *alterna, peltata, novemlobata: Lobis exterioribus majoribus, magis angulatis; Nervis totidem ab umbilico ad loborum apices excurrentibus; obtuſiuſcule inæqualiter ſerrata, reticulato venoſa, utrinque lævia, ſupra glabra, diſco extrorſum verſa.*

Hæc *ante explicationem plicata, ſerraturis glanduloſis.*

Petio-

❧ XI. SKETCHES

325. SKETCHES contain the history of a plant, for instance the *names* (VII), *etymologies* (234–42), *classes* (II), *characters* (VI), *definitions* (VIII), *varieties* (IX), *synonyms* (X), *descriptions* (326), *pictures* (332), *locations* (334), and *seasons* (335).

> *The method of demonstration* proposed in [*my*] *Systema naturae* [*ed.*] 6. *p.*222. establishes the order according to which the history of a plant ought to be put together.
>
> A sketch must contain everything that is relevant to the history of the plant, such as its names, marks, appearance, nature, and use.

326. A DESCRIPTION (325) is the natural character of the whole plant, and it should describe all its external parts (80–6).

> A complete description should not be satisfied, in the usual manner, with the *root, stem, leaves* and *fruit-body*, but must take proper notice of the *petioles, peduncles, stipules, bracts, glands, hairs, buds, the foliation,* and *the whole habit.*
>
> RICINUS with shield-shaped, saw-toothed leaves, and petioles bearing glands.
>
> [*My*] *Hortus Cliffortianus* 450.
>
> Root *branchy and fibrous.*
>
> Stem *upright, rounded, green, jointed, hollow and smooth: with sparse lengthwise streaks, and flexible in the upper part, one or two fathoms high.*
>
> Branches *solitary; those in the upper axils of the leaves are similar to the stem and comparatively tall;* those in the lower axils *are shorter, either withering or comparatively late.*
>
> Leaves *alternate, shield-shaped, and nine-lobed: with the outer lobes larger and more angular, and with as many nerves running outwards from the umbilicus to the tips of the lobes; somewhat bluntly and unevenly toothed, veined in a net-like pattern, smooth on both sides, glabrous on the upper side, and turned outwards from the disc.*
>
> The leaves *are folded before their expansion, with glandular serratures.*
>
> Petioles *rounded, smooth, open, sessile, thread-like, as long as the leaves, and open.*
>
> A gland *above the base of the petiole: on the upper side, blunt and solitary.*

A pair of shield-shaped glands *at the tip of each petiole: on the upper side.*
Two opposite glands *at the base of the petiole, on the stem.*

A stipule *opposite the petiole, membranous, glabrous, solitary, surrounding the stem right up to the petioles, concave, pointed and deciduous.*

A peduncle *at the end of each branch, in the direction of the petiole, between the branch and the stipule, upright, bare, and sprinkled with alternate sessile partial umbels.*

The involucre *of the partial umbel has three leaves, is membranous, very small, uneven, and withering.*

> The *lower* partial umbels *have many flowers and are male; the upper ones are fewer, have only one flower, and are female.*

Pedicels *that grow out alternately and have flowers that wither.*

Male flowers *situated on the longer pedicels.*

> The male ones. The perianth of the calyx *with a single leaf, divided into five: with egg-shaped, concave lappets. No* corolla. The filaments of the stamens *various, thread-like, with branches and sub-branches and a comparatively long calyx.* The anthers *almost round, and in pairs.*

> The female ones. The perianth of the calyx *with a single leaf, divided into three, and deciduous, with egg-shaped, concave lappets. No* corolla. The ovary of the pistil *egg-shaped, covered with spines that are awl-shaped and stingless. Three styles, [each] divided into two, upright and spreading, awl-shaped, hairy and purplish.* Stigmata *simple.* Capsule *almost round, with three grooves, vaguely triangular, pointed all round, with three chambers, triply dehiscent, and elastic.* The seeds *solitary, almost egg-shaped, with uneven spots.*

327. The description (326) should portray the parts as compendiously as possible, yet completely, by means of technical terms only, if adequate ones exist, according to their *number, shape, relative size,* and *position.*

> The natural character of a species ought to be prepared in the same way as that of the genus (167), but it may include more accidental features than the character of the genus.

> The characteristic features of the principal description should always be observed in all parts of the plant; they are *(a)* the number, *(b)* the shape, *(c)* the relative size, and *(d)* the position.

> PHARNACEUM, smooth, with peduncles equal [in length] to the leaf. Alsine *Amm. ruth.* 84. *Common at Rostock.*

> Root *a. solitary, b. thread-like, with sub-branches; c. whitish*[1]*, d. perpendicular.*

> Stems *a. several, b. thread-like, rounded, smooth and noded: with bare joints, c. finger-sized, d. fairly upright, later prostrate, e. flesh-coloured and translucent.*

> Branches *a. very scarce, b. with appearance like that of the stem, c. taller than the stem, d. growing out from the nodes.*

Radical leaves *a. numerous, b. linear, perfectly entire, veinless, succulent, convex on the under-side, somewhat blunt at the edge, and slightly pointed at the tip, c. quite often as long as the joints of the stem, d. the lower ones becoming gradually shorter.* Stem leaves *d. in a whorl, a. in groups of three, four or more, b. very much like the radical ones in shape and c. in size.* Peduncles *a. solitary, more rarely in pairs, b. thread-like, c. slightly shorter and narrower than the joints, d. growing out from the node between the leaves, ending at the tip with pedicels, which are a. two or three, b. hair-like and bare, with a single flower at the tip, c. projecting and flowering unevenly, d. jointed at the base, and hanging down when the flowering is finished.* Bracts *a. of the same number as the peduncles, b. egg-shaped, c. very small, d. at the bases of the common peduncle and of the partial ones.* The calyx *of the flowers. The perianth a. with five leaves; b. with leaflets that are almost egg-shaped, concave, c. even, and d. very smoothly[2] united at the base.* Corolla, *none, or one that is united with the calyx (since the perianth is snow-white at the edge and on the inner side, at the time of flowering).* The filaments of the stamens *a. five, b. awl-shaped, c. of the same length as the calyx, d. inserted into the receptacle.* Anthers *a. solitary, b. divided into two at the base, c. very short, d. attached by the division.* The ovary of the pistil *a. solitary, b. egg-shaped and triangular, c. very short.* Styles *a. three, b. thread-like, c. as long as the stamens.* Stigmas *b. blunt.* The capsule of the pericarp *b. egg-shaped, vaguely triangular, c. as long as the calyx, d. covered by the calyx, with three chambers and three valves.* Seeds *a. numerous, b. circular, flattened, surrounded by a sharp edge, and shiny, c. small, d. fixed to the columella.*

328. The description should follow the order of growth.

The order of the description should proceed according to the order of the growth of the parts of the plant.

At the same time, this rule allows some latitude, but not in every respect.

It would be a bad practice to present first the tendrils, next the *peduncles*, then the *glands*, after that the *leaves*, followed by the *stem*, and lastly the *petioles*, in a confused manner.

It is best to follow nature, from the *root* to the *stem, petioles, leaves, peduncles* and *flowers*. TILIA. [My] *Hortus Cliffortianus* 204.

Root: the *descending* stock *straggling, very branchy, rounded, flexible, with a deciduous epidermis:* with radicles *that are capillary and flexible, and have sub-branches.* The ascending stock *tree-like, rounded, very branchy:* with bark that is thick and porous, the stock covered with an epidermis, *which is full of cracks where there were originally streaks, glabrous in the delicate and smooth parts:* the last year's branches being furnished with alternate buds.

Buds: *egg-shaped, prominent, made out of two scales that are alternate, oblong egg-shaped, blunt, rolled together, somewhat fleshy and stipulaceous.*

Stipules: *bud-like, opposite each other, egg-shaped, glabrous, perfectly entire and concave, enveloping the leaves and the stem.*

Stem: *very simple, rounded, somewhat flexible from leaf to leaf, spreading horizontally, smooth, sprinkled with a number of dots that are scattered and indeterminate.*

Leaves: when still tender, they are *folded double, turned towards one side, wrinkled, and villous all round;* when fully grown, they are *heart-shaped, full of nerves and veins, unevenly saw-toothed, pointed, glabrous on the upper side, sprinkled with hardly visible hairs, and on the underside they have tufts in the axils of the larger vessels.*

Petioles: *slightly rounded, smooth, shorter that the leaves, mostly growing in two rows, with spaces, which are shorter than the leaves, placed between them.*

Peduncles: *solitary, at the sides of the petioles, relatively long and thread-like, with the tip divided into three: those at the sides being divided into three, with single flowers at the ends; with up to seven* florets, *of even height.*

The bracts: *lanceolate, slightly blunt, white-coloured, perfectly entire, as long as the peduncles, joined to the peduncles from the base to the middle of their length.*

The perianth of the flower: *divided into five, concave, yellowish in colour, about the same size as the corolla, deciduous.* The petals of the corolla *five, oblong, blunt, scalloped at the top, whitish with a tinge of yellow.* The filaments of the stamens *very numerous: thirty or forty, bristling, inserted into the receptacle, as long as the corolla.* Anthers *almost round.* The ovary of the pistil *almost spherical and hairy.* Style *cylindrical, as tall as the stamens.* Stigma *blunt, with five corners.* Pericarp *leathery, spherical, with five chambers and five valves, and dehiscent at the base.* Seeds *solitary and almost round.*

Cotyledons: *divided into five, with the end ones and the middle one comparatively long.*

329. **The description should set forth the separate parts of the plant in separate paragraphs.**

The parts of the plant should appear in the description as distinct as they are in the actual plant.

All the parts of the plant should be indicated in large letters, but the parts of the parts in small letters, different from the common ones.

The resulting advantage is that not only can the parts be more readily found by the reader, but also that any omissions in the description can be more readily perceived.

Nothing is more boring than a very full description that is neither divided by paragraphs nor by large letters, according to the parts of the plant.

PASSIFLORA with three-lobed leaves that are heart-shaped and hairy, and with involucres that are much divided and capillary. *Amoenitates academicae* 228.

Root	*fibrous, annual.*
Stem	*rounded, of a height greater than a man's, frail: sprinkled with hairs that are very much spread out, uneven and white.* Branches above *the axils, separated by spaces, and imitating the stem in structure.*
Leaves	*alternate, separated by spaces as long as the leaves, heart-shaped, vaguely three-lobed, with five nerves at the base, veined: the veins being transparent; perfectly entire, indistinctly ciliate: with bristles that bear glands; sprinkled all over with white, upright hairs; folded into three before they open out.*
Petioles	*rounded, likewise hairy, spread out, shorter than the leaves by half, the upper part slightly flattened; with awl-shaped* glands *which are alternate, upright, and arranged lengthwise on the horizontal part.*
Stipules	*crescent-shaped, embracing the stem, persistent, uncovered and ciliate: with awl-shaped* glands, *as long as the stipules.*
Tendrils	*thread like, spread out, in the axils, hairy, rolled together and rolled back spirally above the middle, longer than the leaves.*
Peduncles	*in the axils, beside the tendrils, solitary upright, rounded, shorter than the petioles, hairy, terminating in involucres.*
Involucre	*with three leaves, linear, divided like a feather, ciliated with awl-shaped glands, persistent, with single flower.*
Flower	*sessile, with a five-leaved calyx, hairy, with a sharp point* etc.

330. **A description may be longer or shorter than it should be: either is wrong.**

A description turns out to be too *long,* when the green colour of the herbage, the dimensions of the parts, and such-like matters, which are very liable to variation, are set forth with diffuse rhetoric.

Descriptions turn out to be *shorter* than they should be, when particular features and essential parts of the herb, even though very small, such as stipules, bracts, and glands, are omitted.

Let the description of LINUM serve as an example;

A. *Dodoens'* very short and incomplete description.
Roots *slender.* Culms *or shoots thin and round.*
Leaves *oblong, narrow and tapering.*

Flowers *at the tops of the shoots, beautiful, blue.* Vessels *small, round and perfectly circular.* Seed *somewhat oblong, smooth, glabrous, and shiny,* [turning] *from tawny to crimson.*

B. A very long, unnecessary, and futile description.

Roots *narrow, subdivided, hidden within the earth.*

Stems *upright, round, green, two or three feet tall, branchy:* with branches *that are shorter than the stem by half.*

Leaves *narrow, green, pointed, very numerous, an inch long, diverging from the stem at an acute angle, fixed by the base, not tomentose or villous; the upper leaves are only half an inch long and four lines wide; the lower ones are three lines wide, but the final ones hardly attain two lines in width.*

Peduncles *simple, an inch or* $1\frac{1}{2}$ *inches long, hardly half a line in thickness.*

Flowers *at the tops of the branches, large, spread out* etc. as on page 145.

C. The natural and legitimate description.

Root	*simple, vertical, flexible and pale: with capillary* radicles *at the sides.*
Stem	*simple, vertical, rounded and thread-like.*
Leaves	*alternate, sessile, lanceolate, vaguely three-nerved, tapering, somewhat upright, glabrous on both sides: those under the axils slightly larger.*
	Branches *in the axils of the topmost leaves upright, furnished with smaller leaves; withering* rudiments *of branches in the axils of the lower leaves.*
Peduncles	*opposite the leaves and longer than them, thread-like, bare, ending with a solitary* fruit-body, *which has been described in 'the genera'.*
Cotyledons	*about four, opposite each other in the form of a cross, the two lower ones being almost egg-shaped and double the width of the upper ones.*

331. Measurement of size in plants is most conveniently taken from the [human] hand.

The measurement of plants according to a geometrical scale, and that very accurate, for describing the parts of plants, was introduced by Tournefort; he was followed by his satellites, with the result that the essence of descriptions consisted in very accurate geometrical measurement.

I do not at all doubt that it is very well known to anyone experienced in botany that plants vary in the length and width of their parts, more than in anything else whatsoever.

I very rarely allow any measurement, except that which is proportional between different parts of the plant, where this or that part is longer or shorter, wider or narrower, than another.

But if indeed a measure or scale is to be adopted, I do not judge that it is necessary for such a scale to be carried around by the botanist; but measurement should be made by the hand or stature of a man.

A HAIR is the diameter of a single hair.

A twelfth part of a line.

A LINE is the length of a lunula[3], extending from the root of a nail towards the nail (but not in the thumb).

One line of Paris measure.

A NAIL is the length of a nail.

Six lines or half an inch.

A THUMB is the length or diameter of the utmost joint of the thumb.

One Paris inch.

A PALM is the diameter of four transverse parallel fingers, excluding the thumb.

Three Paris inches.[4]

THREE-QUARTERS [OF A FOOT] is the space extending between the tip of the thumb and the little finger.

Nine inches.

A SPAN is the space extending between the tip of the thumb and the index finger.

Seven inches.

A FOOT is measured from the joint of the elbow to the base of the thumb.

Twelve inches.

A CUBIT, from the joint of the elbow to the utmost tip of the middle finger.

Seventeen inches.

AN ARM, from the armpit to the tip of the middle finger.

Twenty-four inches.

A FATHOM, six foot, or the height of a man, is the measurement between the hands when spread right out.

Six feet.

332. Pictures should be drawn in the natural size and position.

The ancients' pictures present the tallest trees and the smallest Alsines as of the same size: they generally depict prostrate and creeping plants as upright; this should be carefully avoided.

In the case of the largest plants, when the [actual] size cannot be depicted, it is best to offer a small branch, and put the whole plant on a small scale next to the part, as Ehret does with Napaea pl. 8, Hibiscus pl. 6, Verbena pl. 14, and Martynia pl. 1.

Basic figures, section 11, notably Plumier's, with their very easy elaboration, are the best representatives of plants.

Woodcut figures, once the most commonly used, notably Rudbeck's, competed with copper plates, yet made botanical works less costly; now they have become antiques, and an occasion of expense to botanists.

A draughtsman, an engraver, and a botanist are equally necessary to produce a praiseworthy figure; if any of these is at fault, the figure turns out to be flawed.

For this reason, the botanists that have also practised the arts of drawing and engraving have left the most outstanding figures.

333. The best pictures should show all the parts of the plants, even the smallest parts of the fruit-body.

The most numerous and outstanding differences, which do most to distinguish a species, lurk in the *smallest* parts, especially those of the fruit-body.

Hairs, glands, stipules, stamens, and pistils, which were excluded from the ancients' figures, must never be omitted in a picture, if it is to turn out satisfactory.

334. The native locations of plants relate to *region, climate, soil*, and *ground*.

The foundation of horticulture depends on the location of plants, from which the rules and principles of the art should be formed; I have demonstrated this in *Acta Stockholmensia*[5] 1739, p. 1.

Miller's 'The Gardeners Dictionary' sets forth the special culture of each particular plant; but this method of horticulture would be too diffuse and burdensome, in going through all the species of plants that have been discovered.

The habitations of the plants are obvious from their location; and it is established from them whence plants are to be obtained for herbaria, gardens, medicine, and household use.

The REGION should set forth the realm, the provinces, the districts, and precisely the particular places where the rarest plants [grow].[6]

CLIMATE has three dimensions, the latitude, longitude, and altitude of a location.

Latitude is either north or south of the equator, each divided into 90 degrees.

Longitude commonly reckons 360 degrees from the island of Hierro in the Canaries to the same.

Altitude is the vertical measurement of the land from the sea to the topmost peaks of the alps. It is measured with a barometer, which goes down according to the height of the vertical location; and this is its scale by English measure:

Inch				Fathoms	
	–	½		"	63
"	1			"	133
		½		"	208
"	2			"	290
		½		"	377
"	3			"	471
		½		"	570

continued

"	4		"		676
		½	"		787
"	5		"		905
		½	"		1028
"	6		"		1150
		½	"		1293
"	7	–	"		1435

Vaillant was one of the first to take notice of climate in the native locations of
plants, but he regarded only the latitude; for example,

Panax with leaves in sets of three and five, in latitude 45. 46. It is obvious, from the
places far apart that produce widely differing plants, that latitude alone is not
enough; much less longitude.

The latitudes of *Rome* in Italy, *Peking* in China, and *New York* in America, are situated
in almost the same degree north.

Rome 41°:51´. Peking 39°:55´. New York 41°:0´.

Likewise *Palestine* and *Florida* in the north, and *South Africa* and *Chile* in the
south.

The Cape of Good Hope, 34°:15´ and *Jerusalem* 31°:40´ almost agree in latitude, but
produce very different plants.

The North Cape (Nordkapp), *Uppsala, Rome,* and the *Cape of Good Hope* are of the same
longitude, with very different plants.

Altitude is more in accordance with the habitations of plants.

The water-plants of India often agree with those of Europe, such as *Utricularia,*
Drosera, Aldrovanda, Nymphaea, and *Sagittaria.*

The *alpine* plants of Lapland, Greenland, Siberia, Switzerland, Wales, the
Pyrenees, Olympus, Ararat, and Brazil, though widely separated, are often
the same.

Unless I am mistaken, the earth's strata above the water table are everywhere the
same: the uppermost is *rock,* next *schist,* then *marble,* after that *schist,* and
lastly *flint,* [my] Wästgöta-resa 77.

If a meadow a little above sea-level is crowded with very prolific meadow plants,
and if the ground that is next to it and further from the sea is raised above it,
soon other wild plants will cover the latter; there are examples of this in *Skånska
resa.*

THE SOIL relates to the nature of the land.

THE SEA is full of salt water, and conceals plants without roots, which must be fed
through pores and do not tolerate the cold.

Fucus	*Najas*	*Ulva*
Zostera	*Ceratophyllum*	*Spongia*

The sea-SHORES consisting of sand or shingle impregnated with salt, and exposed
to the waves and winds, sustain peculiar plants.

Salicornia	Crambe	Glaux
Salsola	Beta	Triglochin
Cakile	Atriplex	Statice
Anastatica	Eryngium	Isatis
Seriphium	Samolus	Hippophaë
Arenaria maritima	Centaureum minus	

SPRINGS gush with water, which is cold and absolutely pure.

Mnium [my] Flora Suecica 913.	Fontinalis	Angelica
Montia	Hippuris	Petroselinum
Beccabunga		

RIVERS carry water that is pure, rather cool, and agitated by movement.
  Potamogeton   Sparganium   Ranunculus capillaris.
The **BANKS** of rivers and lakes are hidden under the water in the winter.

Phalaris Flora Suecica 48	Hydrocotyle	Alsima
Lycopus	Limosella	Lysimachia
Scutellaria	Ranunculus Flora Suecica 459.	
Lythrum		
Eupatorium		

LAKES are full of pure water and possess firm bottoms.

Isoëtes	Lobelia Dortmanni	Arundo
Subularia	Spongia	Scirpus
Plantago monanthes	Nymphaea	Elatine minima

PONDS and ditches are full of slimy bottoms and still water.

Chara	Typha	Cicuta
Stratiotes	Butomus	Sium
Elatine verticillata	Sagittaria	Phellandrium
Vallisneria	Hydrocharis	Rumex britann[ica]
Callitriche	Nymphoides	Ranunculus [flora suecica] 457

MARSHES are full of loose muddy earth and water, and dry out in the summer.

Carex	Sceptrum carolinum	Myrica
Menyanthes	Scheuzeria	Calla

PEAT marshes are full of earth mixed with Sphagnum, covered with tubers, surrounded by slimy, deep water.

Sphagnum	Ledum	Lichen Flora Suecica 936.
Scirpus Flora Suecica 42.	Tetralix	
Eriophorum	Andromeda	
Rubus Flora Suecica 413.	Oxycoccus	

FLOODED locations are filled with water in the winter; in the summer they are putrid and dried out, and sometimes bathed with showers.

Peplis	*Cypripedium*	*Saccharum*
Bidens	*Tamarix*	*Oryza*
Filago Flora Suecica 676.		

I regard as DAMP locations those that are spongy, suffer from putrid water, are hated by farmers, not being suitable for producing corn or hay, and are noted for their peculiar plants.

Primula Flora Suecica 162.	*Pedicularis*	*Ulmaria*
Valeriana Flora Suecica 31.	*Anthericum ossifragum*	*Comarum*
Cynosurus Flora Suecica 82.	*Potentilla fruticans*	*Pinguicula*
Aira Flora Suecica 71.	*Vaccinium* 312	*Burmannia*
Cardamine Flora Suecica 562.		*Selinum palustre*

ALPS are the highest mountains, which penetrate the second region of the air and are deprived of trees, with their topmost peaks covered by perennial snow, and their valleys filled with peaty earth.

Dryas	*The Alpine Pediculares*	*Crocus*
Sibbaldia	*The Alpine Ranunculi*	*Soldanella*
Diapensia	*Arbutus* Flora Suecica 340.	*Betula nana*
Azalea	*Alchemilla* digitata	*Silene* Flora Suecica 368.
The Andromedas	*Rumex* Flora Suecica 294.	*Veronica* Flora Suecica 13.
	Bartsia 575.	*Viola* Flora Suecica 720.
	Thalictrum Flora Suecica 455.	*The Gentianas* Flora Suecica 201 and 204.

CRAGS consist of rocks or abrupt, very dry walls.

Aira Flora Suecica 64.	*Polygonatum*	*Capparis*
Aloë	*Asclepias*	*Cymbalaria*
Mesembryanthemum	*Melica*	*Clinopodium*
Sedum	*Origanum*	

MOUNTAINS and hills are gravelly, dry and barren, and hardly admit any water.

Asperula cynanch[um]	*Jasione*	*Gnaphalium*
Oreoselinum	*Carlina*	Red Vacciniums[7]
Arnica	*Cneorum*	*Veronica major*
	Lithospermum	*Melampyrum* Flora Suecica 510.

PLAINS are sunny, exposed to the winds, dry and rough.

Artemisia campestris	*Buphthalmum*	*Gentianella*
Pulsatilla	*Adonis lutea*	*Daucus*
Stellaria	*Medicago falcata*	*Bellis*
Thesium	*Uva ursi*	*Locusta*
Draba	*Myosotis*	*Echium*

WOODS are shady and full of gravelly, barren earth.

Hypnum Flora Suecica 872	The Pyrolas	Empetrum
Linnaea	Trientalis	Erica
Sibthorpia	Pulsatilla sylvatica	Melampyrum Flora Suecica 514.
		Solidago

GROVES are at the roots of mountains, between glades, and are covered with spongy earth; they are always shady, and continually exhale slightly moist air, which is very little exposed to the winds; they sustain spring plants, which do not tolerate cold or heat.

Lathraea	Hepatica	Actaea
Martynia	Anemone	Asarum
Dentaria	Pulmonaria	Galanthus
Adoxa	Orobus vernus	Leucojum
Oxalis	Paris	Agrimonoides
Fumaria bulbosa	Daphne	Amomum
Lunaria	Prenanthes	Struthiopteris
Impatiens	Mercurialis	Convallaria verticillata
Cardamine Flora Suecica 561.	Allium ursinum	Galeobdolon
Milium	Chrysosplenium	Alsine Fl. 371.
Epimedium	Asperula odorata	Sanicula
Melampyrum Fl. 512.	Ranunculus ficaria	Cyclamen

MEADOWS abound with herbs, and consist of low-lying plains and valleys.

a. Fruitful meadows	b. Dry meadows	c. Somewhat moist meadows
Lotus	Briza	Alopecurus Fl. 52.
Trifolium rubrum	Lagopus	Succisa
Scorzonera	Hypochoeris	Lychnis palustris
Campanula	Lilium convallium	Geum palustre
Millefolium	Viscaria	Fritillaria
Rhinanthus	Bistorta sobolifera	Thalictrum Flora Suecica 453.
Linum catharticum	Agrimonia	Opulus
Lathyrus luteus	Helianthemum	Frangula
Melampyrum Flora Suecica 513.	Geranium sanguineum	Dulcamara
Galium luteum	Arnica	Clymenum parisiense
Ranunculus acris		Parnassia

PASTURES differ from meadows, in that they are more barren, drier and more gravelly

Tormentilla	Euphrasia	Ranunculus lanceolatus
Pimpinella	Brunella	Poa Flora Suecica 78.
Sagina		

ARABLE consist of fallow fields.

Aira Flora Suecica 72.

Aphanes

Anagallis

Myosurus

Rumex hastata

Campanula Rap[vnculoidas]

Fumaria vulgaris

Erysimum Flora Suecica 555.

Tribulus

Scandix

Myosurus

Anthemis

Chamaepithys

Leontice

Lathyrus tuberosus

Ononis

Thlaspi

FIELDS possess of prosperous, worked ground.

Chrysanthemum segetum

Ranunculus echinatus

Melampyrum segetum

Galeopsis

Sinapis arvensis

Napus

Hypecoum

Agrostemma

Cyanus

Delphinium

Vicia cracca

Androsace major

Convolvulus

Helxine scandica

Bromus Fl. 84

Lolium annuum

Triticum Fl. 105

Stachys palustris

Lycopsis

Conium

Sison

Anethum

Nigella

Psyllium

Adonis bulbosa

The TURNING SPACES or edges of the fields are regarded as manured meadows.

Festuca Flora Suecica 91

Lolium

Scabiosa

Cichorium

Anchusa

Cerinthe

Heliotropium

Orchards[8]

Aegopodium

Chaerophyllum

Chelidonium

Ground CULTIVATED in gardens, which is worked, mixed, and very fertile, encourages plants that are disliked by gardeners, when they abound among the vegetables.

Urtica annua

Alsine vulgaris

Chenopodium

Aparine

Aethusa

Sonchus

Euphorbia annua

Lamium

MUCK-HEAPS are built up from the dung of animals.

Urtica major

Persicaria

Asperuga

Xanthium

Blitum

Ricinus

Datura

RUBBISH-HEAPS are near to buildings, dwellings, roads, and streets.

Hyoscyamus	*Solanum*	*Cynoglossum*
Absinthium	*Ballote*	*Lappula*
Chenopodium 209.	*Morrabium*	*Caucalis*
Plantago	*Cardiaca*	*Polygonum*
Verbascum	*Erisymum* Fl. 554.	

The kinds of GROUND that favour plants are *earth, sand, clay, and chalk.*
Very fine EARTH is the principal food of plants, as *Kylbelius* observed.
Therefore most plants delight in earth, and this accords with the experience of
gardeners.
SAND is dry, crumbly, and parched.

a. Drifting sand; [my] Systema Naturae 2.
   Arundo [my] Flora Suecica 102
   Elymus Fl. 106.
   Carex Fl. 749.

b. Floury sand; S.N. 1.
   Erica
   Pinus
   Iberis Flora Suecica 536

c. Common sand.
   Asparagus
   Scleranthus
   Peganum
   Ornithopus
   Ulex
   Ceratocarpus

d. Shingly, mountain sand.
   Herniaria
   Digitalis
   Acinos
   Serpylum
   Androsace Flora Suecica 160.
   Radiola

CLAY is oily in wet weather, but is hardened in dry weather.

Thlaspi Flora Suecica 531.	*Medicago* Flora Suecica 621.	*Horminum glutinosum*
Papaver	*Tragopogon*	*Anthyllis*
Rhoeas	*Blattaria*	
Persicaria amphibia		

CHALK is found in the driest, most arid hills.

Hippocrepis	*Verbena*	*Reseda vulgaris*
Onobrychis	*Campanula* [my] Hortus Cliffortianus 4.	*Cheiranthus luteus*
Trifolium capitatum asperum		

So, by mere inspection of the plants, the earth and soil beneath can be discerned.
   Potentilla vulgaris, when it wears green in its leaves, indicates clay in the hidden
   ground.

Melampyrum,	Flora Suecica	510,	indicates	mountainous places.
"	"	511,	"	fields.
"	"	512,	"	groves.

continued

"	"	513,	"	meadows.
"	"	514,	"	woods.
Pedicularis,	Flora Suecica	504,	"	damp places.
Aira,	Flora Suecica	71,	"	peat.

335. The seasons of growth, germination, leafing, flowering, watching, fruiting, and leaf-shedding indicate the climate.

GERMINATION is the season in which the seeds entrusted to the earth push it out, for the growth of the cotyledons; for this, see page 125.
[Nicolaus] Winckler C. K. *Chronica Herbarum.*

Days		
	3.	*Napus, Eruca, Blitum.*
	4.	*Anethum.*
	5.	*Lactuca.*
	6.	*Raphanus, Cucumis.*
	7.	*Hordeum.*
	8.	*Atriplex.*
	15–20.	*Faba.*
	19–20.	*Cepa.*
	40–50.	*Apium.*

Rudbeck. propag. plant. 54.

Days		
	1.	*Milium, Triticum.*
	3.	*Faba, Sinapis, Rapa.*
		Spinachia, Phaseolus.
	4.	*Lactuca.*
	5.	*Cucumis, Cucurbita.*
		Nasturtium.
	6.	*Beta.*
	10.	*Brassica.*
	30.	*Hyssopus.*
	40.	*Petroselinum.*
Year	1.	*Persica, Amygdalus.*
		Juglans, Castanea, Paeonia.
	2.	*Cornus, Avellana.*

LEAFING is the season of the summer in which the species of plants unfold their first leaves, for example.

1748, at Uppsala.			1749, at Uppsala.			1750, at Landskrona.[9]		
April	28.	*Ribes* Fl. 195.	April	29.	*Ribes* 195.	Febr.	26.	*Grossularia.*
May	4.	*Padus* 396.	May	2.	*Syringa* H.6.1.	March.	8.	*Sambucus.*
	10.	*Euonymus* 133.		3.	*Betula* 776.		9.	*Ribes rubrum.*

continued

11.	*Spiraea* Hort. 1.	4.	*Padus* 396.		10.	*Philadelphus.*
12.	*Crataegus* 7.	6.	*Fagus* 785.		11.	*Syringa.*
	Populus 819.	8.	*Tilia* 432.		12.	*Rosa.*
13.	*Sambucus* 150.	10.	*Quercus* 784.		13.	*Oxyacantha.*
17.	*Quercus* 784.	12.	*Carpinus* 786.		14.	*Lonicera.*
18.	*Fraxinus* 830.				15.	*Ulmus.*
					16.	*Cerasus, Euonymus.*[10]
					17.	*Malus, Alnus.*
					18.	*Corylus, Opulus.*
					19.	*Betula, Sorbus.*
					20.	*Salix viminea.*
					21.	*Salix caprea.*
					24.	*Carpinus.*
					25.	*Salix minor.*
					26.	*Prunus.*
					27.	*Populus nigra.*
					28.	*Pyrus.*
				April.	1.	*Rhamnus cathartica.*
					2.	*Vaccinia nigra. [the black Vacciniums]*[11]
					5.	*Frangula.*
					7.	*Aesculus.*
					11.	*Juglans.*
					12.	*Tilia.*
					20.	*Fagus.*
					22.	*Populus trem[ula].*
					25.	*Acer.*
				May	5.	*Quercus.*
					15.	*Fraxinus.*

FLOWERING is the time of the month in which the species of plants show their first flowers, for example: 1748, at Uppsala.

APRIL.				
17.	*Hepatica* [my] Flora Suecica 445.		29.	*Empetrum* 832.
			30.	*Anemone* 450.
18.	*Fumaria* 585.	**MAY.**		
22.	*Tussilago* 680.		1.	*Ranunculus* 460.
23.	*Daphne* 311.		2.	*Tussilago* 683.
24.	*Pulmonaria* 156.		3.	*Lathraea* 518.
25.	*Draba* 523.		4.	*Myrica* 817.
26.	*Ornithogalum* 270.		5.	*Viola* 718.
27.	*Viola* 716.		6.	*Primula* 161.
28.	*Pulsatilla* 446.		7.	*Glechoma* 483.

continued

8.	Betula 776.			Lotus 609.
9.	Caltha 473.			Trifolium 615.
10.	Oxalis 385.			Ranunculus 469.
11.	Vaccinium 313.			Chaerophyllum 243.
12.	Fraxinus 830.		26.	Triglochin 299.
13.	Viola 710.		27.	Pinus 788.
14.	Androsace 160.			Juniperus 824.
15.	Draba 526.		28.	Potentilla 419.
16.	Leontodon 627.			Cynoglossum 154.
17.	Saxifraga 350.			Hyoscyamus 184.
	Orobus 595.			Erysimum 558.
18.	Adoxa 326.		29.	Berberis 290.
	Alchemilla 135.			Syringa 6.1.
19.	Chelidonium 430.		30.	Ledum 341.
	Fragaria 414.			Vaccinium 312.
	Convallaria 274.			Asclepias 200.
20.	Fritillaria 811.			Sorbus 400.
	Cynosurus 82.			Geranium 571.
21.	Actaea 431.			Dentaria 565.
	Menyanthes 163.			Ranunculus 472.
	Paris 325.			
22.	Primula 162.	JUNE.		
23.	Convallaria 273.	1.		Geum 423.
	Trientalis 302.			Gnaphalium 671.
	Orobus 596.			Pyrola 334.
	Lonicera 192.			Thymus 477.
24.	Pyrus 130 1.			Potentilla 415.
	Pyrus 130. 2.			Bryonia 790.
25.	Statice 253.			Nymphaea 426.
	Polygala 586.	2.		Anchusa 153.
				etc. etc.

The various kinds of *Carduus* do not flower until the solstice is over.

Parnassia is the herald of hay-making.

Colchicum is the messenger of fall and frost.

The WATCHES of plants are finished at fixed times of day, at which the flowers open, display, and close their flowers each day.

Flowers that keep these fixed times of unfolding and closing are called *solar*: these are of three kinds.

1. *Solar* flowers that are *affected by weather*: they keep the hour of unfolding rather inaccurately, and open earlier or later on account of shade, moist or dry air, or greater or less atmospheric pressure.

2. *Tropical* flowers open in the morning, and close each day before the evening; but the hour of unfolding is earlier or later as the length of the day increases or decreases; and so they observe Turkish or unequal hours.

3. *Equinoctial* flowers open at a certain definite hour of the day, and for the most part also close up at a fixed hour each day: these observe European or equal hours.

The *watches* of solar flowers are very well known.

	Flora Suecica and *Hortus Upsaliensis*			rises,			sets.			
1.	Leontodon	627.	Taraxacum	5.	6.	—	8.		9.	
2.	— —	628.	Taraxaconoides	4.		—	———		3.	
3.	— —	629.	Chondrilloides	7.		—	———		3.	
4.	Hypochaeris	631.	pratensis	6.		—	———		4.	5.
5.	— —	1.	hispida	7.	8.	—	———		2.	
6.	— —	2.	Chondrilloides	9.		—		12.	1.	
7.	Hieracium	635.	Pilosella	8.		—	———		2.	
8.	— —	637.	Pulmonaria	6.	7.	—	———		2.	
9.	— —	639.	fruticosum	6.		—	———		5.	
10.	— —	1.	latifolium	7.		—	———		1.	2.
11.	— —	3.	rubrum	6.	7.	—	———		3.	4.
12.	Crepis	640.	tectorum	4.	5.	—	10.	12.		
13.	— —	1.	alpina	5.	6.	—	11.			
14.	— —	6.	rubra	6.	7.	—	———		1.	2.
15.	Picris	1.	magna	4.	5.	—	12.		9.	
16.	Sonchus	642.	repens	6.	7.	—	10.		12.	
17.	— —	643.	laevis	5.		—	11.		12.	
18.	— —	644.	lapponicus	7.		—	12.			
19.	— —	1.	belgicus	6.	7.	—	———		2.	
20.	Lactuca	1.	sativa	7.		—	10.			
21.	Scorzonera	3.	tingitana	4.	6.	—	10.			
22.	Tragopogon	48.	luceum	3.	5.	—	9.		10.	
23.	— —	3.	Columnae	5.	6.	—	11.			
24.	— —	4.	Dalechampii	6.	7.	—	12.		4.	
25.	Lapsana	1.	Rhagadiolus	5.	6.	—	10.		1.	
26.	— —	4.	Rhagadioloides	7.	8.	—	———		2.	
27.	— —	3.	glutinosa	5.	6.	—	10.			
28.	Cichorium	650.	scanense	4.	5.					
29.	Nymphæa	427.	alba	7.		—	———		5.	
30.	Calendula	712.	arvensis	9.		—	———		3.	
31.	— —	2.	africana	7.		—	———		3.	4.
32.	Papaver	4.	nudicaule	5.		—	———		7.	
33.	Hemerocallis	a.	fulva	5.		—	———		7.	8.
34.	Convolvulus	2.	rectus	5.	6.					
35.	Malva	4.	helvula	9.	10.	—	———		1.	
36.	Alyssum		Alyssoides T.	6.	8.	—	———		4.	
37.	Anthericum	267.	album	7.		—	———		3.	4.
38.	Arenaria	376.	purpurea	9.	10.	—	———		2.	3.
39.	Anagallis	169.	rubra	8.						

continued

40.	— —	1.	*caerulea*	7.	8.			
41.	Portulaca	1.	*hortensis*	9.	10.	—	11.	12.
42.	Dianthus	7.	*prolifer*	8.		—	———	1.
43.	Mesembryanthemum	1.	*barbatum*	7.	8.	—	———	2.
44.	— —	2.	*crystallinum*	9.	10.	—	———	3. 4.
45.	— —	10.	*neapolitanum*	10.	11.	—	———	3.
46.	— —	5.	*linguiforme*	7.	8.	—	———	3.
				☽		☽		☾

A floral *clock* should be formed from the following table, after the flowers affected by weather and the tropical[12] ones have been excluded: these are dealt with elsewhere.

☽	☾		☽	☾	
3.	—	Tragopogon *luteum* 22.	—	—	Lactuca *sativa* 20.
.4	—	Leontodon *Taraxonoides* 2.	—	—	Calendula *africana* 31.
4.5	—	Picris *magna* 15.	—	—	Nymphæa *aloa* 29.
—	—	Cichoreum *scanense* 28.	—	—	Anthericum *album* 37.
—	—	Crepis *tectorum* 12.	8	—	Hypochaeris *hispida* 5.
.6	—	Scorzonera *tingitana* 21.	—	—	Lapsana *Rhagadioloides* 26.
5.	—	Sonchus *lævis* 17.	—	—	Mesembryanthemum *barbatum* 43.
—	—	Leontodon *Taraxacum* 1.	9	—	Hieracium *Pilosella* 7.
—	—	Crepis *alpina* 3.	—	—	Anagallis *rubra* 39.
—	—	Tragopogon *Columnae* 23.	—	—	Dianthus *prolifer* 42.
—	—	Lapsana *Rhagadiolus* 25.	—	8.9	Leontodon *Taraxacum* 1.
—	—	*glutinosa* 27.	9	—	Hypochaeris *Chondrilloides* 6.
—	—	Convolvulus *rectus* 34.	10	—	Malva *helvula* 35.
6.	—	Hypochaeris *pratensis* 4.	—	—	Arenaria *purpurea* 38.
—	—	Hieracium *fructicosum* 9.	—	—	Mesembryanthemum *crystallinum* 44.
—	—	— — *Pulmonaria* 8.	—	10	Lapsana *glutionosa* 27.
—	—	Crepis *rubra* 14.	—	—	Lactuca *sativa* 20.
—	—	Sonchus *repens* 16.	—	—	Scorzonera *tingitana* 21.
.7	—	—— *belgicus* 19.	10.	11.	Mesembryanthemum *neapolitanum* 45.
7.	—	Hieracium latifolium 10.	—	11	Crepis *alpina* 13.
—	—	Sonchus lapponicus 18.	—	—	Tragopogon *Columnae* 23.
—	1.	Hypochaeris *Chondrilloldes* 6.	—	12.	Sonchus *laevis* 17.
—	—	Malva *helvula* 35.	—	12.	—— *lapponicus* 18.
—	—	Dianthus *prolifer* 42.	—	3	Mesembryanthemum *neapolitanum* 45.
—	2	Hieracium *latifolium* 10.	—	—	—— *linguiforme* 46.
—	—	Crepis *rubra* 14.	—	.4	Hieracium *rubrum* 11.
—	2-	Hypochaeris *hispida* 5.	—	—	Mesembryanthemum *crystalinum* 44.
—	—	Hieracium *Pulmonaria* 8.	—	—	Calendula *africana* 31.
—	—	Sonchus *belgicus* 19.	—	—	Anthericum *album* 37.
—	—	Lapsana *Rhagadioloides* 20.	—	4	Alyssum *Alyssoides* 36.

continued

—	—	Mesembryanthemum *barbarum* 13.	—	-5	Hypochaeris *pratensis* 4.
—	3	Arenaria *purpurea* 38.	—	5.	Hieracium *fruticosum* 9.
—	3.	Leontodon *Chondrilloides* 3.	—	—	Nymphæa *alba* 29.
—	—	Calendula *arvensis* 30.	—	7.	Papaver *nudicaule* 32.
			—	.8	Hemerocallis *fulva* 33.

Calendula africana Hortus Upsaliensis 274 n. 2 is subject to watches between the sixth and seventh hours of the morning, watching till the fourth hour of the afternoon, if the day's weather is dry; but if it does not take up its watches, or does not open its flowers by the seventh hour of the morning, rain will fall on that day, according to a consistent rule; but it does not readily teach us how to avoid showers caused by thunder.

If *Sonchus sibiricus* closes at night, the next day will generally be fine; but if it stays awake through the night with the flower open, the following day will generally be rainy.

We have touched on plants that *sleep* through the night in Section 133.

FRUITING comprises the time at which plants scatter ripe seeds.

HORDEUM [BARLEY], with all its florets hermaphrodite, and seeds coated with bark. Hortus Upsaliensis 22.

1732 [in] Lapland	1750 at Uppsala	Ripens	
Sown *May* 31.	Sown *March* 6.	*Uppland*	110
Mown *July* 28.	Mown *August* 4.	*Skåne*	90
Ripened in 58 days.	*Ripened* in 155 days.	*Lapland*	60

Barley ripens faster in Skåne in the south and in Lapland in the north, than in Uppland which is in between.

LEAF-SHEDDING is the season of autumn, when the trees shed their leaves, and by this they indicate the passage of autumn and winter, which follows.

Fraxinus [ash] is one of the first to shed its leaves, and one of the last to produce leaves.

The first flowers of *Colchicum* should be observed.

The time of THRIVING comprises the years in which the plants are living: the years are easily reckoned from the concentric circles or resinous rings in a stock that has been cut down.

Most trunks are reckoned by their resinous or internal rings.

A *Quercus* [oak] of Öland, dating from 1581. was 260 years old. Ölandska resa 68.

A *Pinus* [pine] of Varmland, dating from 1337, was 409 years old. Wästgöta resa 247.

The ages of Pinus, Cedrus [cedar], Malus [apple], Pyrus [pear] etc., etc., are reckoned by the previous year's branches.

A *chronicle of the winters,* harder or milder, is provided by the internal rings in most trees, especially the oak.

Hitherto, the botanists engaged in the identification of plants, which are extremely numerous, and who have been overwhelmed by the variety of objects, have not been able to organize observations like those of the astronomers, even though they pursue objects in a lower plane; nevertheless, by their observations they will provide something of much greater use to the public.

Floral calendars should be completed every year in every province, according to the leafing, flowering, fruiting, and leaf-shedding, with simultaneous observations of the climate, so that it may be ascertained how the regions differ among themselves.

Floral clocks under any climate are to be worked out according to the watches of the plants, so that anyone can make a calculation of the hour of the day without a clock or sunshine.

Expressive maps that indicate the region, climate, and ground everywhere.

The climate, which results from the progression of the year according to the leafing and leaf-shedding, and from which come the extremes of heat and cold in a place, must be carefully noted by the botanist.

The botanical thermometer will be ours, with *freezing* point as 0, and the temperature of boiling water as 100.[13]

Virginian plants are *autumn ones*, and they flower most happily in September and October, when the summer is most agreable to them in their own country; whereas in Sweden the seeds ripen with great difficulty.

South African plants are *winter ones*, and their flowers are brought out by agreeable heat even at midwinter, when summer reigns in our country; at other times, they would be propagated in vain.

All *Alpine* plants are *spring ones*, since winter succeeds spring in the Alps, and there is hardly a taste of summer, so that they flower and fruit very rapidly.

Twice-bearing plants are those which which flower twice a year, to wit in spring and fall, according to the custom in their countries, as all *Indian* plants do between the tropics.

Cold plants can hardly bear the thirtieth degree of heat.

Alpine	*German*
Siberian	*Netherlandish*
Canadian	*English*
	French north of Paris.

Temperate plants cannot bear winters of 28 degrees of cold.

Southern French	*Italian*
Portuguese	*Syrian*
Spanish	

Warm plants can tolerate the fortieth degree of heat, but are extinguished by the tenth degree of cold.

East Indian	*Egyptian*
South American	*Of the Canaries.*

Cold plants in the heat grow vigorously at first, but soon become flabby and die.
Warm plants in the cold cease to grow, shed their leaves, and become barren.

XII VIRES.

336. Vires plantarum a *Fructificatione* (86) desumat Botanicus (7), observato Sapore (365), Odore (362), Colore (364) & Loco (357).

Auctores sententiæ fuere *Hermannus*, R. J. *Camerarius*, *Hoff-mannus*, & alii.

Dissertatio de *Viribus plantarum* a F. Hasselquist Upsaliæ 1747. propugnata, hoc de Viribus Caput explicat.

Falsæ sunt veterum Theoriæ de viribus plantarum ex *Astrologia* §. 47. *Signatura* §. 47. *Chemia* §. 48.

Probant Ordines naturales veritatem aphorismi. §. 77.

337. Plantæ, quæ *Genere* conveniunt (165), etiam virtute conveniunt; quæ *Ordine* Naturali (77) continentur, etiam virtute propius accedunt; quæque *Classe* naturali congruunt, etiam viribus quodammodo congruunt.

Genere convenientes easdem possident vires:

Convolvuli; *Scammonium*, *Mechoacanna*, *Turpethum*, *Soldanella*.

Allii: *Moly*, *Porrum*, *Cæpæ*, *Victorialis*.

Lauri: *Cinnamomum*, *Malabathrum*, *Cassia*, *Camphora*, *Sassafras*, *Benzoë*.

Euphorbiæ: *Esula*, *Cataputia*, *Tithymalus*.

Artemisiæ: *Abrotanum*, *Absinthium*, *Cina*, *Seriphium*.

Ordine naturali & viribus conveniunt.

Columniferæ: *Malva*, *Althæa*, *Alcæa*, *Gossypium*.

Scitamina: *Zingiber*, *Cardamomum*, *Galanga*, *Zedoaria*, *Costus*, *Grana paradisi*, *Curcuma*.

Orchideæ: *Orchis*, *Satyrium*, *Serapias*, *Epidendrum*.

Multisiliquæ: *Pæonia*, *Aquilegia*, *Aconitum*, *Delphinium*, *Nigella*, *Helleborus*, *Ranunculus*, *Pulsatilla*.

Contortæ: *Apocynum*, *Cynanche*, *Asclepias*, &c.

CLAS-

❧ XII. POTENCIES

336. The botanist (7) should extract the potencies of plants from the fruit-body (86), noticing the taste (365), smell (362), colour (364), and location (357).

> The authors of [this] opinion were *Hermann, R.J. Camerasius, Hoffmann,* and others.
> The dissertation *De viribus plantarum,* which was defended by F. Hasselqvist at Uppsala in 1747, explains this heading concerning potencies.
> The ancients' theories about the potencies of plants, derived from *astrology* Section 47, *symbolism* Section 47, or *chemistry* Section 48, are false.
> The natural orders prove the truth of the aphorism.

337. Plants that agree in genus (165) also agree in effective properties: those that are included in one natural *order* also approximate to each other in effective properties; and those that are of the same natural *class* are also to some extent of the same potencies.

> Those that agree in GENUS posses the same potencies.
> > The Convolvuli: *Scammonium, Mechoacanna, Turpethum,* and *Soldanella.*
> > The Alliums: *Moly, Porrum, the Caepas,* and *Victorialis.*
> > The Lauruses: *Cinnamomum, Malabathrum, Cassia, Camphora, Sassafras,* and *Benzoë.*
> > The Euphorbias: *Esula, Calaputia,* and *Tithymalus.*
> > The Artemisias: *Abrotanum, Absinthium, Sina,* and *Seriphium.*
> These agree in natural ORDER and in potencies.
> > Plants bearing columns: *Malva, Althaea, Alcaea* and *Gossypium.*
> > Delicacies: *Zingiber* [ginger], *Cardamomum, Galanga Zedoaria, Costus, Grana paradisi,* and *Curcuma.*
> > Orchids: *Orchis, Satyrium, Serapias,* and *Epidendrum.*
> > Plants with many siliques: *Paeonia, Aquilegia, Aconitum, Delphinium, Nigella, Helleborus, Ranunculus,* and *Pulsatilla.*

Twisted plants: *Apocynum, Cynanchum, Asclepias,* etc.
These are of the same CLASS, and the same potencies.

Grasses,	*With three kernels,*	*Butterfly-shaped,*
Compound plants,	*With runners,*	*With loments,*
Umbellate plants,	*Kitchen vegetables,*	*Siliquous,*
Ferns,	*Whorled.*	

338.   The leaves of GRASSES (77:14) form gladsome pastures for farm animals and beasts of burden; the smaller seeds are food for birds, the larger for humans.

The leaves of grasses are the principal sustenance for herbivorous animals.
The comparatively small seeds of *Phalaris, Panicum,* and *Milium* are very acceptable to sparrows and chickens.
Cereals are the comparatively large grass-seeds, which form the daily bread of humans:
Oryza [rice], *Triticum* [wheat], *Secale* [rye], *Hordeum* [barley], *Avena* [oats], *Milium* [millet], *Panicum* [panic-grass], *Holcus* [mouse-barley], *Zizania* [tares], and *Mays* [maize]: excepting perhaps only *Lolium* [darnel], unless it is skilfully prepared.

339.   Stellate [plants] (77:44) are diuretic.

Examples: *Rubia, Asperula, Aparine,* and *Galium.*

340.   Rough-leaved [plants], which are kitchen-garden plants, more or less, are mucilaginous and glutinous.

Kitchen-garden : *Anchusa* [bugloss], *Borrago* [borage], etc.
Mucilaginous: especially the root of *Consolida major,* which is the chief among glutinous [objects].

341.   Sallow [plants] (77:33) are suspect plants.

Stinking: *Solanum, Hyoscyamus, Nicotiana, Atropa, Mandragora,* and *Datura.*
Maddening and narcotic: *Atropa, Mandragora, Nicotiana, Hyoscyamus, Melongena* and *Lycopersicon,*[1] with our people.
Entirely corrosive: *Capsicum*

342. UMBELLATE [PLANTS] (77:22) are spicy, warming, and laxative in dry [locations]; in wet [locations] they are poisonous: their potency is in the root and seeds.

> Plants that grow in a wet location are also poisonous: *Cicuta, Oenanthe, Sison, Phellandrium,* and *Apium palustre.*
>
> Those that evacuate sweat, urine, wind, menses, or milk are *Assa foetida, Levisticum, Angelica, Imperatoria, Pimpinella, Peucedanum, Opoponax, Galbanum, Carvi, Cyminum, Daucus, Meum,* and *Foeniculum.*

343. The roots of [a PLANT] WITH SIX STAMENS (68) are edible or harmful, according to their taste and smell.

> Roots without smell are edible: *Martagon,* the *Tulipas,* and the *Ornithogali .*
> *Gloriosa, Scilla, Hyacinthus, Anthericum, Leucojum, Narcissus,* and *Corona imperalis* are POISONOUS, in accordance with their venomous smell.

344. TWO-HORNED [PLANTS] (77:24) are astringent, but the acid berries are edible.

> Astringent are: *Erica, Pyrola,* and *Vaccinium,* but most of all *Uva ursi.*
> Edible acid berries, in this kind, are: *the Vacciniums, the Myrtilli, the Oxycocci,* the *Uvae ursi, the Arbuti,* the *Guajacanas,* and *the Melastomas.*

345. The pulpy fruit of [a PLANT] WITH TWENTY STAMENS (68) is edible.

> Pulpy fruits in this class are:
> with pomes 77:37, *the Malums [apples], the Pyrums [pears], the Granatums [pomegranates], the Crataegi [hawthorns], the Mespilums [medlars], the Sorbi [services]* and *the Ribes [currants].*
> Thorny 77:35. *The Cynosbatons [dog-thorns], the Rubums (brambles), and the Fragums [strawberries].*
> With stone-fruits 77:38. *the Amygdalums [almonds], the Persicums [peaches], the Prunums [plums], the Armeniacums [apricots],* and *the Cerasums [cherries].*
> Woody 77:39. *Guajava* and *Eugenia [cloves].*

346. [A PLANT] WITH MANY STAMENS (68) is generally poisonous.

> Plants with many siliques 77:23. *Aconitum, Anthora, Aquilegia, Staphisagria, Delphinium, Helleborus, Apium risus*[2], *Clematis, Pulsatilla,* and *Paeonia.*

The Rhoeases [poppies] 77:30. *Papaver*, *Chelidonium* [celandine] and *Actaea* [bane-
berry].
Others: *Euphorbia* (spurge), *Cambogia*, and *Pergamum*.

347. WHORLED [PLANTS] (77:58) are fragrant, affecting the nerves,
laxative, and evacuant: their leaves are potent with effective
properties.

The most fragrant: *Marum [cat-thyme], Dictamnus [dittany], Origanum [wild
marjoram], Majorana [marjoram], Ocimum [basil], Pulegium [pennyroyal], Horminum
[wild sage],* and *Sclarea* [clary].
Fragrant plants are laxative by acting on the nerves. They evacuate wind, menses,
milk, and sperm.

348. SILIQUOUS [PLANTS] (77:57) are watery, sour, incisive, detergent,
and diuretic; their effective properties are diminished by drying
out.

Since the effective properties are diminished by drying out, for that reason they
should be applied fresh by physicians.
Incisive [substances] reduce viscous ones by their sourness, and so they dissolve
oedemas and cold tumours: *Cochlearia*, *Armoracia*, *Sisymbrium*, etc.

349. [PLANTS] BEARING COLUMNS (77:34) are mucilaginous, lubricant,
blunting, and softening.

They lubricate by smearing sour substances with bird-lime, in cases of *cough*,
strangury, nephritis, colic, and *excoriation*, and so relieve *pain*.
They soften by mollifying.

350. The leaves of BUTTERFLY-SHAPED [FLOWERS] are edible by beasts
of burden and farm animals, and their seeds, which are edible by
various animals, are floury and flatulent.

*The Fabas [beans], the Vicias [vetches], the Pisums [peas], the Phaseoli [runner beans], the
Cicers [chick-peas],* and *the Lenses [lentils]* are flatulent and edible.
*Trifolium [clover], Medicago [medick], Trigonella, Hedysarum, Vicia [vetch], Lotus [bird's-
foot trefoil],* and *Lathyrus [vetchling]* supply most excellent nourishment for
animals.

351. [A PLANT] WITH UNITED ANTHERS (68) belonging to the compound flowers (77:21) is completely accepted in medicine, but is in general bitter.

> Bitter: *Eupatorium [hemp agrimony], Tanacetum [tansy], Balsamita, Santolina, Absinthium [wormwood], Abrotanum, Artemisia [mugwort], Ageratum, Matricaria [mayweed], Chamomilla [chamomile], Cotula, Acmella, Auricula muris [mouse-ear], Taraxacum [dandelion], Cichoreum [succory], Carduus mariae, Carlina [carline thistle], Carduus benedictus.*

352. ORCHIDS (77:4) are aphrodisiac.

> The Americans' *Vanilla*, the orientals' *Salep*, and the Europeans' *Satyrium* are among the aphrodisiacs.

353. CONIFERS (77:15) produce resin and they are diuretic.

> They all produce resin, and are therefore also evergreen.
> The diuretic ones, which turn urine violet, are *Terebinthina, Juniperus [juniper], Sabina [savin], Olibanum, Pinus [pine], Abies [fir]*, and *Cupressus [cypress]*.

354. THE CRYPTOGAM [CLASS] (68) quite often includes vegetables that are suspect.

> *Ferns* (77:64) are disagreeable on account of their strong smell.
> *Mosses* (77:65) likewise have a disagreable smell.
> Very few *Algae* are edible, and most are cathartic.
> *Funguses* are dubious food, according to Pliny.

355. Plants with flowers that have the nectary separated from the petals (110) are generally poisonous.

> Separate nectaries are observed in *Aconitum [monk's-hood], Helleborus [Christmas rose], Aquilegia [columbine], Nigella [love-in-the-mist], Parnassia [grass of Parnassus], Epimedium, Clutia, Kiggellaria, Hyacinthus [bluebell], Stapelia, Asclepias, Mirabilis [jalapa],*[3] *Nerium, Narcissus [daffodil], Zygophyllum, Dictamnus [dittany]* and *Melianthus;* all these are poisonous.

356. [Plants] with MILKY juice are generally poisonous, except for those with demi-florets (77:21a).

> Most [plants] with milky juice are poisonous.

Twisted 77:20	Poppies 77:30	With three kernels 77:47	Various
Rauwolfia	*Papaver*	*Euphorbia*	*Rhus*
Thevetia	*Argemone*	*Cambogia*	*Ficus*
Cerbera	*Chelidonium*	*Dalechampia*	*Acer*
Plumieris	*Bocconia*	*Jatropha*	*Melia*
Tabernaemontana	*Sanguinaria*	*The Agarici.*	
Periploca			
Apocynum			
Cynanchum			
Ceropegia			
Asclepias.			

> [Plants] with demi-florets (77:21.a.) and milky juice are hardly poisonous: *Prenanthes, Chondrilla, Hieracium, Crepis, Hypochaeris, Picris, Hyoseris, Leontodon, Tragopogon,*[4] and *Lactuca*; but even in the case of Lactuca, the venomous prickly species are extremely poisonous.
> Bell-shaped plants (77:32) are sometimes poisonous, such as *Lobelia*; and sometimes not exactly harmful, such as *Campanula*.

357. A *dry* location makes plants relatively tasty, a *moist* one makes them relatively tasteless, and a *wet* one quite often makes them corrosive.

> Most water plants are sour and corrosive:
>> Ranunculus, Calla, Nymphaea, Sium, Phellandrium, Cicuta, Persicaria, Armoracia, and Sisymbrium.
>> Most spring plants are also sour, from this cause.
> Plants that grow in a most soil are relatively tasteless, like most *kitchen vegetables*.
> The most outstanding *spices* arise from a dry soil: *Cinnamomum*
>> *Rosmarinus, Thymus, Salvia, Origanum, Clinopodium, Hyssopus, Lavandula, Melissa,* and *Caryophyllus.*
> The strongest-tasting *spices* are dry, and plants that have been dried out turn out to have a stronger flavour.
> Seasonal fruits are rather tart in damp and shady locations, and sweeter in those that are dry and exposed to the sun.

358. The QUALITIES of the plants in which the potencies subsist are indicated by the *taste*, the *smell*, and the *colour*.

> The outward senses are the natural instruments with which the qualities of the plants are investigated.

Things without taste or *smell* hardly exercise any medicinal potency.

Plants with very strong tastes and *smells* always possess very great potency.

If the taste and smell are eliminated in plants, their potency is also rendered ineffective, as in the burnt tartar and magisteries of *Galla, Arum, Jatropha*, and *Elaterium*.

The potencies that are produced from something tasty are best investigated by chewing.

359. *Tasty* and *sweet-smelling [plants]* are wholesome; *nauseous* and *strong-smelling* ones are poisonous.

> These notions are written into the senses of all animals, according to the texture of each body.
>
> Strong-smelling plants are unwholesome: *Funguses, Cotula, Sambucus, Actaea, Aconitum, Helleborus, Veratrum, Asarum, Anagyris, Solanum, Datura, Nicotiana, Hyoscyamus, Tagetes, Cassia, Stachys, Doronicum, Colocynthis, Coriandrum, Buxus, Cynoglossum, Juglans,* and *Opium.*

360. OPPOSITE qualities produce an opposite effect.

> Opposite diseases indicate an opposite cause, which produces an opposite effect.
> It follows that diseases are cured by diseases.

361. All plants act, either on the *nerves* by means of something smelly, or on the *fibres* by means of something tasty, or on *fluids* by means of either.

> *Tasty* things never act on the nerves, nor do *smelly* ones act on the muscular fibres.
> *Fluids* are altered by tasty things, and are evacuated by both tasty and smelly things.

362. *Ambrosial*[5] things are restorative, *fragrant* things are exciting, *spicy* things are stimulating, *noisome* things are stupefying, and *nauseous* things are corrosive.

> AMBROSIAL components are active, such as *Ambra, Moschus, Zibethum, Asperula, Abelmosch, Geranium moschatum, Malva moschata, Melium, Aira.*
>
> FRAGRANT things recommend themselves by their very agreeable smell.
> The flowers of *Crocus, Cheiranthus, Polianthes, Jasminum, Lilium, Tilia,* and *Viola.*
> Herbs: *Lavandula, Thymus, Majorana, Ocimum, Origanum, Satureja, Melissa,* and *Marum.*
>
> SPICY PLANTS usually agree in smell and taste.
> *Cinnamomum, Laurus, Sassafras, Camphora, Macis, Cardamomum, Caryophyllus, Myristica, Acorus, Ammi, Angelica,* and *Citrus.*

The following are notable STRONG-SMELLING VEGETABLES:
Garlic-scented: *Allium, Cepa, Porrum, Alliaria, Scordium, Petiveria, Assa foetida.*
Goat-scented: *Orchis, Vulvaria, Hieracium foetidum,* and *Geranium robertianum.*
NOISOME VEGETABLES are chiefly remarkable for their disagreeable smell.
Stachys foetida, Cotula foetida, Tagetes, Opium, Cannabis, Ebulus, and *Anagyris.*
NAUSEOUS VEGETABLES, which, if they are swallowed, are spewed back by nature.
Veratrum, Helleborus, Convallaria, Asarum, Nicotiana, and *Colocynthis.*

363. TASTY things act on fluids and solids.

Sweet	sweetening and fattening,	
	incisive and corrosive	*sour.*
Fatty	blunting and mollifying,	
	thickening and astringent	*styptic.*
Acid	cooling and thinning,	
	balsamic and tonic	*bitter.*
Viscous	mucilaginous and lubricant,	
	penetrating and detergent	*salty.*
Wet	cleansing and moistening,	
	absorbent and drying	*dry.*

Opposite QUALITIES
Wet and *dry*
fatty and *styptic*
acid and *bitter*
sweet and *sour*
salty and *viscous.*

Similar QUALITIES
Wet and *viscous*
sweet and *fatty*
acid and *salty*
sour and *bitter*
dry and *styptic.*

OPPOSITE POTENCIES in fluids.
Cleansing and *absorbent*
cooling and *balsamic*
sweetening and *incisive*
blunting and *thickening*
mucilaginous and *penetrating.*

OPPOSITE POTENCIES in solids.
Moistening and *drying*
thinning and *tonic*
fattening and *detergent*
mollifying and *astringent*
lubricant and *corrosive.*

Examples of	wet:	*water,*	dry:	*flour*
	viscous:	*gum,*	sour:	*mustard*
	fatty:	*oil,*	salty:	*salt*
	sweet:	*sugar,*	bitter:	*bile*
	acid:	*vinegar,*	styptic:	*gall.*

364. *Pale* COLOUR indicates something tasteless, *green* something raw, *yellow* bitter, *red* acid, *white* sweet, and *black* indicates something disagreeable.

> YELLOW [VEGETABLES] are bitter: *Gentiana, Aloë, Chelidonium, Curcuma*, and *yellow flowers*.
> RED means acid:
>> Berries of *Oxycoccus*, of *Berberis, Ribes, Rubus*, and *Morus*.
>> Acid pomes of *Hippophaë, Sorbus, Rosa*, and *Cerasus*.
>> Herbs that turn red towards the autumn:
>> *Acetosa, Oxalis, Lapathum sanguineum*, and *Brassica*.
> GREEN means raw: *leaves and unripe fruit*.
> PALE means tasteless:
>> *Lactuca, Cichoreum*, and *Asparagus*.
> WHITE means sweet: *Ribes, Rubus*, sweet *Malums*, white *Prunums*.
> BLACK [VEGETABLES] are disagreeable and quite often poisonous:
>> Berries of *Atropa, Actaea, Goriaria, Solanum, Tinus, Empetrum*, and *Padus*.
> The discoverers of *acid* are the blue or violet colours, for example:
>> The colours of Croton, of Viola, and of other vegetables.
>> Blue colours turn ruddy if acid is poured on, and green if alkali is poured on.
> Tournefort used blue paper, which was changed by pouring on it the juice of a plant and indicated its quality, acid or alkaline.

365. The ECONOMIC use of plants is of great utility to the human race.

> The plants generally used for *bread, food, drink, buildings, utensils, tools, works, and dyes* should be properly observed by the botanist, in all kinds of locations, and should be described by him.
>> [Our] Flora Oeconomica has gathered together some examples of this kind.
>> Our Öländska, Gothlandska, Wäst-göta and Skånska *resor* [*journeys*] have provided various examples.
> The use of plants for the *feeding* of animals should be carefully investigated.
>> [Our] *Pan Suecicus* has laid the foundation of this.
> The use of plants for *the Universal Economy of Nature* should be assiduously researched.
>> [Our] *Oeconomia naturae* (an academic dissertation) has sought out several examples of this kind.

In natural science
the elements of truth
ought to be confirmed by observation.

PLATE I

SIMPLE LEAVES

Fig.
1. Orbiculatum [circular]
2. Subrotundum [almost round]
3. Ovatum [egg-shaped]
4. Ovale or *Ellipticum* [*elliptical*]
5. Oblongum [oblong]
6. Lanceolatum [lanceolate][1]
7. Lineare [linear]
8. Subulatum [awl-shaped]
9. Reniforme [kidney-shaped]
10. Cordatum [heart-shaped]
11. Lunulatum [crescent-shaped]
12. Triangulare [three-cornered]
13. Sagittatum [arrow-headed]
14. *Cordato-sagittatum* [*heart-shaped arrow-headed*]
15. Hastatum [halberd-shaped]
16. Fissum [split]
17. Trilobum [three-lobed]
18. Praemorsum [premorse, or bitten off]
19. Lobatum [lobed]
20. Quinquangulare [five: cornered]
21. Erosum [erose, or gnawed]
22. Palmatum [palm-shaped]
23. Pinnatifidum [split like a feather]
24. Laciniatum [slashed in strips]
25. Sinuatum [with recesses]
26. *Dentato-sinuatum* [*toothed, with recesses*]
27. *Retrorsum sinuatum* [*with recesses, turned back*]
28. Partitum [cloven]
29. Repandum [bent back]
30. Dentatum [toothed]
31. Serratum [saw-toothed]
32. *Duplicato-serratum* [*doubly saw-toothed*]
33. *Duplicato-crenatum* [*doubly scalloped*]
34. Cartilagineum [cartilaginous]
35. *Acute crenatum* [*pointedly scalloped*]
36. *Obtuse crenatum* [*bluntly scalloped*]
37. Plicatum [pleated]
38. Crenatum [scalloped]
39. Crispum [curly]
40. Obtusum [blunt]
41. Acutum [pointed]
42. Acuminatum [tapering]
43. *Obtusum acumine* [*blunt with a tapering point*]
44. Emarginatum acute [sharply notched]
45. *Cuneiforme emarginatum* [*knotched, wedge-shaped*][2]
46. Retusum [retuse]
47. Pilosum [shaggy]
48. Tomentosum [tomentose]
49. Hispidum [prickly]
50. Ciliatum [ciliate]
51. Rugosum [wrinkled]
52. Venosum [with veins]
53. Nervosum [with nerves]
54. Papillosum [pimply]
55. Linguiforme [tongue-shaped]
56. Acinaciforme [sabre-shaped]
57. Dolabriforme [chopper-shaped]
58. Deltoides [deltoid]
59. Triquetrum [three-edged]
60. Canaliculatum [grooved]
61. Sulcatum [furrowed]
62. Teres [rounded]

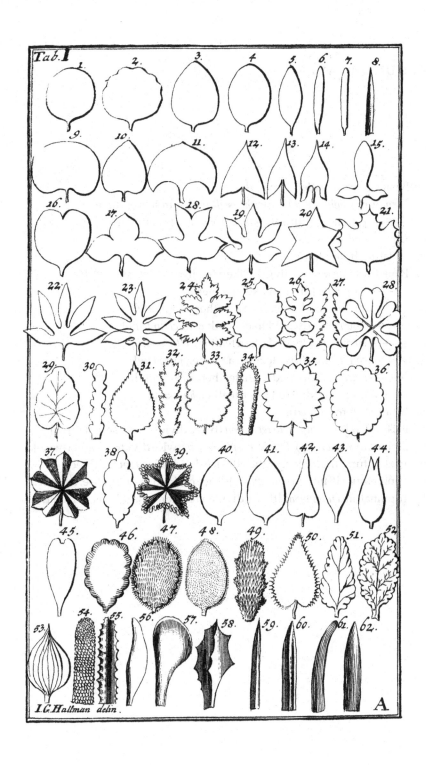

PLATE II

COMPOUND LEAVES

Fig. 63. Binatum [with leaflets in pairs].

64. Ternatum *foliolis sessilibus* [*with sessile leaflets* in sets of three].

65. ---------- *petiolatis* [with *leaflets on petioles* in sets of three].

66. Digitatum [fingered].

67. Pedatum [foot-like].[3]

68. Pinnatum *cum impari* [divided like a feather, *with an odd leaflet*].

69. ---------- *abruptum* [divided like a feather, *broken off*].

70. ---------- *alternatim* [divided like a feather, *alternately*].

71. ---------- *interrupte* [divided like a feather, *interruptedly*].

72. ---------- *cirrhosum* [divided like a feather, *with tendrils*].

73. ---------- *conjugatum* [divided like a feather, *joined up*].

74. ---------- *decursive* [divided like a feather, *decurrently*].

75. ---------- *articulate* [divided like a feather, *jointedly*].

76. Lyratum [lyre-shaped].

77. BITERNATUM, *Duplicato ternatum* [twice divided into sets of three].

78. Bipinnatum (*Sauvag.*), *Duplicato-pinnatum* [twice divided feather-wise].

79. Triternatum, *Triplicato-ternatum* [thrice divided into sets of three].

80. Tripinnatum [*Sauvag.*] *sine impari* [thrice divided feather-wise, *without an odd leaflet*].

81. Tripinnatum *cum impari* [thrice divided feather-wise *with an odd leaflet*].

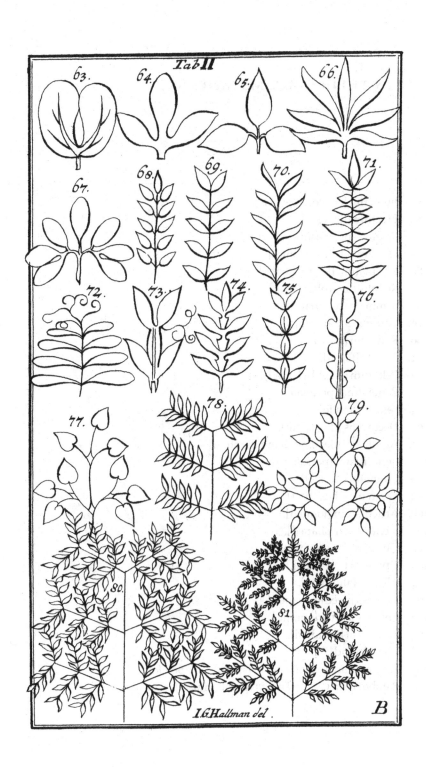

I.G.Hallman del.

PLATE III

LEAVES IN THEIR CIRCUMSTANCES

Fig. 82. Inflexum [curved inwards].
     83. Erectum [upright].
     84. Patens [spread out].
     85. Horizontale [horizontal].
     86. Reclinatum [leaning back].
     87. Revolutum [rolled back].
     88. Seminale [seminal].4
     89. Caulinum [cauline].
     90. Rameum [rameal].
     91. Florale [floral].
     92. Peltatum [shield-shaped].
     93. Petiolatum [petiolate].
     94. Sessile [sessile].
     95. Decurrens [decurrent].
     96. Amplexicaule [embracing the stem].
     97. Perfoliatum [perfoliate].5
     98. Connatum [united].
     99. Vaginans [sheathing].
   100. Articulatum [jointed].
   101. Stellata [star-shaped].
   102. Quaterna [in sets of four].
   103. Opposita [opposite].
   104. Alterna [alternate].
   105. Acerosa [needle-shaped].6
   106. Imbricata [overlapping].
   107. Fasiculata [bundled].
   108. Frons [frond].
   109. Spathulatum (*Sauv.*) *Folium* [spatula-shaped *leaf*].
   110. Parabolicum (*Sauv.*) *Folium* [parabolic *leaf*].

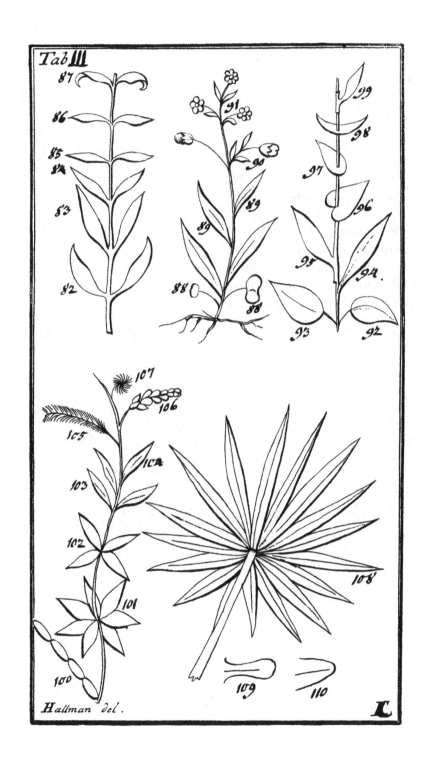

PLATE IV

STEMS

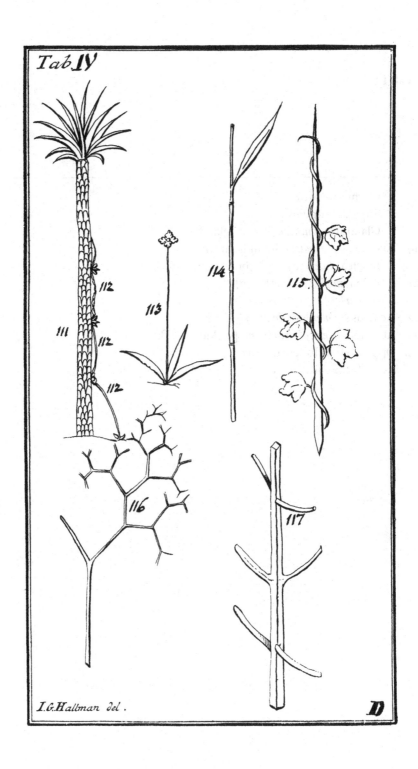

PLATE V

SUPPORTS

Fig. 118.   *a.* Cirrhus [tendril].

          b. Stipulae [stipules].

          c. Glandulae *concavae* [*concave* glands].

    119. Glandulae *pedicellatae* [glands on *pedicels*].

    120. *a.* Bractea, *diversa a foliis b.* [bract, *distinct from the leaves, b.*].

    121. *a.* Spina *simplex* [*simple* thorn].

          b. ---- *triplex* [*triple* thorn].

    122. Aculeus *simplex* [*simple* prickle].

    123. Aculeus *triplex or Furca* [*triple* prickle or *fork*].

    124. Opposita folia [opposite *leaves*].

          a. Axilla [axil].

PLATE VI

ROOTS

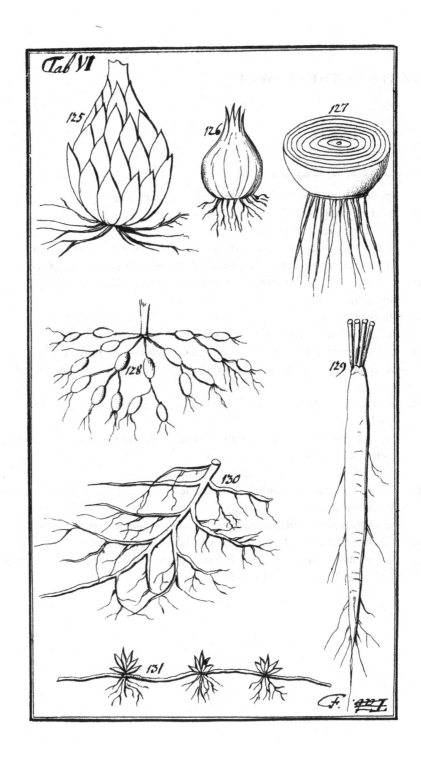

PLATE VII

THE PARTS OF THE FLOWER

Fig. 132. Spatha, Calyx, *Narcissi* [spathe and calyx *of Narcissus*].

133. Spatha, Spadix [spathe and spadix].

134. *a.* Gluma *calyx. b. Arista* [*calyx, a.* husk; *b.* awn].

135. a. *Umbella universalis. b. partialis* [*a.* complete umbel; *b.* partial umbel].

  c. Involucrum universale. *d.* partiale [*c.* complete involucre; *d.* partial involcure].

136. *c.* Calyptra; *b.* Operculum. *a.* Capitulum [*c.* hood; *b.* lid; *a.* head].

137. Amentum [catkin].

138. Strobilus [cone].

139. Fungi *a.* Pileus. *b.* Volva. *c.* Stipes [fungus's *a.* cap; *b.* volva; *c.* stipe].

140. *a.* Receptaculum commune nudum [*a.* uncovered common receptacle].

141. Receptaculum commune paleis *imbricatum* [common receptacle covered with *overlapping* stubble].

142. *Corollae monopetalae a.* Tubus. *b.* Limbus. [*a.* tube; *b.* selvage, *of corolla with a single petal*].

143. Flos. *a.* Germen. *b.* Stylus. *e.* Stigma. *d.* Filamenta [flower, *a.* ovary; *b.* style; c. stigma; *d.* filaments].

  c. Antherae. *f.* Petala [*e.* anthers; *f.* petals].

144. *Corollae polypetalae. a.* Ungues; *b.* Laminae [*a.* narrow bases; *b.* blades, *of a corolla with several petals*].

145. Nectarium campanulatum in *Narcisso* [bell-shaped nectary in *Narcissus*].

146. Nectaria cornuta in *Aconito* [horned nectaries in *Aconitum*].

147. Nectarium cornutum *in Calyce Tropaeoli* [horned mectary *in the calyx of Tropaeolum*].

148. Nectaria *in Parnassia* [nectaries in *Parnassia*].

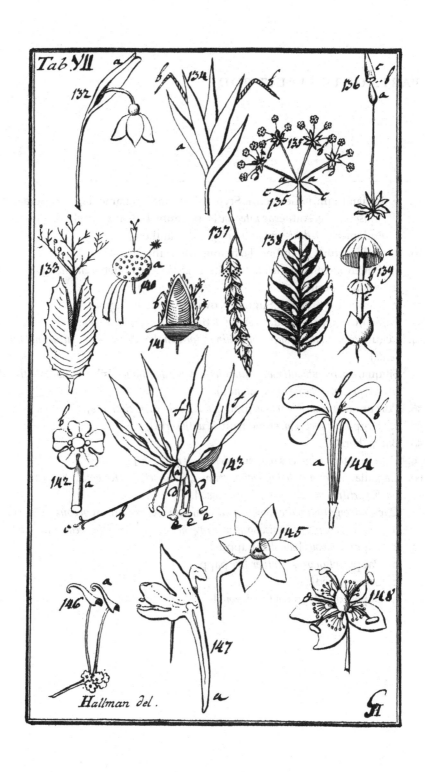

PLATE VIII

THE PARTS OF THE FRUIT-BODY

Fig. 149.   *a.* Perianthium, *b.* Germen, *c.* Stylus, *d.* Stigma, *e.* Filamenta, *f.* Antherae
dehiscentes, g. Antherae *integrae* [*a.* perianth, *b.* ovary, *c.* style, *d.* Stigma,
e. filaments, *f. dehiscent* anthers, *g. entire* anthers].

150.   *a.* Filamentum. *b.* Anthera [*a.* filament, *b.* anther].

151.   *a.* Pollen *microscopio visum. b.* Halitus *elasticus* [*a.* Pollen *seen through a microscope,*
b. elastic vapour].

152.   *a.* Germen, *b.* Stylus, *c.* Stigma [*a.* ovary, *b.* style, *c.* stigma].

153.   Folliculus, *a, Receptaculum seminum* [follicle, *a. receptacle for seeds*].

154.   Legumen, *a.a. sutura superior semina adnectens* [pod, *a.a. upper suture holding onto*
seeds].

155.   Siliqua, *a.b. receptaculi margo uterque, c. Valvula siliquae* [silique *a.b. both the edges of*
the receptacle, c. the lid of the siliques].

156.   Pomum *a. capsula, b. inclusa* [pome, *a. capsule, b. its contents*].

157.   *a.* Drupa, *b. Nucleo* [*a.* stone-fruit, *b. with kernel*].

158.   Bacca [berry].

159.   Capsula *apice dehiscens* [capsule *dehiscent at the tip*].

160.   Capsulae *a. Valvula. b. Dissepimentum, c. Columella. d. Receptaculum* [the capsules *a.*
lid, b. partition, c. columella, d. receptacle].

161.   Capsula *longitudinaliter dissecta, ut seminum* receptaculum *in conspectum prodeat*
[*capsule cut open lengthwise so as to bring the* receptacle *of the seeds into view*].

162.   *a.* Pappus *pilosus* [*hair-like* pappus].

      b. Pappus *plumosus* [*feather-like* pappus].

      c. Semen [seed].

      d. Stipes *pappi* [stipe of the *pappus*].

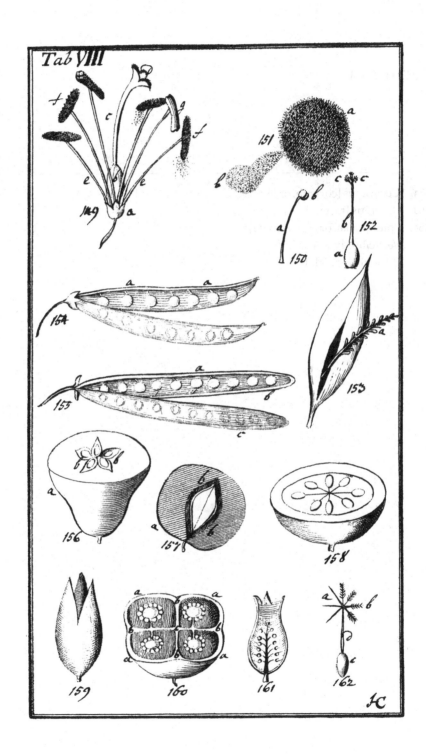

PLATE IX

THE PEDUNCLE

Fig. 163. Corymbus [corymb, or cluster].
  164. Racemus [raceme].
  165. Spica [spike, or ear (of corn)].
  166. Verticillus [whorl].
  167. Panicula [panicle].

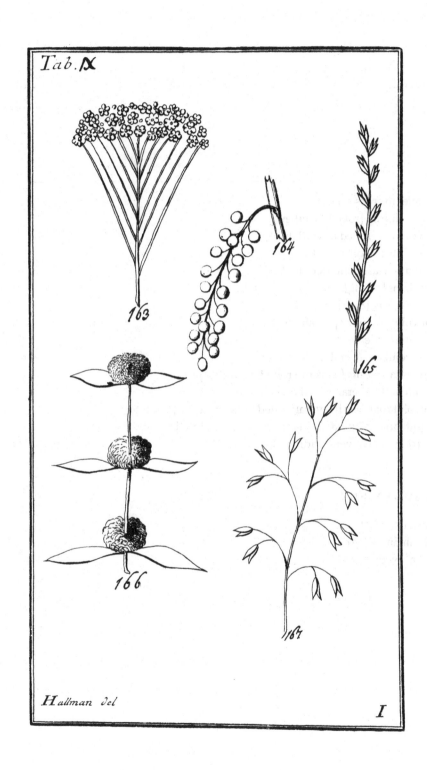

Tab. IX

163

164

165

166

167

Hallman del

I

PLATE X

FOLIAGE

with leaves in cross-section
 1. Convolutum [rolled together].
 2. Involutum [rolled inwards].
 3. Revolutum [rolled back].
 4. Conduplicatum [folded double].
 5. Equitans [riding].[9]
 6. Imbricatum [overlapping].
 7. Obvolutum [wrapped about].
 8. Plicatum [pleated].
 9. Convoluta [(several) rolled together].
10. Involuta *opposita* [(several) rolled inwards, *opposite*].
11. --- *alterna* [(several) rolled inwards, *alternate*].
12. Revoluta *opposita* [(several) rolled back, opposite].
13. Equitantia ancipitia [(several) riding, two edged].[9]
14. --- triquetra [(several) riding, three edged].[9]

MEASUREMENT

P. $\frac{1}{4}$ of a Paris *foot*.
A. $\frac{1}{4}$ of an English *foot*.
S. $\frac{1}{4}$ of a Swedish *foot*.

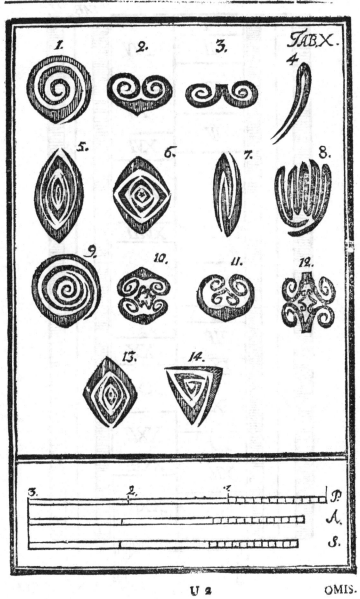

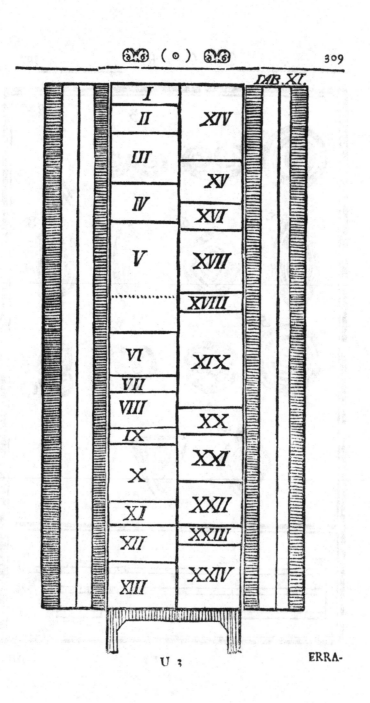

MEMORANDA
(see pp. 6 and 350)

THE BEGINNER

He should make all the *parts of a plant* well-known to himself.

He should learn to identify *the commonest plants* from the appearance of the vegetation.

He should himself *collect*, dry, and glue on to sheets of paper the larger plants, as many as he can.

He should learn to distinguish the principal *parts of the fruit-body*.

He should make himself familiar with the *classes* and *orders* of the system, and place the simpler and more obvious flowers in them.

He should frequently be present at *demonstrations* in a garden.

He should have thoroughly studied the *technical terms* according to their definitions.

He should examine about fifty *genera* that are known to him, with reference to [my] *Genera plantarum* collating the *fruit-bodies* with their characters.

By the same method, he should by his own effort prepare about fifty *generic characters*, and correct them with reference to *Genera plantarum*.

He should prepare about sixty *descriptions* of species by technical rules, beginning with the simplest plants, and going on to the more difficult ones; and these should be corrected by the professor.

He should himself *examine* species of plants not known to him, according to the keys, characters, and definitions of the system.

He should have a proper understanding of the *principles* and *foundations* of botany.

He should make himself familiar with the *literary history* of botany; and the authorities on species of plants should be consulted in the first place.

He should become accustomed to referring[1] to the *synonyms* used by the authorities, going back to the discoverers.

He should describe the medical and economic *uses* of the plants.

THE HERBARIUM[2]

Plants that have been collected by a given method (p. 18) into a herbarium made from living plants, which some call a *hortus siccus* (dry garden), are arranged according to a system, so that they may be promptly produced.[1]

The following is the way to set up a herbarium according to the sexual system; others may arrange it on the same principle, according to any other system, observing what should be observed.

A wooden CUPBOARD, which can be closed by two long folding doors, nicely corresponding to a vertical partition.

The *dimensions* of the cupboard on the inside should be accurately determined.

Height, 7 ½ Paris feet from top to bottom.

Width, 16 inches, besides the partition.

Depth, 1 foot, to wit from front to back.

The cupboard should be *divided* by the partition into two equal parts, from front to back.

The compartments should be separated horizontally, by shelves 6 lines thick.

The spaces for the classes should be arranged according to the amount of species, as indicated by the numbers in each of the classes.

Class	1,	2	inches	Class	14,	10	inches
"	2,	3	"	"	15,	5	"
"	3,	6	"	"	16,	4	"
"	4,	5	"	"	17,	8	"
"	5,	14	"	"	18,	2	"
"	6,	6	"	"	19,	12	"
"	7,	2	"	"	20,	3	"
"	8,	3	"	"	21,	5	"
"	9,	2	"	"	22,	5	"
"	10,	7	"	"	23,	2	"
"	11,	3	"	"	24,	7	"
"	12,	5	"	"			
"	13,	6	"				

This cupboard has *room* for 6,000 dried plants, glued to sheets of paper.

If the folding doors are *marked* with the numbers and names of the genera, with the spaces[3] on the shelves corresponding exactly, and linden bands are kept between the spaces, enclosing the same genera and themselves marked with the numbers of the genera, then any plant can be pulled out and produced[1] without delay.

BOTANIZING

Botanical OUTINGS[a] are arranged differently by different people; with us, the following [arrangements] are usual.

Very light and very loose CLOTHING, proper to botanists,[b] (where circumstances permit), and the most appropriate for the business.[4]

INSTRUMENTS: books: *Systema* naturae, and the *Flora* and *Fauna* of the region. A *microscope*, a botanical *needle*, a botanical *knife*, and *black lead*.

A Dillenian *case*,[c] and bound *paper*.

A *box* with pins for insects.

The SEASON is from the time of the trees' leafing onwards, except the dog days, until they shed their leaves.

[a] Called *herborisations* by the French.

[b] The *clothing of the herborisant*, besides linen, should be a short coat; very thin breeches extending from the hypochondria to the heels; smooth[5] shoes; a hat with a very large brim, or else a sunshade, so that he shall not be tired by the way, the warmth, heat, or sweat.

[c] A *Dillenian case* <*vasculum*>: a semi-cylindrical copper bowl, 9 inches long, with a proper lid, an opening large enough for a hand, and the side slightly concave rather that flat, to fit the thigh, for plants that have been gathered and watered, to keep them alive till the evening.

In any week, twice in the summer, once in the spring.

From 7 in the morning till 7 in the evening.

There are RULES for those who come late, depart early, or are absent.

And for division [of labour], lunch at 2, rest at 4, and for a secretary.

The ROUTE; meeting place, dispersed walking, halts.

2 $^{1}/_{2}$ miles at the most.

GATHERING: *plants*, especially flowering plants, *mosses*, etc.

Insects, amphibians, fish, and *small birds* killed with shot.

Rocks, minerals, fossils, and above all *soils*.

A single DEMONSTRATION should be made by the professor, lasting not longer than half an hour.

All the natural objects that have been gathered should be announced by *numbers* from books.

The essential *characters* of the genus and the species.

Peculiarities to be observed in an object.

The *economic* use, especially the *medical* one.

USES: the collected objects give pleasure at CHRISTMAS, facilitate memory, and give an impression of their habit and nature.

In one day, as many objects are offered as at other times on days equalling the number of the members of the company.

Arguments are collected by the secretary, to be written up for later use.

THE GARDEN

The botanical garden, called a PARADISE, should support a large number of plants.

GREENHOUSES, with walls containing windows facing towards the south, should protect the plants from the severity of the winter.

A HOT-HOUSE with three stoves, benches, a hot-bed, and windows precisely adjusted, for *wild* plants. Thermometer, 12–36°.

A *warm house* with two stoves, and benches arranged in steps, for *tropical*[6] and *succulent* plants. Thermometer, 4–12°.

A *cool house* with a single stove, very spacious, for demonstrations in the summer, in the winter for *cultivated* plants. Thermometer, 2–10°.

Behind these, a *museum*, which should be oblong, tall, and narrow.

Animals in glasses filled with spirits of wine.

Minerals in a case suitable for keeping them in.

The houses of the *gardener* and the *professor* should be close by.

SUMMER FRAMES, built in the form of very small greenhouses, should protect the plants from harm in the summer.

A STEAM-PIPE, protected by a covering consisting of a horse's paunch should promote the propagation of seeds, in pots.

A BALCONY, with benches arranged in steps, will provide sunshine for the pots containing plants; it should be shut up at night, by means of windows; for *tropical* plants.

Separate AREAS should be enclosed by quickset hedges, shaped by topiary work.

The PERENNIAL area should be very spacious, and divided into level[7] hot-beds, which are renewed annually; and it should systematically support *domesticated* perennial herbs, or even *common* ones.

The ANNUAL area should be as spacious as the preceding, and divided into two newly worked level parts, with hot-beds assigned to the plants by numbers.

The SOUTHERN area should receive the *cultivated* plants from the cool house in the summer, from the leafing of the oak till the flowering of the Colchicum.

The SPRING area should be furnished with a wall to the north, for *climbing* and *very delicate* plants.

The AUTUMN area is for *comparatively rare* plants, which should be protected in winter by moss or leaves.

WATERWORKS should be dug out for plants, and should be surrounded by growing turf.

One like a RIVER (if circumstances permit) should carry water that is agitated by motion.

One like a LAKE should contain deep stagnant water.

One like a MARSH should let in water, when it is flooding as a result of showers, through its muddy bottom.

PLANTATIONS OF TREES should protect the sides from storms, without causing harm by their shade; they should be arranged in *quincunxes*[8], *walks*, *arbours*, and *pergolas*.

The places for *tools*, *wood*, and *manure* should be OUT OF SIGHT.

The USES of a living plant can be demonstrated by the professor.

The comparison of the [several] species is of the first importance.

Expensive journeys are avoided, when more exotic plants are found in one garden than can readily be found growing naturally throughout all Europe.

TRAVELLING

The *starting-point* must be to marvel at all things, even the most commonplace.

The *means* is to commit to writing things that have been observed, and are useful.

The *end* must be to depict nature more accurately than anyone else.

The JOURNEY: measurement of the daily route through districts, villages, etc.

GEOGRAPHICAL [features] *mountains*, *ridges*, *rivers*, *lakes*, and *cities*.

PHYSICAL: *waterfalls*, *springs*, and *mineral waters*.

Soil that is relatively uncommon or peculiar to a place.

MINERALOGICAL: *earths*, peculiar *rocks*, various *minerals*.

Metals for *mining*, processes of *smelting*, the economics of *metallurgy*.

BOTANICAL: native *plants*: according to the place, the time, and the name given by the inhabitants.

The *rare* ones should be described, drawn, and coloured.

* Wild plants *are those that must be kept indoors throughout the year*; cultivated plants, *are those that go outside in the summer*; domesticated plants *are those that can stand our winter climate*.

ZOOLOGICAL: descriptions of the rarer *animals*, their shapes, food, and habits.

> The food of *insects*, their transformations,[9] and the harm that they do.
>
> The fins of *fishes* and the scales of *serpents* should be counted
>
> The wings and tails of *birds*, and the structure of *worms* should be examined.

ECONOMIC: private, its[10] foundation.

> *Architecture* of buildings and fences.
>
> *Agricultural* tools, methods and seasons.
>
> *Horticulture*, how it is done, and with what materials.
>
> *Monotrophaea*[11], what, and of what kind.
>
> The cultivation and condition of *meadows*.
>
> The sources of the staple *diet* throughout the year.
>
> The technique of *cookery*, and the *cuisine*.
>
> *Bread-making*, and the preparation of *drink*.
>
> *Trades* and *manual* operations.

Peculiar CUSTOMS in social intercourse, *speech*, and *clothing*.

> *Festive customs* at weddings and funerals.
>
> *Superstitious customs*.

ANTIQUITIES: *ruins, barrows*, and *runes. Historic* and *legendary* [features.]

The REWARD: natural knowledge of plants, animals, and minerals, influence in the system of the world, and usefulness to the human race.

THE BOTANIST

The character of a BOTANIST should be very clear to the beginner, so that he may distinguish true *authors* from learned *compilers*.

The TRUE BOTANIST advances the science of botany everywhere.

The crude showman contributes nothing to the growth of science.

1. The true botanist arranges his plants *systematically*;
   And he does not enumerate them out of order.
2. He acknowledges the *fruit-body* as the starting point, in his theoretical arrangement (152);
   And does not change the arrangement according to the herbage (79).
3. He accepts the natural *genera* (162);
   And does not create erroneous genera because of an aberrant feature of a species (170).
4. He sets forth the separate *species*;
   And does not fabricate false ones from varieties (283).
5. He puts the *varieties* back into their species (317);
   And does not allow them to walk in steps with the species.
6. He seeks out and selects the most important *synonyms*;
   And is not satisfied with just any nomenclature that he comes across.
7. He should investigate the characteristic *definitions*;
   And he does not prefer worthless names to the true ones (319).
8. He applies himself to removing *plants* of no fixed abode to genera;
   And does not regard the rarer plants that he comes across with fleeting glances.

9. He compendiously (327) presents *descriptions* that include the essential definitions;
   And does not trumpet the natural structure (93) with a rhetorical speech (199).
10. He carefully examines the *smallest* parts (326);
    And does not undervalue the things that are most illuminating.
11. He elucidates the plants everywhere by his *observations*;
    And is not satisfied with a doubtful name.
12. He observes all things that are peculiar with his own *eyes*.
    And does not compile his own [notes] solely from the authorities

MASTERS in the art climb up the temple of FLORA by this ladder.

The chief contemporary masters will readily grant me this, for instance:

The Jussieus,	*Gronovius,*	*Royen,*
Gesner,	*Haller,*	*Gmelin,*
Burman,	*Wachendorff,*	*Ludwig,*
Guettard,	*Dalibard,*	*Gleditsch,*
Celsius,	*Browallius,*	*Lecheus,*
Kalm,	*Clayton,*	*Colden,*
Mitchell,	*Monti,*	*and others.*

VEGETABLE TRANSFORMATION

The *root* reaches downwards; the *herbage* upwards.

All *herbage* is an extension of the medullary substance of the root.

The *flower* is the ending of every plant, *fertilization* is its distinction, and the *seed* is its beginning.

The *ending of a plant* is the same, whether it is life continued in a *bud* or propagated in a *flower*.

The *bud* protects the ends of a plant from the severity of the climate, and the *seeds* protect its beginnings.

The *bud* is the winter abode of the herbage, the *seed* is its box.

The *bud* preserves the embryos between scales, the *seed* does so between teguments.

Perennial herbs are *propagated by bud* no less than trees.

Southern trees cannot be *northern ones*, on account of the buds.

A perennial plant possesses a duplicated *flower*: one is the glorious *predecessor*; the other, the invisible *successor*.

The bud is pregnant either with *little buds* or with *flowers*, or both.

Quite often, either a flower or a bud forms the ending of the *plumule*.*[12]

The origin of the *flowers* and the *leaves* is the same.

The origin of the *buds* and the *leaves* is the same.

A bud consists of the rudiment of leaves.

Stipules are appendages of the leaves.

A *perianth* is made out of the rudiments of leaves, fused together.

If nourishment is channelled to the scales of a catkin, the florets are destroyed and the scales turn into *leaves*.

If the nourishment is channelled to the florets of the catkin, the leaves become *calyces*.

Luxuriant vegetation produces leaves by continuing after the flowers.

Meagre vegetation produces flowers at the ending, after the leaves.

Buds hardly *burgeon* unless the tree has shed its leaves, or the leaves have undergone a change caused by the climate.

Buds quite often destroy the flowers of the following year, if the tree loses its leaves in the summer.

Buds turn out to be *early* by a whole year, if the tree loses its eaves in the summer.

A *flower* growing from a bud is a whole year earlier than the leaves.

Amygdalums [almonds] sown in the hot-house produce alternate leaves; but [they are] crowded together at a certain distance above the root, on a rather swollen stem, at a certain location on the bud, when a more rapid forward thrust has overcome delay and recalled the next year to the present one.

✿ ADDENDA[13]

Fullness.

The *infilling of* XERANTHEMUM *by means of paleas* is special and peculiar, as in *Jasione*.

The *paleas* are discs that are elongated and spread out, with the appearance and structure of scales, counterfeiting the ray of the calyx.

The *florets* are hardly bigger than natural ones, and therefore shorter than the paleas.

The *female rays* are barren! with an uncovered *style* which is as long as the paleas, and a simple *stigma*.

The *hermaphrodite dises* have a shorter *style*, and a *stigma* that is split in two.

> *Note:* the florets of a simple *flower*, that next to the calyx, are hermaphrodite; so it is peculiar that the full flower adopts a female ray.

RIBES inerme [thornless], with oblong flowers. [my] *Hortus Cliffortianus,* 269.

> The *ovary* is excluded from the full flower.
>
> The *calyx* is divided into five at the base, bent back open and coloured, with linear lappets.
>
> Instead of a *corolla* there is a triple series of petals, very similar to the calyx.
>
> Instead of *stamens and pistils* there is a pointed cone, divided into five.

HYACINTHUS with a panicle of many leaves, sanesius. *Colonna, Ecphrasis* 2, *p.* 10, *pl.* 12.

> The *spike* is expanded into a crowded *panicle*.
>
> All the flowers seem to consist [entirely] of *styles*, though even these are not very visible, without any corolla, stamens, or ovary, or any other rudiment of flowers, often twenty on the same peduncle.
>
> The *scales* are very small and are alternate, being placed under every style: a new pedicel seems to grow forth from each scale; so what could be more unnatural?

❧ OMISSA

Page Paragraph

Page	Paragraph			
166		Kæmpferia	- -	Knoxia.
		Nyctanthes	- -	Ophioxylum.
		Ophiorhiza.		
		Phœnix	- -	Prasium.
		Stœbe	- -	Staehelina.
180	16	Rubia	- -	from redness <rubedo>.
		Salvia	- -	from saving <salvare>.
		Panicum	- -	from bread <panis>.
191	-	Veronica	- -	Borrago Corago.
				Cuscuta κασυτας.
				Tanacetum Athanasia [immortality].
198	-	- -		Narcissus from Narce [numbness] Pl.
200		Anemone	- -	from ανεμος wind.
		Cratægus	- -	from κρατοςstrength
		Cardamine	- -	from καρ [sic: κηρ, κεαρ] heart, δαμαω tame.
		Doronicum?	- -	from δωρον gift, νικη victory.
		Fagus	- -	from φαγειν eat.
		Lotus	- -	from λωτος sweet.
		Ocymum	- -	from ωκεως (germinating) quickly.
		Scandix	- -	from σκανδιξ shepherd's needle.
		Polypremum	- -	from πολυς much, πρεμνος trunk.
15	1748.	Lecheus	- -	1749. Dalibard. 1750. Schiera.

❧ ERRATA

Page	19	Section	14.	Variora	read	rariora.
	20	"	16.	primarii	"	primariae.
	40	"	77:2	Coix	"	Cycas. Caryota.
	47		55.	Glycine	"	Astragalus.
			57.	Bunias	"	delete, at end.
			59.	Carniolaria	"	Craniolaria.
	49		68.	Adoxa		delete.
			"	Calaumaria	"	Craniolaria.
	53	"	80.	Truni	"	Trunci.
	65	"	86:7	Valva	"	Volva.
	72	"	93.	longis	"	longae.
	80	"	102.	Melica	"	Melia.
	92	line	29	plena	"	quasi plena.
	117			sinistrum	"	dextrum.
	118			hygrometica	"	hygrometrica.
	[?]			qua	"	quo.
	133	[section]	170	Ananas	"	Ananas T.
		"	"	Melocactus	"	Melocactus T.
	134	"	"	Bernhardia A.	"	Bernhardia H.
	136	"	177	sed	"	nec.
		"	"	Flores	"	Florem.
	139	"	183	superstruuntur	"	superstructa.
		"	"	Dracunculus	"	Dracunculoides.
	158	"		*Murucuja*	"	Murucuja.
	163	"		Lupularia	"	Lunularia.
	165	"		Ceratocarpus D.	"	Ceratocarpus B.
	167	[column]	2	Pontederia	"	Sterculia.
	166	"	[4]	Glycine	"	Barleria.

Page	[column]	[4]				
166	"	"	Tetragonia D.	"	Pothos.	
173			Cisso-Ampelos	"	CissAmpelos.	
196			Plematis	"	Clematis.	
			Ceriploca	"	Periploca.	
199			prafu	"	supra.	
200			αμαμελις	"	αμαμηλις	
205	[Paragraph]	3	Hydrophyllum	"	Ceratophyllum.	
212	[section]	247	pinguenda	"	pingenda.	
213	"	257	irregula	"	irregularis.	
220			283	"	273	
235			174	"	274	
240	"	280	nequeant	"	nequeunt.	
295	line	-	caniculares	"	tropici.	

❧ NOTES

TITLE-PAGE

1. Of St. Petersburg.
2. L'Académie des Sciences.

TO THE BOTANICAL READER

1. That is, *Philosophia Botanica*.
2. Latin *in ipsa herba*.
3. Linnaeus' *Species plantarum* was first published at Stockholm in 1753.

INTRODUCTION

1. Linnaeus uses the neuter plural of the adjective, *naturalia*, without any substantive noun; but one such is required in English.
2. That is, the 6th edition (1748).
3. Sections 8 and 9 are actually on page 210.
4. That is, *Acta Holmensia* (the Transactions of the 'Collegium curiosorum', later the Royal Academy of Sciences, Stockholm).
5. Here *corpora* 'bodies' appears in the text.

CHAPTER I: THE LIBRARY

1. Ovidio Montalbani.
2. Bock (French, Le Bouc).
3. Wieland.
4. L'Écluse.
5. Joachim Kammerer, the younger.
6. Bergzabern.

7. Johannes Bodaeus à Stapel.
8. Bachmann.
9. Pompey was defeated in Pharsalia, 48 BC, Antony at Actium, 31 BC. Several of the dates in this section are inaccurate.
10. Perhaps an error for Abenguefi*l*.
11. See p. 329 and Plate XI p. 328.
12. Garden at Eltham belonging to James Sherard, brother of William.
13. Botanic garden at Leiden.
14. Botanic garden at Amsterdam.
15. Latin *Adonides*: it is 'Adonides' the French version of 1788; and 'Adonistas' in the Spanish version of *Fundamenta Botanica*, 1788. The cult of Adonis, the lover of the goddess Aphrodite, originated in Phoenicia (Lebanon), and was very popular in the Hellenistic world: a feature of it was the 'Adonis garden', with a statue of him surrounded by plants in flower. In Linnaeus' time, hot-houses for exotic plants were called 'Adonis houses': (W. Blunt, *The Compleat Naturalist*, p. 103).
    Many of the dates are incorrect. The first university garden was founded at Pisa in 1543, followed by Padua in 1545; the Oxford garden was founded in 1621.
16. *Amoenitates exoticarum* (delights in exotic e.g. Japanese plants).
17. The British claimed the right to fell timber in the bay of Campeche; this was one of the causes of the war of Jenkins's ear.
18. Linnaeus presided when his pupil Johan Gustaf Wahlbom responded with this dissertation, later published Stockholm in 1746.
19. Linnaeus regards the *Philosophia* as a revised edition of the *Fundamenta*: see his preface 'To the Botanical Reader'.
20. Latin *critici*.
21. Or 'bell-glass'.
22. Latin *ablactatio* (literally 'weaning').
23. Latin *signatores*; French (1788) *'faiseurs de talismans'*.

CHAPTER II: SYSTEMS

1. These strange terms, (some of which cannot be rendered in English by a single word), were invented by Wachendorff; the most obscure are given as follows in Quesné's French version of 1788:
    4, à étamines en nombre double des pétales, ou des divisions de la corolle (with twice as many stamens as petals); *6, à étamines réunites par les filets et*

formant un cylindre (with stamens joined by fillets, and forming a cylinder); 9, à anthèrers plus nombreuses (with anthers more numerous); 13, à étamines séparées des pistils sur le même pied (with stamens separated from the pistils on the same base); 14, à étamines séparées des pistils sur des pieds différents (with stamens separated from the pistils on different bases).

2. Assuming that *cynaro* is for κιναρο-; if it is for κυναρο-, it should be rendered 'dogthorn-headed'.

3. The god of the underworld.

4. Read 'Cycas | Caryota' for *Coix*, in accordance with a correction made in the Errata (p. 339)

5. A kind of grapevine, according to Lewis and Short; but the genera listed mostly belong to the Primulaceae.

6. Read 'Astragalus', for *Glycine*: see the Errata.

7. *Bunias* should be deleted, since it already appears above in the list.

8. To be deleted: see Errata.

9. Read 'Isoetes' for *Calaumaria*: see Errata.

10. The significance of the letters A-H is not obvious.

CHAPTER III: PLANTS

1. For fuller details concerning the terms used to describe the leaves, see the descriptions of the plates, (pp. 308–12).

2. *Repandus*, originally bent back' or 'turned up'; here used of a leaf with a wavy edge.

3. *Papillosus*, literally 'with nipples'.

4. The text mistakenly has 82:V and 82:IV for 82:E and 82:D.

CHAPTER IV: THE FRUIT-BODY

1. These numbers do not refer to other sections, but to other sentences in section 88.

2. In the Errata, *longis* is altered to 'longae'; but the former appears to be correct.

3. Perhaps *viviparus*, applied to vegetables, refers to seeds that sprout before sowing, like malt.

CHAPTER V: SEX

1. Reading 'lĕvius' (lighter) for *laevius* (smoother).
2. *Thalamus* and *aulaeum:* to be consistent with what follows, they should refer to parts of an animal body; this is rectified in the ensuing sentences.

CHAPTER VI: CHARACTERS

1. Ariadne provided the Athenian hero Theseus with a thread, which he unwound on his way into the Cretan labyrinth; this enabled him to find his way out by the same route.
2. Transactions of the French Academy of Sciences.
3. See Plate X.
4. That is Class XV (*Icosandria*).
5. That is Class XIV (*Didynamia*).
6. That is Class XII Order III (*Icosandria Polygynia*).
7. Or 'lentil-shaped'; *lenticularis*.
8. That is Class XV (*Tetradynamia*).
9. That is Class XII (*Icosandria*)
10. Here rendering *definitio* (not *differentia*): see Ch. VIII.
11. That is Class V (*Pentandria*) and Class XIX (*Syngenesia*).
12. That is, Class V, Order I (*Pentandria Monogynia*).
13. Reading 'propius' for *proprius*.
14. That is Class IV, Order I (Tetrandria Monogynia).
15. That is Class V, Order I (Pentandria Monogynia).
16. *Lapis Lydius*: a stone from Mount Tmolus (Boz Dağ in Lydia, Asia Minor (Turkey); used in assaying gold and silver.
17. This is a mixed metaphor. For Ariadne's thread, see note 1 to this Chapter. Gordius, the peasant King of Phrygia in Asia Minor (Turkey), dedicated his wagon to Zeus, and it was preserved at Gordium, the capital; the pole was fastened to the yoke by a very complicated knot, and it was prophesied that whoever untied it should rule all Asia. Alexander the Great severed the knot with his sword.

CHAPTER VII: NAMES....

1. St John's Gospel xx 17.

2. The flowers of the coltsfoot appear before the leaves.

3. St Matthew's Gospel ix 5.

4. St Matthew's Gospel viii 31–2.

5. That is, the spear used to pierce his side: St John's Gospel xix 34.

6. Now called English *'Lady's* slipper' etc. in English.

7. *Cypris* (from Cyprus) is another for Venus.

8. These names are perhaps the titles of collections.

9. *Pistillum,* rendered 'pistil' in botanical contexts.

10. That is, neither Latin nor Greek.

11. From the old English Christian name Osmund.

12. From αθανασια (immortality): see Omissa, (p. 338).

13. Vitex, Abraham's balm or *agnus castus*

14. The dew resembles the effects of distillation, practised by the alchemists.

15. Surely from Proserpina, the goddess of the underworld.

16. The names are arranged in two alphabetical series; perhaps they were taken from two different sources.

17. In section 238, Nicot appears to be acknowledged as the discoverer of Nicotiana (tobacco).

18. *Gall(ia) aequinoct(ialis);* in the French (1788) 'France équinoctiale'; perhaps this refers to tropical colonies, such as Cayenne and Martinique.

19. The *h* is inorganic; the name is from αμαραντος (unfading), and has no connexion with ανθος (flower).

20. See note 7.

21. Here Linnaeus has singular and plural.

22. The usual meaning of αιθουσα is a 'corridor' or 'verandah'.

23. *Vella* Gal: if 'Gal.' stands for Galen, it is strange that no Greek form is given. According to Pliny, (22, 25, 75 para. 158) 'Vela' is Gallic for 'Erysimon': perhaps 'Gal.' is for 'Gallic'.

24. *Ervum* is accidentally repeated.

25. 'Phellandrion', with Greek ending, appears in Pliny, 27.12.101 section 126.

26. *Hydrophyllum* appears in the text but is corrected in the Errata.

27. There is confusion between οθοννα (the plant) and οθονη ('fine linen').

28. The Latin words consist of half a hexameter line, followed by a whole line. Cf. Horace *Ars Poetica* l. 97.

29. See section 69 (Wachendorff), 249 and 295.

CHAPTER VIII: DEFINITIONS

1. *Differentia* is rendered as 'definition' rather than 'difference'.
2. From the Dalmatian island Mljet (Melita, Meleda); they were lap-dogs, quite unlike the modern Dalmatian. The island is often confused with Malta: see note 3.
3. In the French version of 1788, the list is *de Malthe* (Maltese), *spagneuls* (spaniels), *à poil ras* (with shorn hair), *mâtins* (mastiffs), *Turcs* (Turkish), and *barbets* (water spaniels).
4. Here the definition is not given in italics, perhaps erroneously.
5. Latin *si fides habenda rebus*; French (1788) 'pour déterminer sa croyance'.
6. *Gallia Narbonensis*, with its capital at Narbonne, was the Roman province that comprised the whole of southern France.
7. Evidently a proverbial phrase. The motto of the Elzevir family is 'Ne extra oleas.'
8. *Heri*, literally 'yesterday'. Were the dog-trials perhaps held on day before Linnaeus delivered lecture?
9. Now known as 'mangetout'.
10. Juvenal X 356.
11. Linnaeus here treats *Acetosa* and *Lapathum* as generic names, though subordinate to *Rumex*. Three of the species of the genus *Rumex* are acetosa, alpina, and hydrolapathum.
12. *Hoc opus, hic labor est*: Virgil, Aeneid VI 129.
13. See note 9 to Chapter VII.
14. These invented words can be rendered as follows, though the precise application of the descriptive adjectives is sometimes doubtful.
    Alsine, mouse's ear, apple-flower.
    Androsace, (obscure; perhaps corrupt).
    Auricula muris (mouse's ear), white mountain flower.
    "   "   "   small white mountain flower.
    Betonica, mountain tooth-shaped root; and white flower.
    Campanula, alpine, ox-tongue leaves.
    "   , beautiful white flower.
    "   , autumn, greater mountain.
    "   , small mountain, autumn.
    Carduus, spineless.
    "   , white wool head.
    Condrilla, small apples, many stems.
    Corruda, white stem, inclined globe.

Cynogossum, alpine ground, red flower.
Doronicum, mountain, many twigs, apple flower.
Gentianella, spring, stem, mountain.
Glycyrrhiza, long root, much divided.
Hieracium, long narrow leaf.
   " , flat flower.
Jacea, beautiful mountain flower.
   " , mountain rock, single stem.
   " , mountain, stemless, purple.
Ground, meadow, *Montis Ceti*
Nardus, mountain, bare stem.
Plantago, inclined flat leaves.
Polygonum, Alsine-like flower.
Pulsatilla, flower somewhat apple-like.
Quercus, maritime, flat-leaves.
Racemus marinus, hammer-shaped.
Tanacetum, without smell.
Thlaspi, mountain-stem crowned with leaves.
Trachelium, alpine, pyramid-shaped.
   " , mountain, (uncertain), thick leaves.
Tulipa, flower very deeply incised.
   " , sharp, middle, flower.

15. *Finium Regendorum*: Justinian, Codex (Code of Laws), III 39.
16. Latin *calyx*, here not used in the usual botanical sense.
17. From Daedalus, the architect of Cretan labyrinth, or maze.
18. The positions of the adjectives are unnatural in English.
19. The effect of these suffixes, attached to the second of the conjoined (or disjoined) adjectives or substantives, cannot be precisely reproduced in English.
20. That is, *tertia* (third): see sect. 263.
21. In A, the lappets appear to be a feature of the leaves; in B. and C., a separate part of the plant.
22. These forms refer to the three grammatical genders.

CHAPTER IX: VARIETIES

1. 'Bon chrétien'; the other varieties are mostly obsolete.

2. Ερυσιβη (Erysibe) is the usual classical form (Plato and Aristotle).

3. The Latin *puniceus* and *phoeniceus* are two forms of the same word, (the latter being nearer to the original Greek φοινικεος); they refer to the famous 'Tyrian purple' dye, made in Phoenicia (Lebanon) from shell-fishes of the genera Murex and Thais. It appears that *puniceus* refers to the red colour from M. trunculus and T. haemastoma, and *phoeniceus* to the deep blue-violet from M. brandaris. See Stearn, *Botanical Latin*, pp. 230–2.

4. There is a noteworthy lack of varieties of blue and green. In ancient times these colours were not easily reproduced: A. E. Kober (1932), quoted by Stearn, *Botanical Latin*, p. 232.

CHAPTER X: SYNONYMS

1. *Evolvi*, literally 'be unrolled', originally referring to papyrus or parchment rolls; here used of finding the appropriate passage in a book.

2. Neither Greek nor Latin.

CHAPTER XI: SKETCHES

1. *Albida* seems to be a mistake, as c. should indicate relative size, not colour.

2. *Laevissime* is perhaps an error for 'lĕvissime' (very lightly).

3. The crescent-shaped white mark at the base of a finger-nail.

4. The 'hand', in the measurement height of a horse, is 4 English inches, or about 3 $^1/_3$ Paris inches. See Plate X.

5. See note 4 to the Introduction.

6. The publication of these sites would hardly be recommended nowadays.

7. For example, cranberry: see note 11.

8. *Pomaria*: the significance of this word here is not obvious.

9. Landskrona is about 4° south of Uppsala.

10. *Euon[ymus]*: the *u* appears as a *v*, perhaps the choice of Linnaeus' informant at Landskrona.

11. For example, whortleberry: see note 7.

12. Reading 'tropici' for *caniculares*, as in the Errata.

13. This centigrade scale is commonly attributed to Celsius; but he made freezing point 100° and boiling point 0°. The present usage is due to Linnaeus, and he rightly claims it as 'ours': (see Blunt, *The Compleat Naturalist*, p. 117).

CHAPTER XII: POTENCIES

1. These are the aubergine and the tomato, not now regarded as toxic.
2. The name contains a pun: the genus *Apium* = celery; *apium* (genitive plural) = bees', *risus* (nominative) = laughter.
3. Also known as 'marvel of Peru' or 'belle de nuit'.
4. When cultivated, it is known as 'salsify'.
5. From ambrosia, the food of the gods.

THE PLATES

1. Lanceolate: like a small lance. The precise meaning of this term varies among different authors: Stearn, *Botanical Latin*, p. 316, citing Alphonse de Candolle, *La phytographie*, pp. 198–200.
2. That is, with wedge-shaped notches.
3. Also described as *ramosum* (branchy), (section 83, No. 100), Stearn, *Botanical Latin*, p. 462, renders it 'palmate ... with the lateral lobes themselves divided'; this is clearly shown in the figure.
4. That is, a cotyledon.
5. That is, with a stem passing through the leaf.
6. The figure does not agree with the descriptions given in section 83 (no. 11) and section 277. Perhaps the artist Hallman took *acerosus* in the classical sense, 'chaffy'
7. That is, branching into two at every division.
8. That is, with branches resembling arms.
9. See section 163: VI. Stearn, *Botanical Latin*, p. 407, defines *equitans* (riding) as 'conduplicate and overlapping in two ranks, the base of the folded outer leaf clasping the base of the leaf opposite it'.
10. Plate XI, which has no explanation, refers to the herbarium, section 11 and p. 359.

The MEMORANDA endorsed on the reverse side of the sheets containing the explanations of the plates.

1. *Evolvere*: see note 1 to Chapter X.
2. See section 11 and Plate XI.
3. Assuming that *loculis* is an error for 'loculamentis', since it is followed by the neuter relative pronoun *quae*: see Preface p. XI
4. The recommended garments do not appear in the illustrations on p. 164 of Blunt's *The Compleat Naturalist*.
5. Or perhaps 'light' if 'lĕves' is read for *laeves*.
6. *Aethiopicae*: that is, from Africa south of Libya and Egypt.
7. Or 'equal' (in size): *aequales*.
8. Squares consisting of 5 trees, one at each corner and one in the middle.
9. For instance, the change from a caterpillar to a butterfly.
10. It is not clear what this refer to.
11. Cf. Greek μονοτροφεω, 'to eat but one kind of food' (Strabo, Geographica, p. 154 in Casaubon's edition).
12. The asterisk does not appear to refer to the footnote on page 335, (which is related in sense to the lines immediately above it). But in the 4th edition (Vienna, 1770), the footnote is preceded by an asterisk.
13. Presumably to be added at the end of Chapter IV.

✻ APPENDIX I

THE LINNAEAN CLASSES AND ORDERS

There are 24 classes, as follows:

I. Monandria, with one stamen.

II. Diandria, with two stamens.

III. Triandria, with three stamens.

IV. Tetrandria, with four stamens of equal length (see XIV).

V. Pentandria, with five stamens.

VI. Hexandria, with six stamens, either equal or else three long and three short (see XV).

VII. Heptandria, with seven stamens.

VIII. Octandria, with eight stamens.

IX. Enneandria, with nine stamens.

X. Decandria, with ten or eleven stamens.

XI. Dodecandria, with ten to nineteen stamens fixed to the receptacle.

XII. Icosandria, with twenty or more stamens inserted into the calyx.

XIII. Polyandria, with twenty or more stamens fixed to the receptacle.

XIV. Didynamia, with four stamens, two long and two short (see IV)

XV. Tetradynamia, with six stamens, four long and two short (see VI).

XVI. Monadelphia, with united filaments in one set.

XVII. Diadelphia, with united filaments in two sets.

XVIII. Polyadelphia, with united filaments in three or more sets.

XIX. Syngenesia, with united anthers.

XX. Gynandria, with stamens inserted into the pistil.

XXI. Monoecia, with male and female flowers on the same plant.

XXII. Dioecia, with male and female on different plants.

XXIII. Polygamia, with male, female, and hermaphrodite flowers distributed in various combinations.

XXIV. Cryptogamia, with the fruit-body concealed.

The classes are divided into *orders*.

In classes I–XIII, the orders are defined by the number of pistils: *monogynia* with one, *digynia* with two, etc.

In classes XIV and XV, the orders are described according to the form of the seed-vessel, or its absence.

In classes XVI–XVIII and XX–XXII, the orders depend on the number of stamens: *monandria*, etc., (as in classes I–XIII).

The orders of class XIX are defined by the arrangement and fertility of the florets.

In class XXIII there are three orders, *monoecia*, *dioecia*, and *trioecia*, with various combinations of male, female, and hermaphrodite flowers.

Class XXIV contains the orders of *filices* (ferns), *musci* (mosses), *fungi* (funguses), etc.

These *Linnaean* orders are distinct from the *natural* orders, for which see Appendix II.

APPENDIX II

THE NATURAL ORDERS

1. Piperitae [pungent].

2. Palmae [palms].

3. Scitamina [delicacies].

4. Orchideae [orchids].

5. Ensatae [sword-shaped].

6. Tripetalodeae [with three petals].

7. Denudatae [uncovered].

8. Spathaceae [with spathes].

9. Coronariae [coronary].*

10. Liliaceae [lilies etc.].

11. Muricatae [muricate].†

12. Coadunatae [united].

13. Calamariae [reeds etc.].

14. Gramina [grasses].

15. Coniferae [conifers].

16. Amentaceae [with catkins].

17. Nucamentaceae [with fir-cones].

---

* That is, suitable for making garlands.
† Rough like the shell of a murex (see Chapter IX, note 1).

18. Aggregatae [clustered].

19. Dumosae [bushy].

20. Scabridae [somewhat rough].

21. Compositi [compound].

22. Umbellatae [with umbels].

23. Multisiliquae [with many siliques].*

24. Bicornes [two-horned].

25. Sepiariae [hedge-plants].

26. Culminiae [with upward projections].

27. Vaginales [scabbard-shaped].

28. Corydales [crested].

29. Contorti [twisted].

30. Rhoeades [poppies].

31. Putaminea [with stones in the fruit].

32. Campanaceae [bell-shaped].

33. Luridae [sallow].

34. Columniferi [bearing columns].

35. Senticosae [thorny].

36. Comosae [with top-knots].

37. Pomaceae [with pomes].

38. Drupaceae [stone-fruit].

39. Arbustiva [woody].

40. Calycanthemi [with calycine flowers].

* A silique is similar to a leguminous pod, but with seeds attached to both sutures.

41. Hesperideae* [citrous].

42. Caryophyllei [pinks etc.].

43. Asperifoliae [rough-leaved].

44. Stellatae [starry].

45. Cucurbitaceae [marrows gourds, etc.].

46. Succulentae [juicy].

47. Tricocca [with three kernels].

48. Inundatae [aquatic].

49. Sarmentaceae [with runners].

50. Trhilatae [with three hila].

51. Preciae, (see Chapter II, note 5).

52. Rotaceae [wheel-shaped].

53. Holeraceae [culinary].

54. Vepreculae [briars].

55. Papilionaceae [butterfly-shapped].

56. Lomentaceae [with loments].

57. Siliquosae [with siliques].†

58. Verticillatae [whorled].

59. Personatae [mask-like].

60. Perforatae [perforated].

61. Statuminatae [ribbed].

62. Candelares [taper-like].

63. Cymosae [with cymes].

*From the garden of the Hesperides (nymphs), in the far west, where the 'golden apples' grew.

†see note to No. 23

64. Filices [ferns].

65. Musci [mosses].

66. Algae.

67. Fungi [funguses].

68. Indeterminate.

APPENDIX III

ABBREVIATIONS OF NAMES OF BOTANISTS AND OTHER AUTHORITIES

A. *Amman;* (in section 241) *Aristotle.*

Act. Haffn. *Acta Haffniensia (Copenhagen Transactions).*

Act. Stockh. *Acta Holmensia (Stockholm Transactions).*

Arist. *Aristotle.*

Ath., Athen. *Athenaeus.*

B. *Boerhaave.*

Brom. *Bromel.*

Buxb. *Buxbaum.*

C. *Cesalpino.*

C.B. *Caspar Bauhin.*

Clus. *Clusius (de l'Écluse).*

Comm. *Commelin.*

Condam. *Condamine.*

D., Dill. *Dillenius;* (in section 241) *Dioscorides.*

Dalib. *Dalibard.*

Dod. *Dodoens.*

Fabr. *Fabregou (-gow).*

G. *Gmelin.*

Gal.? *Galen.**

* See note 23 to Ch. VII.

Gron. *Gronovius.*

Guett. *Guettard.*

H. *Haller; Houstoun;* (in section 241) *Hippocrates.*

Herm. *Hermann.*

Hor. *Horace.*

I.B. *Jean Bauhin.*

Isn. *Isnard.*

J. *Jussieu.*

J.B. *Jean Bauhin.*

K?, Kn., Kt? *Knaut.*

Kr., Kram. *Kramer.*

L. *Linnaeus.*

Ld., Lind.

Lob. *Lobel.*

Loes. *Loesel.*

Lon. *Lonitzer.*

M. *Monti.*

Malpigh. *Malpighi.*

March. *Marchant.*

Mich. *Micheli.*

Mitch. *Mitchell.*

Mor., Moris. *Morison.*

O.?; (in section 241) *Ovid.*

P. *Plumier;* Pontedera; (in section 241) *Pliny.*

Park. *Parkinson.*

Pk?,. Plk., Plukn. *Plukenet.*

Pl. Plum. *Plumier.*

Pont. *Pontedera.*

Pt., *Petit.*

R. *Ruppius.*

Raj., Rj?, *Ray (Rajus)*

Riv. *Rivinus (Bachmann).*

Rudb. *Rudbeck.*

S?

Sloan. *Sloane.*

S.S.?

Sauv. Sauvag. *Sauvages.*

T. *Tournefort;* (in section 241) *Theophrastus).*

Tab. *Tabernaemontanus (Bergzabern).*

Th., Theophr. *Thephrastus.*

Tourn., Tournef. *Tournefort.*

Trag. *Tragus (Bock).*

V. *Vaillant;* (in section 241) *Virgil.*

Valent. *Valentini.*

Volk. *Volkamer.*

Wachend. *Wachendorff.*

Weh. *? Wehle ? Wehme*

Wehl. *? Wehle*

Wehm. *? Wehme*

There is some ambiguity; for instance, it appears from the list of genera (in section 209), that H. may stand for Houstoun of Haller, and P. for Plumier or Pontedera.

Most of the abbreviations of the titles of the works cited are obvious; it may be noted that lugdb. stands for Lugduno-Batavus (of Leiden). Sometimes rhut. (rather than ruth.) stands for Ruthenus (Russian).

APPENDIX IV

SELECTIVE LIST OF LINNAEUS'S WORKS UP TO 1753

System Naturae (The System of Nature), Leiden, 1735; (3rd edition, 1740–4; 6th, 1748; 12th, 1766–8).

Bibliotheca Botanica (The Botanical Library), Amsterdam, 1736.

Fundamenta Botanica (The Foundations of Botany), Amsterdam, 1736.

Musa Cliffortiana (Clifford's Banana), Leiden, 1736.

Critica Botanica (Botanical Taxonomy), Leiden, 1737.

Flora Lapponica, (Flora of Lapland), Amsterdam, 1737.

Genera Plantarum (The Genera of Plants), Leiden, 1737.

Hortus Cliffortianus (Clifford's Garden), Amsterdam, '1737' (1738).

Classes Plantarum (The Classes of Plants), Leiden, 1738.

Ficus, ejusque Historia naturalis et medica (The Fig, and its history in nature and medicine); Linnaeus, president; dissertation by Cornelius Hegardt; Uppsala, (1744).

Oratio de Telluris habitabilis Incremento (Lecture about the increase of the habitable land), Leiden, 1744.

De Peloria (An Abnormality);* Linnaeus, president; dissertation by Daniel Rudberg; Uppsala, 1744.

Öländska och Gothländska Resa (The Journey to Öland and Gotland), Stockholm and Uppsala, 1745.

Fauna Suecica (The Fauna of Sweden), Stockholm, 1745.

* An epigenetic mutation of *Linaria*.

Sponsalia Plantarum, (The Betrothals of Plants); Linnaeus, president; dissertation by Johan Gustav Wahlbom; Stockholm, 1747.

Wästgöta Resa (The Journey to Västergötland), Stockholm, 1747.

Flora Zeylandica (The Flora of Ceylon), Stockholm, 1747.

Hortus Upsaliensis (The Uppsala Garden), Stockholm, 1748.

De Oeconomia Naturae (On the Economy of Nature); Linnaeus, president; dissertation by I. J. Biberg; Uppsala, 1749.

Amoenitates Academicae (Academic Delights; that is, theses), Stockholm and Leipzig, 1749–(90).

Semina Muscorum detecta (The Seeds of Mosses Discovered); Linnaeus, president; dissertation by Peder Jonas Bergius; Uppsala, 1750.

Skånska Resa (The Journey to Skåne), Stockholm, 1751.

Species Plantarum (The Species of Plants), Stockholm, 1753.

✤ APPENDIX V

SELECTIVE LIST OF OTHER BOTANICAL WORKS

Jean Jacques Ammann, *Stirpium rariorum in Imperio Rutheno sponte provenientium Icones et Descriptio* (Pictures and Descriptions of the rarer Plants that are found growing naturally in the Russian Empire), St Petersburg, 1739.

Caspar Bauhin, *Catalogus Plantarum circa Basileam sponte nascentium* (Catalogue of the Plants that grow naturally around Basle), Basle, 1622.

Jean Bauhin and J. H. Cherler, *Historia Plantarum universalis*, (The Universal History of Plants), Yverdon, 1650–1.

Herman Boerhaave, *Index Plantarum quae in Horto Academico Ludguno-Batavo aluntur*, (List of the Plants that are grown in the Leiden University Garden), Leiden, 1720.

Johann Jakob Dillenius (Dillen), *Catalogus Plantarum sponte circa Gissam nascentium* (Catalogue of Plants growing naturally around Giessen), Frankfurt and Main, 1719.

Joachim Jung, *Isagoge Phytoscopia* (Introduction to the Observation of Plants), Hamburg, 1678.

Pietro Antonio Micheli, *Nova Plantarum genera* (New Genera of Plants), Florence, 1729.

Philip Miller, *The Gardeners Dictionary*, London, 1731.

Charles Plumier, *Nova Plantarum Americanarum Genera* (New Genera of American Plants), Paris, 1703.

John Ray, *Catalogus Plantarum Angliae et Insularum Adjacentium* (Catalogue of the Plants of England and the adjacent Islands), London, 1670.

John Ray, *Historia Plantarum* (The History of Plants), London, 1686.

John Ray, *Synopsis methodica Stirpium Britannicarum* (A Methodical Conspectus of British Plants), London, 1690.

John Ray, *De variis Plantarum, methodis dissertatio brevis* (A Short Dissertation on the various ways of Studying Plants), London, 1696.

Giulio Pontedera, *Anthologia* (The Study of Flowers), Padua, 1720.

Augustus Quirinus Rivinus (André Bachmann), *Introductio generalis in Rem Herbariam* (A General Introduction to Botany), Leipzig, 1690–9.

Adriaan van Royen, *Florae Leydensis Prodromus* (Introduction to the Flora of Leiden), Leiden, 1740.

Heinrich Bernhard Ruppius, *Flora Jenensis* (Flora of Jena), Frankfurt am Main, 1718.

Joseph Pitton de Tournefort, *Élémens de Botanique* (Elements of Botany), Paris, 1694.

Joseph Pitton de Tournefort, *Histoire des Plantes qui naissent aux environs de Paris* (History of the Plants that grow in the neighbourhood of Paris), Paris, 1698.

Joseph Pitton de Tournefort, *Institutiones Rei Herbariae* (Elements of Botany), Paris, 1700. (The Latin version of the French of 1694.)

Joseph Pitton de Tournefort, *Corollarium Institutionum Rei Herbariae* (Supplement to Elements of Botany), Paris, 1703.

Joseph Pitton de Tournefort, *Relation d'un voyage du Levant*, Paris, 1717. (Translated by John Zell as *A voyage into the Levant*, London, 1718.)

APPENDIX VI

PRINCIPAL SOURCES USED FOR THIS EDITION

Wilfrid Blunt, *The Compleat Naturalist*. London, Collins, 1971.

Ian Garrard and David Streeter, *The Wild Flowers of the British Isles*. London, Macmillan, 1983.

V. H. Heywood and others (eds.), *Flowering Plants of the World*, Oxford University Press, 1978.

Albert M. Hyamson, *A Dictionary of Universal Biography*, 2nd edn., Routledge and Kegan Paul, London 1951.

C. A. Johns, *Flowers of the Field*, 29th edn., ed. G. S. Boulger, London, Society for Promoting Christian Knowledge, 1899.

D. J. Mabberley, *The Plant Book*, 2nd edn. Cambridge University Press, 1997.

William T. Stearn, *Botanical Latin*, 4th edn., Newton Abbot, David and Charles, (first published 1966).

Richard Walker, *The Flora of Oxfordshire*. Oxford, Henry Slatter, 1833.

INDEX OF CONTENTS AND TERMS

Figures in **bold** indicate major references.
Figures in *italic* refer to Plates.

❧ INDEX OF GENERA

Figures in **bold** indicate major references.

Printed in the United States
By Bookmasters